Modeling Dynamic Systems

Series Editors

Matthias Ruth
Bruce Hannon

Springer
New York
Berlin
Heidelberg
Barcelona
Hong Kong
London
Milan
Paris
Singapore
Tokyo

MODELING DYNAMIC SYSTEMS

Modeling Dynamic Biological Systems
Bruce Hannon and Matthias Ruth

Modeling Dynamic Economic Systems
Matthias Ruth and Bruce Hannon

Dynamic Modeling in the Health Sciences
James L. Hargrove

Modeling and Simulation in Science and
Mathematics Education
Wallace Feurzeig and Nancy Roberts

Dynamic Modeling of Environmental Systems
Michael L. Deaton and James J. Winebrake

Modeling and Simulation in Science and Mathematics Education

With 124 Illustrations and a CD-ROM

Wallace Feurzeig   Nancy Roberts

Editors

Springer

Wallace Feurzeig
BBN Technologies
10 Moulton Street
Cambridge, MA 02138
USA

Nancy Roberts
Lesley College
29 Everett Street
Cambridge, MA 02138-2750
USA

Series Editors:
Matthias Ruth
Center for Energy and
  Environmental Studies
  and the Department of Geography
Boston University
675 Commonwealth Avenue
Boston, MA 02215
USA

Bruce Hannon
Department of Geography
220 Davenport Hall, MC 150
University of Illinois
Urbana, IL 61801
USA

Library of Congress Cataloging-in-Publication Data
Modeling and simulation in science and mathematics education /
    Wallace Feurzeig, Nancy Roberts (editors).
        p.   cm. — (Modeling dynamic systems)
    Includes bibliographical references and index.
    ISBN 0-387-98316-3 (hc : alk. paper)
        1. Science—Study and teaching (Secondary)   2. Science—Computer
    simulation.   3. Mathematics—Study and teaching (Secondary)
    4. Mathematics—Computer simulation.   5. Computer-assisted
    instruction.   I. Feurzeig, W.   II. Roberts, Nancy, 1938–   III. Series.
    Q181.M62   1999
    501'.13—dc21   99-10025

Printed on acid-free paper.

Production coordinated by Diane Ratto, Michael Bass & Associates, and managed by Lesley
Poliner; manufacturing supervised by Jacqui Ashri.
Typeset by G&S Typesetters, Austin, TX.
Printed and bound by Hamilton Printing Co., Rensselaer, NY.
Printed in the United States of America.

9 8 7 6 5 4 3 2 1

ISBN 0-387-98316-3 Springer-Verlag New York Berlin Heidelberg   SPIN 10636235

This book is dedicated to National Science Foundation program officers Andrew Molnar, Nora Sabelli, and Beverly Hunter, whose encouragement, guidance, and support has made possible the project work reported here. They have consistently promoted the development of computer simulation and modeling methodology for science education.

"The challenges before those of us interested in educational reform, and in educational and cognitive research, are to explore the pedagogical implications of this methodology that has revolutionized science, to adopt and modify it, to collaborate with scientists thinking about the science that will be taught in the 21st century, and experiment with how to best teach it to students with diverse learning styles, in such a way that the goal of general scientific literacy can be reached." (Nora Sabelli, 1994, Interactive Learning Environments, Vol. 4, No. 3, 195–198, Ablex Publishing Corporation, Norwood, NJ.)

Series Preface

The world consists of many complex systems, ranging from our own bodies to ecosystems to economic systems. Despite their diversity, complex systems have many structural and functional features in common that can be effectively simulated using powerful, user-friendly software. As a result, virtually anyone can explore the nature of complex systems and their dynamical behavior under a range of assumptions and conditions. This ability to model dynamic systems is already having a powerful influence on teaching and studying complexity.

The books is this series will promote this revolution in "systems thinking" by integrating skills of numeracy and techniques of dynamic modeling into a variety of disciplines. The unifying theme across the series will be the power and simplicity of the model-building process, and all books are designed to engage the reader in developing their own models for exploration of the dynamics of systems that are of interest to them.

Modeling Dynamic Systems does not endorse any particular modeling paradigm or software. Rather, the volumes in the series will emphasize simplicity of learning, expressive power, and the speed of execution as priorities that will facilitate deeper system understanding.

Matthias Ruth and Bruce Hannon

Preface

Computer-based modeling and simulation are becoming increasingly important in the learning and teaching of science because of three converging trends: the adoption of new standards for science education for all citizens, the explosion in scientific information and in its accessibility through computer networks, and the development of powerful modeling tools made possible by the increased computational and multimedia capabilities of our machines. The Science Education Standards of the National Research Council assign a key role for modeling in science education:

As they mature, students become increasingly able to understand invisible conceptual worlds of science, and to build more abstract understandings. They develop manipulative and cognitive skills that allow more complex experimentation and analysis of quantitative as well as qualitative data. In addition to their ability to identify patterns within such data, students become increasingly able to formulate explanations for phenomena in terms of models and theories, many of them mathematically grounded.[1]

The thesis of this book is that the introduction of computational modeling in the precollege science and mathematics curriculum has the potential to significantly improve the quality of education. However, there is a serious lag between these new capabilities and their effective educational implementation and use. This book considers the question: What issues need to be addressed in bringing modeling into the precollege curriculum? The book brings together in-depth discussions of these issues in the context of several major educational modeling projects from across the United States, as well as a major project from England. The accompanying CD-ROM includes most of the software modeling tools and applications described, together with associated documentation. Updates of some of the included software can be obtained from the various projects' websites, which can be accessed through links from the Springer-Verlag Internet site at http://www.springer-ny.com/biology/moddysys/.

---

[1] National Science Education Standards, Washington, DC: National Academy Press, 1996.

In many ways, this book is an outgrowth of two 3-day conferences on the topic "Setting a Research and Planning Agenda for Computer Modeling in the Precollege Curriculum," sponsored by the National Science Foundation (NSF Grant RED-9255877). The conference participants included precollege teachers; university faculty in mathematics, science, and engineering; software developers; educational researchers; and NSF program officers. The conferences had two major purposes: to significantly increase the level of communication among experts in the field and to provide information and guidance to policy makers, researchers, and educators on issues related to the use of models and simulations in schools.

The major conclusions of the conference participants were twofold: (1) From the earliest grades, children can be—and should be—engaged in the process of building, refining, using, and validating increasingly realistic models of natural and social phenomena. (2) Modeling ideas and activities should have a central role throughout the precollege science and mathematics curriculum.

The conferences were motivated by the fact that computer capabilities for representing knowledge, simulating complex phenomena, creating visually rich animated displays, and enabling people more readily to communicate and share information with each other and with their computer tools—provide enormous potential for modeling, in education as well as research. We are beginning to realize and demonstrate some of the educational benefits made possible by the thoughtful use of computer technology. Students can use computers to model systems with complex structures and behaviors in a variety of representations and at different levels of depth and detail. Models can be made transparent and accessible to students for exploration, study, and analysis. Students can use design tools to build their own models. They can access real data through computer sensors that monitor the physical world. Modeling tasks and projects can contribute greatly to motivating an understanding of the ideas and issues at the heart of learning and thinking within all the disciplines. Modeling activities can foster transformational changes in students' experience of science, mathematics, and other subjects, with dramatic learning.

We hope this book gives educators and policy makers valuable guidance in making thoughtful decisions on the use and benefits of modeling in their schools. And we hope it will contribute to the increased adoption of modeling ideas and activities in precollege science and mathematics curricula.

The editors wish to thank Nora Sabelli, Beverly Hunter, and John Richards for their thoughtful advice and guidance in planning the two conferences as well as for many hours, since then, of valuable conversation.

Wally Feurzeig and Nancy Roberts
Cambridge, Massachusetts
March 1999

Contents

Series Preface   vii
Preface   ix
Contributors   xiii
Introduction   xv

Part 1: Modeling Tools for Students and Teachers   **1**

1. *Modeling Clay for Thinking and Learning*
   Jon Ogborn   5

2. *Training System Modelers: The NSF CC-STADUS and CC-SUSTAIN Projects*
   Ron Zaraza and Diana M. Fisher   38

3. *Construction of Models to Promote Scientific Understanding*
   Michele Wisnudel Spitulnik, Joseph Krajcik, and Elliot Soloway   70

4. *A Visual Modeling Tool for Mathematics Experiment and Inquiry*
   Wallace Feurzeig   95

5. *Decentralized Modeling and Decentralized Thinking*
   Mitchel Resnick   114

6. *An Object-Based Modeling Tool for Science Inquiry*
   Eric K. Neumann, Wallace Feurzeig, and Peter Garik   138

Part 2: Model-Based Inquiry   **149**

7. *GasLab: An Extensible Modeling Toolkit for Connecting Micro- and Macro-properties of Gases*
   Uri Wilensky   151

8. *Designing Computer Models That Teach*
   Paul Horwitz   179

9. *Modeling as Inquiry Activity in School Science: What's the Point?*
William Barowy and Nancy Roberts   197

10. *Alternative Approaches to Using Modeling and Simulation Tools for Teaching Science*
Barbara Y. White and Christina V. Schwarz   226

Part 3: Toward Extended Modeling Environments   257

11. *The Developing Scientist as Craftsperson*
Michael Eisenberg and Ann Eisenberg   259

12. *Multisensory Immersion as a Modeling Environment for Learning Complex Scientific Concepts*
Chris Dede, Marilyn C. Salzman, R. Bowen Loftin, and Debra Sprague   282

Conclusion: Introducing Modeling into the Curriculum
Wallace Feurzeig and Nancy Roberts   320

Appendix A. Websites of Contributing Authors   323

Appendix B. Contents of Enclosed CD-ROM   324

Index   325

Contributors

William Barowy
Lesley College
29 Everett Street
Cambridge, MA 02138-2790

Chris Dede
Graduate School of Education
George Mason University
Fairfax, VA 22030

Ann Eisenberg
University of Colorado
Campus Box 430
Boulder, CO 80309

Michael Eisenberg
University of Colorado
Campus Box 430
Boulder, CO 80309

Wallace Feurzeig
BBN Technologies
GTE Internetworking
10 Moulton Street
Cambridge, MA 02138

Diana M. Fisher
NSF CC-STADUS
NSF CC-SUSTAIN Projects
Franklin High School
5405 SE Woodward St.
Portland, OR 97206

Peter Garik
Computer Science Department
Boston University
605 Commonwealth Avenue
Boston, MA 02215

Paul Horwitz
Senior Scientist
The Concord Consortium
37 Thoreau Street
Concord, MA 01742

Joseph Krajcik
1323 School of Education
University of Michigan
Ann Arbor, Michigan 48109-1259

R. Bowen Loftin
Virtual Environment Technology Lab
University of Houston
Houston, TX 77023

Eric K. Neumann
Director of Research
NetGenics, Inc., Suite 1550
1717 East Ninth Street
Cleveland, OH

Jon Ogborn
Professor of Science Education
University of Sussex
Falmer, East Sussex BN1 9RG
United Kingdom

Mitchel Resnick
MIT Media Laboratory
20 Ames Street
Cambridge, MA 02139-4307

Nancy Roberts
Lesley College
29 Everett Street
Cambridge, MA 02138-2790

Marilyn C. Salzman
US West Advanced Technologies
4001 Discovery Drive, Room 370
Boulder, CO 80303
mcsalzm@uswest.com

Christina V. Schwarz
4633 Tolman Hall, Emst
Graduate School of Education
University of California
Berkeley, CA 94720-1670

Michele Wisnudel Spitulnik
1015 Miner
Ann Arbor, MI 48103

Elliot Soloway
1101 Beal Ave.
106 Advanced Technology Lab
University of Michigan
Ann Arbor, MI 48109

Debra Sprague
Graduate School of Education
George Mason University
Fairfax, VA 22030

Barbara Y. White
4633 Tolman Hall, Emst
Graduate School of Education
University of California
Berkeley, CA 94720-1670

Uri Wilensky
Center for Connected Learning and
    Computer-Based Modeling
Curtis Hall, Room 1B
Tufts University
474 Boston Ave.
Medford, MA 02155

Ron Zaraza
NSF CC-STADUS
NSF CC-SUSTAIN Projects
Wilson High School
1151 SW Vermont
Portland, OR 97219

Introduction

From the time of Galileo until fairly recently, there were two complementary ways of doing science: experiment and theory. Now there is a third way. Computer modeling, a child of our time, is a powerful new paradigm, serving as a bridge between the traditional processes of experiment and theory and as a synergistic catalyst to both. The use of computer models as tools for furthering scientific knowledge is fast becoming a key component of current scientific research. We don't know what scientific discoveries will be made in the twenty-first century, but we can be sure that explorations and experiments with computer models will play a paramount role in making major advances possible.

The essence of computer modeling—mathematical experimentation—provides a powerful tool for connecting observed phenomena with underlying causal processes. Indeed, much of our understanding of the workings of the physical world stems from our ability to construct models of it. Models are particularly valuable mental tools, because in simplifying the complexities of the real world, they enable us to concentrate our attention on those aspects of it that are of greatest interest or significance. It has even been suggested that our ability to create, examine, and refine such models is crucial to our understanding of the world and that without this ability, we would literally be unable to "think" as humans.

Computer-based modeling is particularly powerful when it is linked to the ability of the computer to simulate the system modeled and to display its behavior via computer graphics. Modeling provides a central and fundamental tool for describing and exploring complex phenomena. Real-time interactive models with richly animated graphical displays—the same kinds of tools being used to great benefit in scientific research—can be made accessible for use by students. The models and modeling tools that students work with are typically a great deal simpler than those used by scientists, but the fundamental character of the modeling activity is the same.

Visualization is valuable for students for very much the same reasons that it is valuable for researchers. It enables them to observe and study complex processes as these processes are run and to "see" into phenomena that are not accessible to direct observation, thereby enhancing their comprehension of the underlying mechanisms. It can provide insight into the inner workings of a process—not just what happens, but also how and why. Visualization can greatly aid students in understanding the complex dynamic behavior of systems composed of interacting subsystems—in studying reaction-diffusion processes in chemistry, for example, or the dynamics of competition, predation, and adaptation in multispecies population ecology models.

The sciences we need to learn in the twenty-first century will involve extensive use of computational models. Acquiring fluency in model-based inquiry should become an essential goal of precollege science education. Yet the notions and art of modeling are seldom taught in school science or mathematics classrooms today. Experiments, such as those described in this book, by a small number of science educators have been conducted since the arrival of computers, but there has been essentially no significant impact in precollege classrooms. This situation, which has been frustrating to scientists and science educators who work with computer modeling in their laboratories and classrooms and know its great educational potential, was the driving force that led to the NSF-sponsored conferences and to this book.

A number of obstacles stand in the way of effective adoption and integration of modeling activities into the precollege curriculum. Several key issues that were articulated by participants in the two modeling conferences need to be clarified and resolved in order to guide ongoing modeling developments in a coherent and educationally productive fashion. In the following list, these issues are presented as "tensions" between closely related but distinctly different and ostensibly opposed perspectives. The *versus* in the phrases that describe them is not intended to imply an intractable duality. Rather, the contrasting views are highlighted to invite discussion about resolving differences or reconciling perspectives. The tensions identified were those between

- Modeling in science research versus modeling in science education.
  This issue concerns the differences between modeling by experts and modeling by novices, in particular between modeling by scientists and modeling by precollege students. Some participants claimed that, under the guidance of professionals, average high school students can use the same models and supercomputing facilities employed by research scientists. Others insisted that all but the brightest high school students need specially designed modeling tools and applications to introduce them to model-based inquiry methods.
- Learning to use models versus learning to design and build models.
  Participants differed as to whether (and how) students can learn to design and build their own models in addition to working with models provided to them. Some participants were convinced that students have a sufficiently difficult task learning how to conduct model-based investigations with models that are given to them. Others held the constructivist position that students learn the skills of model-based inquiry from engagement in the

process of building models and simulations—indeed, that the process of designing and building models is a natural part of the process of learning to use models as investigative tools.

- Computer-based modeling versus laboratory experimentation.

Computer models are obviously unreal. How is it, then, that they can be credible tools for capturing key aspects of reality? Participants agreed, of course, that computer models are not a substitute for observation of and experimentation with real phenomena. They also agreed that computer modeling adds a valuable new dimension to scientific inquiry and understanding. The key issue is the appropriate relationship between the activities of computer modeling and laboratory experiment, and how these two kinds of activities should be integrated in the science course.

- Visualization of model output behavior versus visualization of model structure and processes.

Computer modeling programs in science research applications typically employ visual representations of the model's behavior—animated displays of the outputs generated in the course of running the model. Some participants believed that this kind of visualization is sufficient for science education. Others argued that an additional kind of visualization facility is needed. In contrast to visualization of the outputs produced by the model, this tool would make possible the visualization of the model processes themselves—the model structures and algorithms—as they interact during the run. The issues here are: when is this kind of visualization useful in model development and model-based inquiry, and what are effective instructional strategies for its use.

This is a select sample, and not an exhaustive list. To understand the educational and technological problems surrounding the effective introduction of modeling in the precollege curriculum, it is crucial that we address these tensions and investigate the associated issues. These need to be resolved in order to inform and help provide direction to science educators, software developers, curriculum designers, and policy makers for the integration of modeling and simulations into the curriculum. In the following chapters, the authors address these issues in the context of their own work as they seek to convince people involved in precollege education that modeling can be a powerful catalyst for improving science education.

The first part of this book, *Modeling Tools for Students and Teachers*, describes several computer modeling languages and illustrates their use in a wide range of elementary and secondary classroom applications. World-Maker, Model-It, LinkIt, STELLA®, and StarLogo are generic modeling tools. WorldMaker and Model-It are useful for introducing modeling to beginning students in terms of qualitative rules describing their objects' actions and relationships. LinkIt, designed for "modeling without mathematics," represents variables and the "semi-quantitative" causal connections between them graphically, without the user's having to specify the form of the mathematical relations. STELLA®, which derives from the system dynamics modeling perspective, uses Stock and Flow diagrams to provide a graphical representation of differential equations. StarLogo makes it possible to model parallel processes that may involve large numbers of objects.

The tools described in these chapters exemplify a rich variety of modeling paradigms and designs. The computer can be a superb tool for constructing and investigating alternative models of real-world systems. Even though this pedagogical use has been the subject of several research projects during the past few years, two circumstances have made progress difficult: the lack of model development tools that are accessible to students with limited knowledge and skill in mathematics, and the lack of a compelling methodology for introducing students to the notions and art of modeling as a significant part of their science activity throughout the secondary curriculum. The authors describe tools and strategies to overcome these difficulties, show examples illustrating their use by students and teachers in a number of domains, and discuss the benefits of their modeling approaches in enhancing mathematics and science learning.

Model-Based Inquiry, the second part of the book, describes four exemplary projects investigating the introduction and integration of modeling ideas and activities into the secondary science curriculum. These chapters focus on tools that embody the knowledge of specific domains such as genetics, physics, and physiology. The authors capture the new capabilities made possible by computer modeling for supporting science inquiry, particularly modeling experiments and model-based reasoning. They discuss the role of model building by students and the interplay between computer modeling and physical experiment.

The last part of the book, *Toward Extended Modeling Environments*, describes two projects centered on the educational application of new and sophisticated modeling technologies. These chapters present two ends of a continuum: from providing science students with a facility for creating tangible and attractive objects generated by mathematical models, to inhabiting a virtual reality environment designed to provide students with the sensation of being present inside the science space of the model.

Part 1

Modeling Tools for Students and Teachers

Each chapter in this section describes a modeling tool designed for precollege students. The tools employ a variety of modeling representations and structures. Their specific capabilities are suited to different kinds of applications and levels of complexity. Despite the authors' differences in approach and focus, they share the conviction that appropriately structured activities, which give students substantial opportunities for designing and building models, can provide powerful science learning experiences.

Modeling Clay for Thinking and Learning describes two modeling tools developed at the University of London by author Jon Ogborn and his colleagues. These tools are designed to make computational modeling a part of curriculum experience for students from the age of about 10 years onward, without requiring conventional mathematical skills. One tool, WorldMaker, enables students to create a discrete model world by placing various types of objects on a grid of cells and specifying simple rules to describe the interactions of the objects. The model is then run to investigate the consequences of these rules over time. The other tool, LinkIt, goes from a world of interacting objects to a world of interacting variables. It enables students to express causal relationships among variables through causal linkage diagrams instead of mathematical equations. The author's research suggests that by developing models with these tools, students come to confront and treat important mathematical ideas and scientific issues. He argues that the activity of building computer models should be an integral part of the school mathematics course, from arithmetic through calculus.

Training System Modelers is a major initiative for developing materials to train teachers to integrate systems-oriented modeling into their curricula. Authors Ronald J. Zaraza and Diana M. Fisher use the STELLA® modeling language, which was developed for creating models based on the system dynamics perspective originated by Jay Forrester. STELLA® models comprise four types of components. Stocks store information or values. Flows change the values of stocks. Converters carry out logical or arithmetic operations. And connectors carry information from one component to another.

The models are developed as visual diagram structures connecting such elements. Models enable users to experiment with a system, study its behavior, and investigate the effect of changes in the model on the system's behavior. Zaraza and Fisher assert that teachers need to build models of the systems they teach to understand fully the dynamics of the systems and the theory underlying them. They have found is that when teachers start with simple models and incrementally embed appropriate new features, they internalize not only the process of model building but also the generalizations encompassed by the underlying theory. The chapter includes a number of teacher and student models across a wide range of applications.

Construction of Models to Promote Scientific Understanding describes the use of Model-It, a software tool designed to introduce beginning students to modeling. The key elements in the models developed with Model-It are objects, their characteristic properties (factors), and the relationships between factors. The use of Model-It is described in connection with student work in a unit on global climate. Student models include factors such as the amount of carbon dioxide emitted by cars, the level of ozone in the atmosphere, and the amount of pollutants emitted by factories. Relationships are typically qualitative statements connecting two factors—for example, as the level of chlorofluorocarbons in the atmosphere increases, the level of ozone in the atmosphere decreases. The chapter describes the classroom context of the iterative model development process, which is highly structured and includes explanation, demonstration, presentation, review, and evaluation. The authors, Michele Wisnudel Spitulnik, Joe Krajcik, and Elliot Soloway, are committed to teaching model construction as a central element in science learning.

In *A Visual Modeling Tool for Mathematics Experiment and Inquiry*, Wallace Feurzeig describes Function Machines, a visual computer language designed to facilitate student work in mathematical modeling. The language employs high-level representations of mathematical processes. A model's computational processes are represented visually as "machines" with inputs and outputs. Machines communicate data to each other graphically via pipes connecting the output of one to the input of another, in data-flow fashion. Students can view a model's inner workings as it runs. At the same time they can view the model's external behavior, the outputs generated by its operation. These dynamic visual representations significantly aid in the understanding of key computational concepts and make complex modeling programs more transparent and accessible. The chapter discusses recent work with Function Machines in mathematics classrooms and gives examples of its use in expressing parallel processes and in addressing complexity.

Decentralized Modeling and Decentralized Thinking discusses modeling activities that enable students to explore the workings of systems involving complex group behavior that seems to be centrally controlled (such as foraging in ant colonies, traffic jams, and the operation of market economies). The models are developed in StarLogo, a modeling environment designed to help students make a fundamental epistemological shift, moving beyond the "centralized mindset" to more decentralized ways of understanding complexity and making sense of the world around them. StarLogo facilitates the programming of massively parallel processes involving a large numbers of objects. Author Mitchel Resnick argues that the process of learning to design and build models will help students develop theories to improve their understanding of the world. He discusses a set of principles that, in his view, should guide applications of computer modeling in science education.

An Object-Based Modeling Tool for Science Inquiry describes OOTLs (Object-Object Transformation Language), a computational environment for modeling situations and phenomena that satisfy mass action laws. The OOTLs equation language describes "well-stirred" systems composed of large numbers of dynamically interacting objects. The many areas of application include epidemiology, population ecology (competition, predation, and adaptation), economics models, physics (gas kinetics), traffic flow, and chemical processes (reaction-diffusion equations). OOTLs provides students with a parser to construct equations describing interactions between symbolic objects shown visually as colored icons. The objects may represent chemical species, gas molecules, or humans. Objects interact with each other at specified rates. The OOTLs equations describe the transformations that result from the object interactions. The authors, Eric Neumann, Wallace Feurzeig, and Peter Garik, give examples of the application of OOTLs and illustrate its use in conjunction with laboratory experiments.

These six chapters present a rich variety of classroom uses of modeling and simulation to help students learn science and mathematics. Chapters 1 and 2 describe modeling tools based on animated causal diagrams. Ogborn introduces LinkIt, a tool that expresses relationships using "semi-quantitative" variables. It helps students gain insight into phenomena that involve causality and feedback. Zaraza and Fisher take their students into the quantitative world by introducing them to system dynamics ideas, using STELLA® as the modeling software. Work with Model-It, the subject of Chapter 3, moves in the opposite direction from STELLA®. It models relationships among objects by using qualitative descriptions. All three chapters present clear examples showing how a student's model reflects the student's level of understanding of a problem, where this understanding is incomplete, and how the student's understanding deepens through model building.

Chapters 4, 5, and 6 present distinctly different approaches to model representation and processing from those described above. In Function Machines, users express mathematical models as functions; these are represented visually as interconnected machine structures. By inspecting the machines as they run, students can better understand a model's structure, processes, and behavior and can more easily design, modify, and extend models. StarLogo postulates a decentralized object-based modeling world within which each object follows its own set of rules. The objects operate independently and, together, generate the observed group behavior as an intrinsically emergent consequence. OOTLs is another object modeling world. In OOTLs, events are conceptualized as interactions among the objects involved in the model processes. The interactions are expressed as rate equations. OOTLs is particularly well suited for modeling dynamic processes in terms of state transitions among the interacting objects. Function Machines, StarLogo, and OOTLs all support parallel processing. The chapters include examples of parallel modeling applications.

1

Modeling Clay for Thinking and Learning

Jon Ogborn

Need Only Experts Apply?

To start, here is a tale of how the obvious isn't always obvious.

It seems obvious and natural to us, because it is what we are all accustomed to, that the education of students in theoretical thinking in science has to progress through the following foot-dragging and painful sequence of stages:

1. Learn some arithmetic (in elementary school)
2. Learn, or fail to learn, some algebra (in high school)
3. Learn some calculus (maybe only at college)
4. Learn (or not) about finite-difference approximations to calculus
5. Use computational models of processes (maybe in graduate school)

A good proportion of the whole population drops out at the first stage, and only the very few who study mathematics, science, or engineering at university get to stage 3 and beyond. Mathematics is generally held to be too difficult for most, and mathematics with calculus and computers is often saved up for university and graduate school, where most of us never arrive. As a result, the population at large has no idea how computers guide spacecraft or predict the course of the economy. Worse, the limitations of computational models are hidden from public view; certainly, observers cannot distinguish those successes and failures of models that depend on the smartness of people from those that depend on the "smartness" of computers. Mathematics that most of us can't do is needed before one can make models with that mathematics. It all seems so blindingly obvious. *Only experts need apply*.

On the contrary, this writer believes that this alleged "natural progression" is very far from obvious and that much of it can actually be reversed. That is, students can use computers to make models and, through that process, learn some mathematics, rather than having to learn mathematics in order to get started. Thus my progression would look something like this:

1. Make some computer models   learn some arithmetic
2. Make some more computer models learn some algebra
3. Analyze some computer models learn some calculus

When one makes models, one begins to think mathematically. When one compares models and analyzes how they work, one can be led to deeper structures—for example, the calculus notion of a rate of change. It might even be that calculus would come to seem *easy*, just a collection of recipes for how things change: "If it's like this now, how will it be in a moment?" At the least, nearly everybody in the whole population would themselves have made a computer model of something and would have had a chance to grasp the seeming paradox that such models can be, all at the same time, wonderful, fascinating, productive of thought, overly simple, misleading, and maybe downright wrong. They might understand that models are mechanized thought, which, like thought itself, is highly valuable and deeply dangerous. A population that believes that thinking is too difficult for it is a population terribly at risk in any world, let alone in our technologically complex one.

The idea is that making models can help make mathematics. Mathematics is *not* needed to make models. However, making models *is* learning, in one special way, to "think mathematically"—in a way that many people *can* do. *Not only experts need apply*.

Such thoughts map out a long educational journey, perhaps too long to travel in a lifetime. But all journeys start with a few steps. And all those steps need to do is to head in the right direction. The foregoing argument is a compass to guide them. You, the reader, may well feel that the projected journey is absurd, ill-conceived, or dangerous. Bear with me. In this chapter I will suggest only a few, I think rather reasonable, first steps. And I will start, not with mathematics or computers, but with people and how people think.

Reflecting about how people think leads me to two propositions or hypotheses. They are the following:

1. *We need computer modeling systems that express their models in terms of objects and the actions of these objects on one another. Given such tools, children as young as 9 or 10 years old can make interesting models and begin to theorize for themselves in a way that many would characterize as "abstract" or "mathematical."*
2. *We need computer modeling systems that allow one to express relationships between things that can reflect "big" and "small," as well as "increasing" and "decreasing," but do not require one to write algebra. Given such tools, many students can think effectively about quite complex systems, even those that involve feedback, and can learn much regular mathematics from them.*

The rest of this chapter provides a basis for these proposals and suggests some first steps toward realizing them in practice.

How People Think

Of course, people *do* think, sometimes rather well. Oddly enough, many of us have been taught to think negatively of the way we think. To describe one crucial aspect of how people—naturally, fluently, and successfully—think, I will start in a perhaps unexpected place, with some rebellious American linguists.

Over recent years, the linguists George Lakoff and Mark Johnson have, through a succession of books and articles (Lakoff and Johnson, 1980, 1981; Lakoff, 1987; Johnson, 1987), shown the fundamentally metaphorical, concrete substance of everyday thought. Mark Johnson, in *The Body in the Mind*, shows how much of our natural spontaneous thinking is founded on imagined bodily action: an idea "strikes" us; an opportunity is "grasped"; a message is "got across"; a value is "tenaciously held"; a chance is "let slip"; a prospect is "in view"; a project "falls by the wayside." Independently, in his *Women, Fire and Dangerous Things*, George Lakoff displays whole systems of concrete metaphor underlying the ways we think: communication as piping through conduits; causation as action making movement or displacement; logic as putting things in containers. Together, in *Metaphors We Live By*, Lakoff and Johnson first set in motion such a program of understanding thinking. As linguists, they are regarded as rebellious because they deny—or ignore—the current consensus among most linguists that language is a matter of arbitrary syntactic structures. Instead, they focus on something else, the material out of which people make their meanings.

A First Pillar: Objects and Events

The key point is that people think through imagined objects and events. Curiously, Lakoff and Johnson never (so far as I can tell) mention a great Swiss predecessor, though one who worked in a very different field—I refer to Jean Piaget. Perhaps they are inclined to dismiss Piaget as the author of a rigid-seeming system of "stages of thought" from preoperational through concrete to formal. Perhaps they know of the many demonstrations that adults, including many university undergraduates, do not much use "formal"(that is, logical) schemes of reasoning. These results are widely, but in my opinion mistakenly, regarded as showing that something is terribly wrong with people's thinking. I believe they show instead something rather natural—indeed obvious: that Lakoff and Johnson are right about how people think.

Piaget's great merit, in my opinion, was that he was the first to characterize—and then to study in delicate and exhaustive detail—the mode of thinking that he called "concrete operational." It is, simply, thinking done using imagined objects and events. Piaget saw it as characteristic of children between about 5 and 15 years of age; later work indicates its fundamental importance in all adult thinking—work to which Lakoff and Johnson have signally contributed while seeing themselves as doing something else: understanding how language works.

Piaget's great failure, again in my opinion, was that he grotesquely overestimated the importance, and misunderstood the nature of another mode of thought that he called by "formal operations," by which Piaget meant reasoning detached from imagined concrete particulars. Like many other philosophers of this century, influenced and impressed by the growth of mathematical logic and its value in addressing questions of the foundations of mathematics, Piaget attributed far too much importance to logic—that is, to the security of reasoning independent of the content of that reasoning. That did not stop him from joking that he personally had never got beyond the stage of concrete reasoning. He just missed the fact that getting beyond that stage is not the point: It is not a stage but an *essential component* of how we all think. What we call formal reasoning is instead just a special and rather unusual case of concrete reasoning. It is "concrete reasoning with symbols." Mathematicians and logicians have become so accustomed to the way some systems of symbols work that they treat them as concrete objects. What are called rules or laws of reasoning are just expressions of what one can do to these symbolic objects and what the objects themselves can do or cannot do. Concrete thinking with imagined real objects and events is founded on exactly the same sort of knowledge: what things can do, can have done to them, and are made of.

Happily, this last formulation of the basis of the way people create meaning—what things can do, can have done to them, and are made of—was formulated by Piaget himself in a book (Piaget and Garcia, 1987) that appeared posthumously: *Towards a Theory of Signification*. It is a view with which Lakoff and Johnson ought cheerfully to concur.

Here, then, after an express whistle-stop tour of some of this century's thought, is a first pillar for understanding spontaneous human thought. It uses imagined objects and events, often metaphorically, basing ideas on what things can be envisaged as doing, having done to them, and being made of. That is, it rests on imagined *action*. It may not be necessary to add that action is what Piaget saw as the whole basis of intelligence.

A Second Pillar: Semi-Quantitative Reasoning

Pick up any newspaper and turn to the business pages. You will find things like this, in today's paper:

The evidence that no head of steam is building amongst manufacturers for large wage rises will be taken by the Government as another indication that inflationary pressures in the economy remain subdued.

The Guardian, February 17, 1997

You will also find "pressures raising prices," "share values falling due to lack of support," "steady drift" up or down, "continuing momentum" for change, and so on. These phrases reflect the way people naturally think about the relationships between one process and another—namely, that they go together or oppositely, or that one pushes another up or down. Reflect on other common-sense things one says: that students would do better if they were more motivated, that one could do more in one's job given more energy or less constraint, that lack of confidence inhibits achievement. Try thinking about gardening: "Rain makes the flowers grow faster." Try thinking about holidays: "lots of sunshine makes us cheerful." All fit the same mold.

The concreteness and basis in action of such thinking are evident. What is not quite so evident is the underlying structure as a kind of natural mathematics. It is based on a kind of variable, such as "amount of happiness" or "amount of stress." These "variables" are not, however, and usually cannot be, quantified—numbers and units cannot be attached to them (though economists sometimes try to do so, as in "liquidity preference"). But they do contain ordering of magnitudes; such a "quantity" is envisaged as being more or being less, as being large or small, as enormous, normal, or essentially nothing. Such "quantities" are also envisaged as changing—as rising or falling, quickly or slowly.

I shall call reasoning of this kind *semi-quantitative*. It has been studied under the label *qualitative reasoning* by a number of people in cognitive science who are interested in giving computers more "natural" modes of thought (Gentner and Stevens, 1983; de Kleer and Brown, 1985; Forbus, 1985; Hayes, 1985). Interestingly, it is far from being as vague as it seems. It obeys a rather well-defined calculus. It can be modeled as "quantities" that are all on essentially the same scale (small, medium, or large) and affect one another through links that determine the "sign" of the effect (the same or opposite in direction—"positive" or "negative"). The effects of one such "quantity" on another are of two broad kinds. One kind occurs when the effect is direct, in a sort of semi-quantitative proportionality: The wealthier you are, the happier you are. The other kind occurs when one "quantity" causes the other to change: The more you eat, the fatter you gradually get. This latter kind of effect is analogous to the calculus relationship between a quantity and what determines its rate of change.

This type of thinking is sufficiently prevalent and natural to have been used for many years, under the name causal-loop diagrams, in the initial design of dynamic models. Causal-loop diagrams are recommended, for example, as a strategy for devising models for the computer modeling system STELLA®, and they figure prominently in primers such as *Introduction to Computer Simulation*, by Nancy Roberts and others (Roberts *et al.*, 1983). Figure 1.1 shows an example.

Two Steps Forward

Following our instincts, and constructing the kinds of arguments above as we

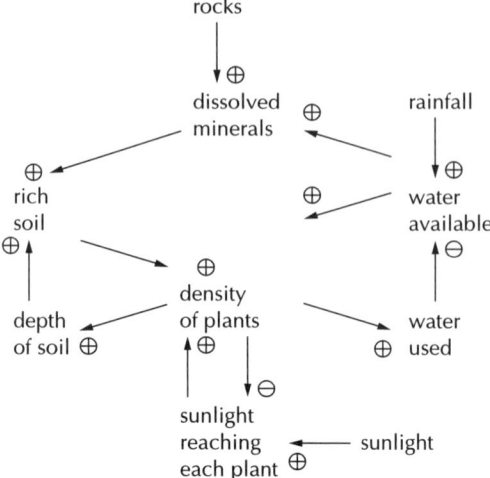

FIGURE 1.1 Forest Development. Adapted from Roberts *et al.*, *Introduction to Computer Simulation*, p. 53.

went along (sometimes to rationalize what we had done and sometimes to think about where to go next), we have, in our research group at the University of London Institute of Education in collaboration with others, constructed two prototypes of computer modeling systems that we believe represent steps forward in using computers to help people think. Both could be called "modeling minus math." But both could also be called "modeling making math."

The first we grandiosely call WorldMaker (Boohan, 1994). It addresses the problem of a modeling system organized around objects and their actions on one another, the first of the themes of natural reasoning discussed above. The second is more prosaically called LinkIt (see Mellar *et al.*, 1994; Kurtz dos Santos, 1995; and Sampaio, 1996). It provides for animated causal-loop diagrams like Figure 1.1, of arbitrary size and complexity, to be drawn on the screen and run to see what happens.

These are only first steps. Both systems have important limitations, which I shall describe insofar as I can see them. And both point to further developments and to problems of incorporating such ideas within the school curriculum. There is much left to do.

WorldMaker: Modeling with Objects and Rules

In WorldMaker, objects interact with one another and with the places where they live, much as things do in the real world. But in WorldMaker, you, the user, create all the objects and places, put them where you like, and tell them

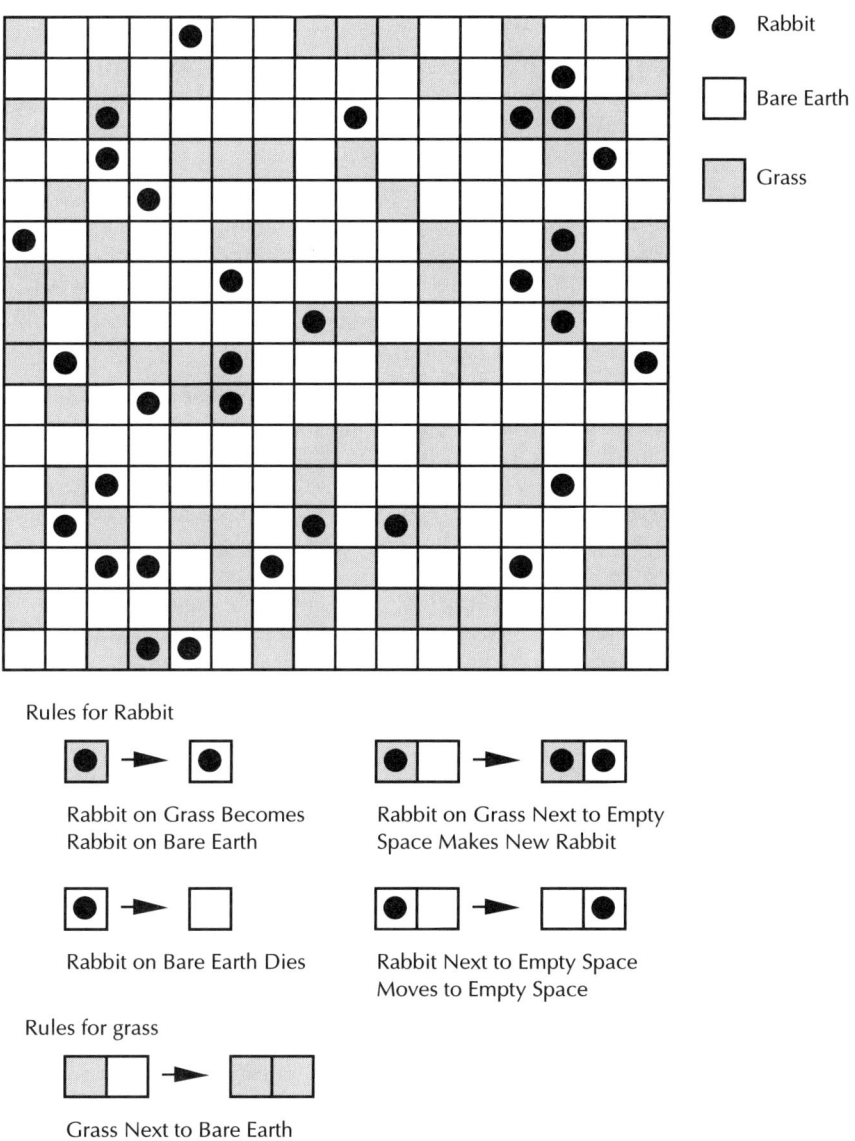

FIGURE 1.2 Rabbits Eat Grass and Breed or Die.

what they can do or have done to them. You might create "people" entering and leaving a supermarket and standing in line at the "checkouts." You might make "beta particles" that are absorbed differently by different materials. You might make "molecules" that move at random. You might make "farmers" who plant "vegetables" and "pests" that eat them. You might make rabbits that eat grass and breed, as in Figure 1.2. Having made such things and told them what they can do, you place them on the screen and set them in action. The result is seldom exactly what you expect and often teaches you something about how the real world works.

Two Students Use WorldMaker

Before describing WorldMaker in detail, let's take a brief look at it in use. The following vignette comes from Nancy Law, who observed two students and their teacher using the system at the University of Hong Kong. They are making a model much like that of Figure 1.2.

Two 17-year-old Hong Kong students decided to use WorldMaker to build an ecological system. They started with a very simple world: an ecology that contains only rabbits and grass. Rabbits eat grass, turning it to bare earth. If they eat they may reproduce, but they may die if there is no grass to eat. Grass can regenerate from bare earth if there is grass nearby. In addition to observing this simple world develop visually as objects on the world grid, they used the built-in graphing module to monitor the changes in the total number of rabbits. They were surprised to see that the total number of rabbits oscillated with a rather stable periodicity and wanted to account for such behaviour. After thinking for a while, one of them said, "Well, this is probably because the rabbits hibernate in winter, so the numbers become smaller in winter." Another said, "The amount of grass would be smaller in winter, so there would be less food." The teacher working with the pair was surprised at their interpretation and queried whether in fact rabbits do hibernate in winter. It is interesting that up to this point, both the students and the teacher were trying to interpret the behaviour of the model not as a consequence of the formal system they had built, but in the context of the actual ecological world that they wanted to model. This kind of response to modeling outcomes is in fact rather common and is frequently reported in other studies. What is interesting in this case is that the students very soon decided that their conjecture was not valid: "We did not set any rules relating to the effect of the seasons, so it cannot be right." One of the students then remembered learning about similar population oscillations in connection with predators and prey. It appears that the transparency of the model helped these students to bridge the gap between the actual world of objects and events that they are familiar with, and the abstract world of models as formal systems. They had also begun to compare models at the level of structure.

From then on, the students' behaviour noticeably changed: they decided to change the icon they used to stand for rabbits from one which looked like a rabbit but did not stand out well against the background to a bright red car which stood out sharply against the various background colours used. The objects and backgrounds had become for them only symbols to be manipulated. Accompanying this change, there was also a clearer focus in their modeling activity: they wanted to look for conditions of ecological equilibrium and ways of minimizing the oscillations they observed. They started by changing the probability settings for the different rules, then tried to add in predators to the ecological system. Cognitively for these students, the model as a representation had moved from a concrete level to an abstract one, thereby laying the groundwork for theorizing to take place.

Nancy Law, University of Hong Kong

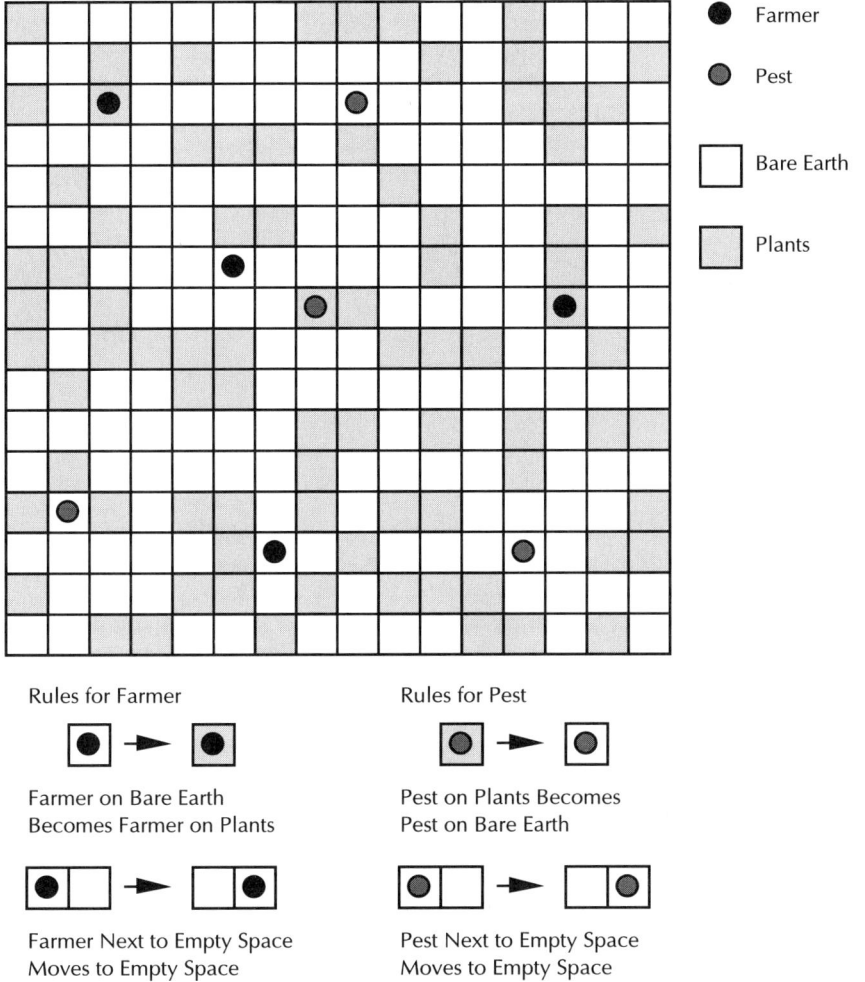

FIGURE 1.3 Farmers and Pests.

WorldMaker with Young Children

We designed WorldMaker to make it accessible to children at as early an age as possible. The following results come from Eleni Maragoudaki (1996), who first showed that it could be used with 9- or 10-year-olds. She used very simple models, including one like Figure 1.3, in which a "farmer" runs around the screen planting "vegetables" on "bare earth," and a "pest" runs around the screen eating the "vegetables."

The question was, whether these young children could understand and follow the operation of the rules? The answer was that they could learn to do so quite quickly, especially if they had a part in making the rules. Could they explain what happens in different circumstances—that with only farmers, the screen fills with plants and that with only pests, it fills with bare earth? They could. Thus we have examples of necessary consequences (that is, mathematical/logical features) that these young children could grasp. They found it harder to see the equilibrium between plants and bare earth with both farmers and pests present. And we noticed the same effect that Nancy Law mentioned in the previous instance: The children were quite likely to import into their explanations things that they knew about the real world but that were not actually represented in the rules. In other words, they set very loose limits on what they supposed the computer might be doing. Of course, these youngsters had yet to learn the idea of a programmed system that does only what its rules say and nothing else. Here they made a start at grasping this idea—one that Nancy Law's older students still had to work out for themselves.

General Description of WorldMaker

Objects in WorldMaker live on a grid of cells on the computer screen, as in Figure 1.2. Each cell can take only one object, but each cell can also have a given background that may affect or be affected by the object on it. A "rabbit" could, for example, change the background it was on from "grass" to "bare earth," simulating rabbits eating grass. Backgrounds can also be affected by neighboring backgrounds. Thus in this example "bare earth" could regenerate to "grass" if there was "grass" adjacent to it.

Objects can move, as long as there is an adjacent empty cell in which to move. Two kinds of motion are possible. In one, an object selects at random from any empty cells next to it and jumps to one of them. In the other, an object has been given or has acquired a "direction" and moves into an empty cell if its direction points to that cell. This second kind of motion allows objects to be steered in motion on the screen—for example, on a background track laid down on the screen or by a background acting like gravity. Colliding objects can be told to exchange directions, so that they bounce off one another or off a wall.

The inspiration for WorldMaker is the cell automaton invented by John von Neumann (for modeling and simulation with cell automata, see Toffoli and Margolus, 1987). But it differs from a cell automaton in having objects whose identity persists as they move around the screen (in a cell automaton, only the cells "exist"). A similar idea has been exploited by Christopher Langton in a system called VANTS (Virtual ANTS).

Objects and backgrounds are told how to behave and interact with rules, all in pictorial form. The rule-pictures show the condition for something to happen—for example, a rabbit on grass or earth next to grass—and they show the outcome—grass changed to earth or earth to grass, in this case. Rules can specify movements of objects, changes to objects (including creating and destroying them), and changes to backgrounds.

There is one absolutely fundamental restricting principle underlying all the rules, and it is that of local action. Objects can affect or be affected only by others in a cell next to them; backgrounds can be affected only by backgrounds next to them; objects can affect only backgrounds they are sitting on; and backgrounds can affect only objects sitting on them.

The icons for objects and backgrounds can be chosen from an available collection, or they can be created or edited by the user via a bitmap editor. The modeler selects among the objects and backgrounds available for use in a given model, adding or removing them from the model as required. In effect, an icon becomes an object or background by having rules given to it. "Worlds"—that is, models—can of course be saved complete with objects, backgrounds, rules, and a screen display of the model in a given state.

Objects and backgrounds are easily and quickly placed as desired on the grid of cells, using simple "paint" tools (pen, fill, outline fill, and random fill). They are erased by using a permanent "empty object" or "blank background."

When the model world is set running, the objects and backgrounds interact according to the rules. The world can be set to run rapidly, slowly, or step by step, so that one can either see a final result quickly or follow carefully just how it arises. The probability that any rule will "fire" when its condition is met can be varied from 100% (always) to 0% (never). Thus the models can be deterministic or probabilistic.

All the objects and backgrounds of a given kind (rabbit, or grass) behave identically in the same circumstances. But of course in a given model, they often find themselves in different circumstances, so that different rules apply or not. For example, some rabbits are on grass with empty cells around them, so they eat and breed. Others are on bare earth far from grass, and they die, while yet others are crowded by other rabbits and have no room to breed. In this way, a few simple objects and backgrounds, with a few simple rules, can yield quite varied and complex behavior.

A graphical plotting module makes it possible to show dynamic scrolling graphs of the total numbers of different objects or backgrounds present on the screen grid.

Examples of WorldMaker Models in Science

Biology and Ecology

Many biological and ecological systems lend themselves naturally to modeling with WorldMaker. One of the simplest examples is the behavior of simple pond organisms, which tend to cluster in nutrient-rich regions and not in nutrient-poor regions, as in Figure 1.4. All that is needed to model this is an object to represent the pond organism and two rules to make it move. If it is on a "nutrient-poor" background, it "jumps" completely randomly to a nearby empty cell, and it does so often (the rule probability is set high). If it is on a "nutrient-rich" background it also jumps at random to a nearby cell, but it does so only rarely (the rule probability is set low). With only these re-

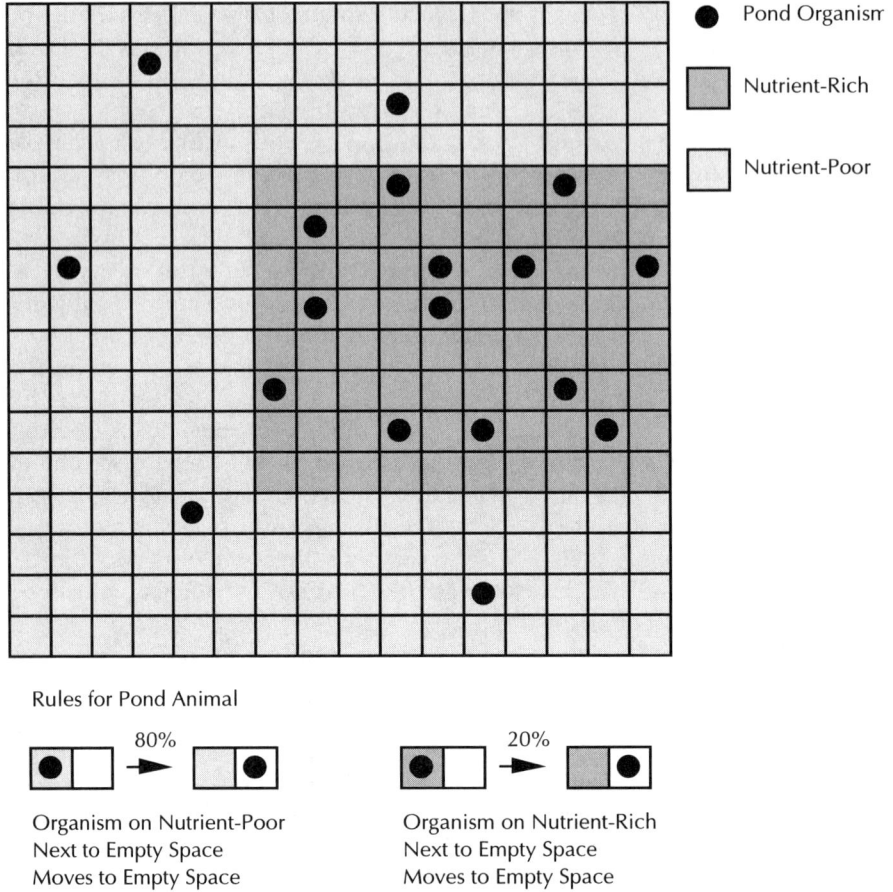

Rules for Pond Animal

Organism on Nutrient-Poor
Next to Empty Space
Moves to Empty Space

Organism on Nutrient-Rich
Next to Empty Space
Moves to Empty Space

FIGURE 1.4 Pond Life.

sources, the behavior of the model is striking. If a block of "nutrient-rich" background is painted anywhere on the screen and the rest is painted "nutrient-poor," the pond organisms very soon congregate mainly in the "nutrient-rich" region. Outside it they skitter about rapidly, but when they happen upon it, by chance they slow down and so have less chance of leaving it again. It's quite a sight to let them settle down—and then to move the "nutrient-rich" region elsewhere and watch them "find" it again.

This model and others like it carry some important biological messages. First, organisms need not have purpose or foresight to behave as though they did. A model of ants forming a line to a morsel of food could deliver the same message. Second, and more generally, purely local rules (acting only between neighboring cells or, in the real world, perhaps only on contact) can have globally patterned consequences. Localized "blind" actions can have global "far-seeing" effects. Let's not forget that our immune systems work like this too.

The rabbits and grass model of Figure 1.2 is another example of local rules (or, in the real world, actions) that have global consequences, in the form of population oscillations. Another, more important global effect is that of dynamic equilibrium. The model of Figure 1.3, which includes a farmer who plants vegetables that are eaten by pests, always arrives after some time at the same ratio between lettuces and bare soil, no matter how much of each we start with. The proportions of each are—with fluctuations—the same; they depend only on the numbers of farmers and pests. But the screen almost never looks exactly the same twice. Nothing decides *where* there will be lettuces or bare soil at a given moment, but the interaction ensures a fixed proportion between their numbers.

A number of "obvious" consequences of these models are, nevertheless, worth showing. If rabbits breed more often than they die, the screen will inevitably fill up with rabbits until no more can be added. Population explosions are a necessary consequence of breeding without checks. A subtler case is that of two competing breeding species each of which kills the other. In the long run—and it may be a very long run—only a single species will be left, but we cannot generally tell in advance which one it will be.

Chemical Change

The difficult idea of dynamic chemical equilibrium is easy to model and illustrate. Take the example of water (H_2O) dissociating into hydrogen ions (H^+) and hydroxyl ions (OH^-). All three objects are given a rule to make them jump at random around the screen. As shown in Figure 1.5, the water molecules are given a rule that converts them into a pair of ions side by side. The two ions follow a rule that if they come together, they can turn into a water molecule.

Starting with only water molecules, there is a net dissociation until ions recombine as often, on average, as they form. The graphs of numbers of particles show that a fluctuating equilibrium has been reached. This model does not include the energy extracted from the surroundings when a water molecule splits up or the energy released when two ions combine, but its effect can be modeled by setting the probabilities of the dissociation and association reactions differently. If water rarely dissociates, but ions readily recombine, we get a balance with few ions, as is indeed the case in real water.

Subtler than this, and very striking, is the case of auto-catalysis. Sequences of reactions can be modeled in which an intermediate species from one reaction is necessary for a later reaction. In this case, the whole reaction proceeds slowly until a number of molecules of the intermediate species have been produced, at which point it dramatically speeds up.

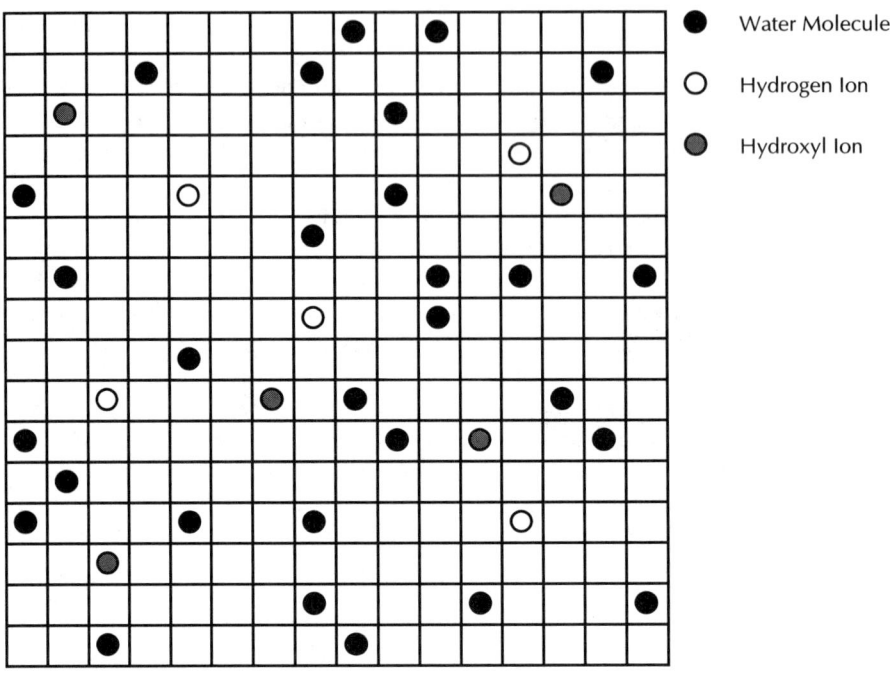

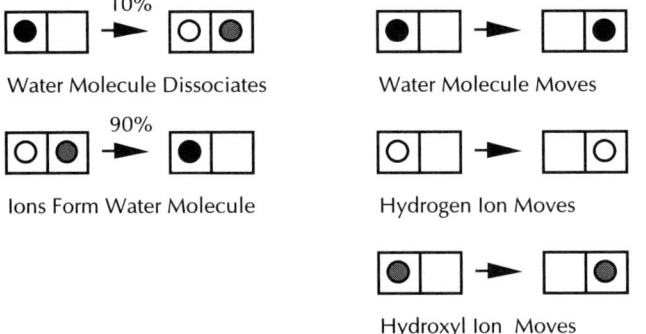

Rules for Water Molecules and Ions

FIGURE 1.5 Simple Chemical Equilibrium.

Physical Changes

Diffusion of randomly moving particles from more to less concentrated regions is very easy to model; it requires just a random jumping rule for a "molecule" and an object to make walls. It is amusing to write one's name in molecules, as IBM is reputed to have done, and watch it dissolve into chaos—not forgetting the common root of the term 'gas' and the Greek word 'chaos'.

The Earth's atmosphere thins out as one goes up a mountain because gravity pulls the molecules of the air down, but this produces a concentration gradient away from the Earth, and the molecules diffuse up this gradient. The actual gradient of pressure (density, concentration) in the atmosphere is a balance between the two. To model it, an "air molecule" needs two rules: one to make it fall and one to make it jump randomly in any direction. The rule for making it fall can be produced by creating a "gravitational field" background and painting that background over the screen with a downward direction. The rule for the molecules is just to acquire this direction and then move along it. If one starts with all the molecules near the ground, they drift upward. If one starts with all of them near the top, they sink down. If one starts with them evenly distributed in height, a graded distribution soon develops. The air can be "cooled" by reducing the probability of random jumping, and the molecules drift down in response, reaching a steeper grading.

Good fun can be had modeling evaporation and rainfall over oceans. A layer of "sea" is needed near the bottom of the screen, with the property of emitting and absorbing vapor molecules. Vapor molecules just jump around at random. The top of the screen can be covered with a "cold" background on which vapor molecules combine to make raindrops. Raindrops fall, and they sweep up vapor molecules near them. With these resources, one can create one's own miniature storms at sea.

Crystallization from a vapor around seed particles on a cold surface is what makes the beautiful frosted patterns on the window on cold winter mornings. Model it with a vapor of randomly moving molecules, adding one stationary "seed." Give the seed the single rule that if a vapor molecule chances to land near it, the vapor molecule turns into a seed particle. What you get is a fractal crystal growing out from the seed, with spidery arms that leave tortuous tunnels between them. This is called diffusion-limited crystallization; the spidery fractal forms because there is a much smaller chance that a vapor molecule will diffuse randomly into the center of the crystal than that it will land by chance near the tip of a branch. Thus branches lengthen, and new small branches keep starting off near the tip of an existing one. Here yet again is a global effect—and one connected to much recent mathematical thinking—arising from a purely local behavior.

Thermal equilibrium—the spontaneous passage of energy from hot to cold—is worth modeling. A simple (if not very realistic) way to do it invokes an idea that Peter Atkins advances in his book *The Second Law*. Pretend that atoms in a solid can have just two energy states, high and low. Make an object for each, and give them the rule that high-energy atoms and low-energy atoms next to one another just exchange energy—high becomes low, and vice versa. There is no need for them to move. With this rule, "energy" from a block of high-energy atoms placed next to a block of low-energy atoms, as in Figure 1.6, will "diffuse" throughout the whole block. Temperature is represented here by the ratio of numbers of high-energy and low-energy particles in a given region. Greater ingenuity is needed to create particles with a whole set of energy levels.

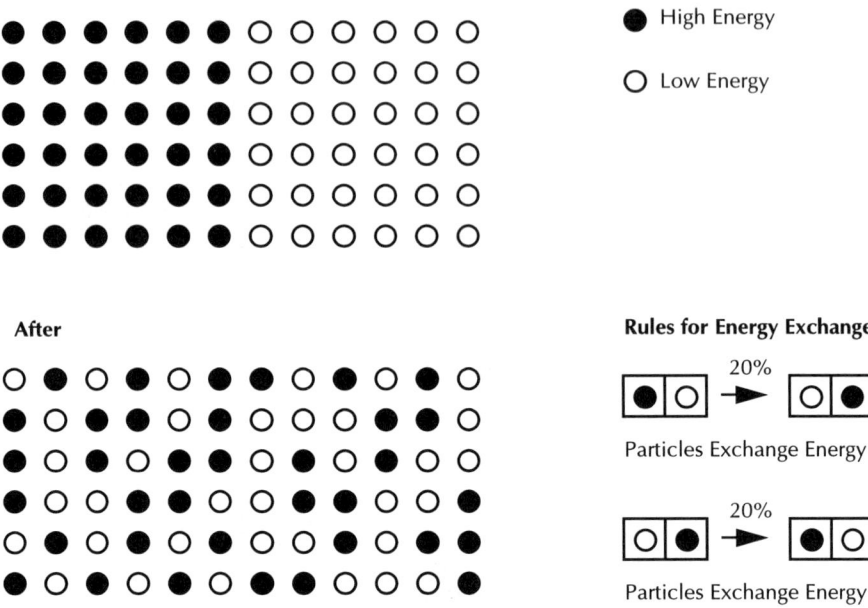

FIGURE 1.6 Simple Model of Thermal Equilibrium.

Radioactive decay chains are rather easy to model. A simple one could use the sequence of rules

$$A \to B$$
$$B \to C$$

with different probabilities assigned to each rule. A screen starting full of the A nuclide builds up a population of B nuclides, which then decay into C. The number of A's declines exponentially; the number of B's rises and then falls; the number of C's rises, slowly at first, then quickly, and then slows down and stops as it reaches a maximum. The relevant differential equations are moderately hard to write and solve; the WorldMaker model is almost trivial to create and easy to understand. Theorems such as that the total number of particles is constant and that the sum of the slopes of the three curves must always be zero, so that the slopes of the A and C curves are equal and opposite where the B curve is a maximum, can be intuited.

Abstract Topics

Mathematicians who have played with cell automata and related systems, in the field of complexity theory, have discovered a range of fascinating phenomena that involve chaotic and self-organizing behavior, much of which is amenable to investigation (and play) with WorldMaker. We will give one example here.

Convert your Neighbor   Be Converted by your Neighbor

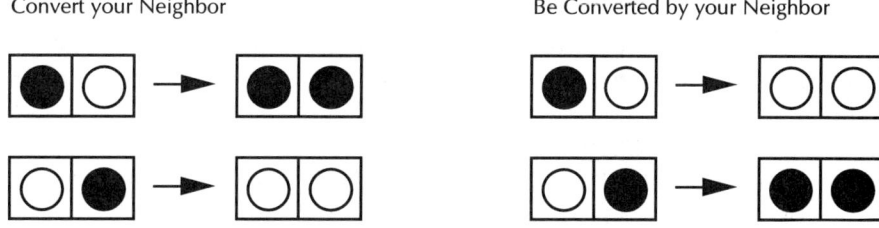

FIGURE 1.7 Ideological Conversion Rules.

Let the problem be called the ideology problem. Suppose the world contains people of two ideologies (political perspectives? religions? languages?) whom we will call greens and reds. Greens try to convert reds to green. Reds try to convert greens to red. There are two ways to express the rules for conversion, as seen in Figure 1.7. They appear very similar but have drastically different effects. Note that both pairs of rules are exactly symmetrical.

If we fill the screen with a random mixture of greens and reds and use the rules whereby each converts the other, with identical probabilities (say 50%), the screen presents an astonishing sight. The random mixture very rapidly "crystallizes" into solid regions that are either wholly red or wholly green. At the boundaries between the two, changes occur and the areas of red and green grow or shrink. In time, either red or green wins completely, but one can never tell in advance which it is going to be. No doubt, if the winners then write history, they will explain how their chance victory was inevitable!

After we watch for a while, the mechanism becomes clear. Any projecting bit of either territory will be lost, because individuals in it are surrounded by more of the opposite kind than of their own kind and hence will more often be converted than get a chance to convert. Straight-sided strips of territory are the least unstable, because excursions or incursions tend to get eaten away. The worst case is that of an island in the middle of hostile territory. This, then, is what we get when the rules make each individual convert others near it to its own cause.

In the second pair of rules, individuals near others of a different kind get converted to that kind, instead of converting them. (See Fig. 1.7) These rules work very differently. Now solid blocks of the two kinds of territory rapidly fragment into a seemingly random pattern, mixing reds and greens close together. This is because, now, any little lack of straightness in a boundary tends to grow: A protrusion of one region into another gives individuals in the invaded region more chances to be converted than there are chances for the invaders to be converted themselves. Again, it just depends on the fact that in the region of a spike of invaders, there are more invaded than invading. Consequently, the pattern on the screen grows as spiky and fragmented as possible, whereas before, it spilled into uniform regions with rather smooth edges.

Now we arrive at the exciting part. What if we run *both sets of rules to-*

gether, all with the same probability? The result is a continually swirling and changing collection of regions of red and green, with constantly shifting boundaries. The system is, as it is said, "on the edge of chaos." Anything can happen.

The model is abstract, but it suggests ways to think about some social systems. Languages may tend to become uniform in small regions because speakers of one language are surrounded by too many speakers of another, who convert them. But languages also fragment, and according to the model, they will tend to do so if people are willing to go along with local innovations of a minority, such as new slang expressions (that is, they are willing to "be converted"). If innovators have a susceptible audience, languages will fragment. If deviants have to conform in order to be understood, languages will tend to cohere across territories. Because both rules probably work at the same time, we can expect languages to be both stable and ever-changing simultaneously. Which is just what they seem to be.

Limitations of WorldMaker

WorldMaker is evidently at its best for studying populations of objects and places, each individual instance of which obeys the same set of fixed rules. Thus it is good for collections of people, animals, plants, molecules, and—perhaps—stars. It completely excludes "action at a distance," so one object can never affect another some distance away from it. Thus it cannot model particles that interact through force fields, for example. A kind of "field" can be established by using the places on which objects move, but its possibilities are limited, though it is easy to use the directions given to background places to make objects tend to move systematically—up or down, for example, or toward or away from some point. But WorldMaker is not adapted to modeling, for example, planets going around stars or systems of stars interacting.

In WorldMaker's present version, objects move only one cell at a time in one time step. Thus there is built into the system a maximum possible speed ('the speed of light', in effect). Newtonian motion cannot be simulated with any exactness; the only way to make an object move more slowly is to reduce the probability of its moving at all, a property that cannot be altered by its interaction with other things. Some features of regular Newtonian types of movement can be modeled, however. If objects are set to move along directions they are given, it is easy to model collisions by having objects that meet exchange their directions of motion. To our initial surprise we found that this gave us "for free" the effect that when one object collides with a stationary row of similar objects, the "direction swap" is passed down the row, and the last object in the row moves off, just as in "Newton's Cradle."

The present system lacks some further features that it may be possible to add in the future. We would like each object to have a clock that counts time steps, so that its rules could depend on the time and thus "older" objects could behave differently from "younger" ones. We would also like an event counter, so that (for example) a rabbit might die only after finding itself on bare earth a number of times in succession. It would also be good if objects could have different "states" that obey slightly different rules. In such a case, a female rabbit might become pregnant after meeting a male, and only female rabbits might give birth (compare the sexless reproduction of Figure 1.2!). Several such improvements, together with others, could be achieved by making the system more "object-oriented" than it is, with generic objects specialized into species each with subspecies, and so on. Rules would then exist at various levels, with (for example) movement rules being rather generic and interaction rules more specific.

Results and Possibilities

Initially, we tried out WorldMaker mainly with rather young students, from 9 or 10 years up to 11 or 12 years, because we were interested in creating a way of making models they could use. Only now are we beginning trials with older students, who range from 15 to 18 years old.

The results with younger students show that they begin to learn to use WorldMaker within half an hour or so but that learning all it can do takes a considerably longer time. The idea of making a model by giving objects rules turns out to be easy to understand, and students rapidly learn to make and interpret rules in pictorial form.

In making models with WorldMaker, both younger and older students sometimes have some trouble deciding which elementary actions available in the system to use to make a model. Where an effect (such as rabbits breeding if they meet and are well fed) has to be made out of several rules, the way forward is often not clear. Simple, direct rules that correspond well to the way the actual situation is imagined (an animal affecting the place it is on; one object affecting another near it) are much easier to arrive at.

We have very often noticed, both in young students and in older ones, a tendency to have an explosion of ideas rather than to run short of them. People like to "improve" a model by adding more and more complexity to it. The idea seems to be that the more different aspects of behavior are taken into account, the more "realistic" the model will be. Readers with some sophistication in modeling will recognize this as a mistake. It is often the case that adding complications adds little or nothing to the overall behavior of a model. Predator and prey populations oscillate in accordance with very simple rules, and the oscillation persists when complexities are added; for example, it matters little whether "reproduction" involves two sexes. This, however, need not be thought of as a problem. That many models produce their behavior from a simple structural core is a lesson to be learned about what makes mathematics valuable—that necessary relationships are generally related to deep structural features.

There is an important aspect of looking "mathematically" at WorldMaker that even the younger students begin to manage. It is easy to have two models that have the same rules and differ only in the names of objects and places. Abstractly, they are the same model, and students can see that they must behave in essentially the same way. This understanding seems to be enhanced by the pictorial expression of rules, in which patterns of similarity are easy to see. Further evidence of such abstract thinking comes from showing students forms of rules without named objects and then asking them to suggest what the rules might describe, which they can quite often successfully do.

It is, however, difficult for younger (and sometimes older) students to grasp that a WorldMaker model "knows" nothing more than the rules it has been given. Young children are quite likely to say that WorldMaker could model a world in which sharks eat fish but to deny that it could model fish eating sharks: "It should know that that is impossible." Such students quite often import known aspects of the real world to explain what happens. For example, if "rabbits" in a model go extinct in interacting with a predator, they may suggest that a disease has killed them. Again, this misconception highlights something useful and general that students need to learn about modeling—that a model contains no more than what has been put into it.

A key issue is thus that of *simplification*. The essential virtue of models is that they represent stripped-down versions of reality, in which only those features essential to producing the effects of interest are retained. This is a lesson that students are slow to learn, because learning it requires a good deal of experience with models. All the more reason for them to get started as soon as possible.

Why Do Buses so Often Come in Threes?

To conclude, here is an example of "mathematical" thinking that we have tried successfully with young students. In cities with public bus services, it is a common complaint that when one waits at the bus stop for a particular bus, one often waits for a long time—after which three buses arrive together. Is there anything that makes this likely?

A simple model is easy to make. A closed loop of "road" carrying directions is drawn on the screen, and buses subject to a rule that makes them move around the road are put on it. They, of course, travel without getting closer together or farther apart. Now we introduce a place from which "people" emerge and a path along which they go to the road. An additional rule now says that if a bus is next to a person, it does not move but instead picks up the person. If people emerge to catch the buses rather infrequently, the buses stay apart, but if many people wait for them, the first bus is held up for several time steps, and the buses behind it catch up. Soon the buses are clustered together on the road.

The problem has been stripped down to essentials. The essentials are that buses move if they can, that buses don't pass one another, and that buses have to wait while there are people there to board. These principles are all that is needed to produce the phenomenon that so many have observed. The natural desire of people to get on the first bus that shows up is enough to produce the undesired effect of bunching the buses. Such simple necessary relations are the essence of mathematical thinking, even if many would not recognize this as "mathematics."

LinkIt: Modeling with Variables and Links

LinkIt takes us from a world made of interacting *objects* to a world made of interacting *variables*. In this new world—a world essential to the scientific imagination—rabbits are replaced by population density, births by fertility rates. In the physical world, an automobile travels from place to place; in this world of variables, it is described by its displacement, velocity, and acceleration.

Such variables are often thought of as affecting one another. The fertility rate increases the population; the death rate diminishes it. The amount of sunlight reaching the earth helps determine its temperature. The depth and volume of water in a bucket increase and decrease together, as do the pressure and temperature of a gas at a constant volume. LinkIt provides for variables to be defined and for such relationships to be established between them. It is not, however, necessary for these relationships to be defined by way of a formula, as one might expect. All one has to do is to draw—and modify the meaning of—links between variables.

A LinkIt Example—the Rain Forest

We can see something of how this works by showing how the rain forest system we first encountered in Figure 1.1 (it is taken from a primer on system dynamics) comes out when translated into LinkIt. Figure 1.8 shows one such translation. Variables are placed on the screen using the mouse, after the box-like variable icon is selected from the icon menu at the top. Links are made by selecting the link-like icon at the left and then clicking, in order, on the variable from which the link is to go and on the variable to which it is to go. Double-clicking on variables or links opens up dialogue boxes in which various parameters are set.

Figure 1.8 has exactly the same "variables" as Figure 1.1, and there are links in exactly the same places. The notation is a little different. In place of "plus" signs by arrows in Figure 1.1, such links carry a symbol with two small arrows pointing in the same direction. This symbol says, "These two variables go together; if the first is large, the other will be influenced to be large." Thus for example, the more the "rainfall" the more the "water available." Other links in Figure 1.8 carry symbols with a pair of opposed arrows. This symbol says, "These variables go oppositely; the more of the first there is, the less of the second there will be." An example is the link from "water used" to "water available." "Water available" is affected by both "rainfall" and "water used" (its value moves with that of the former and in the direction opposite to that of the latter). Therefore, the value it gets must take into account both of these

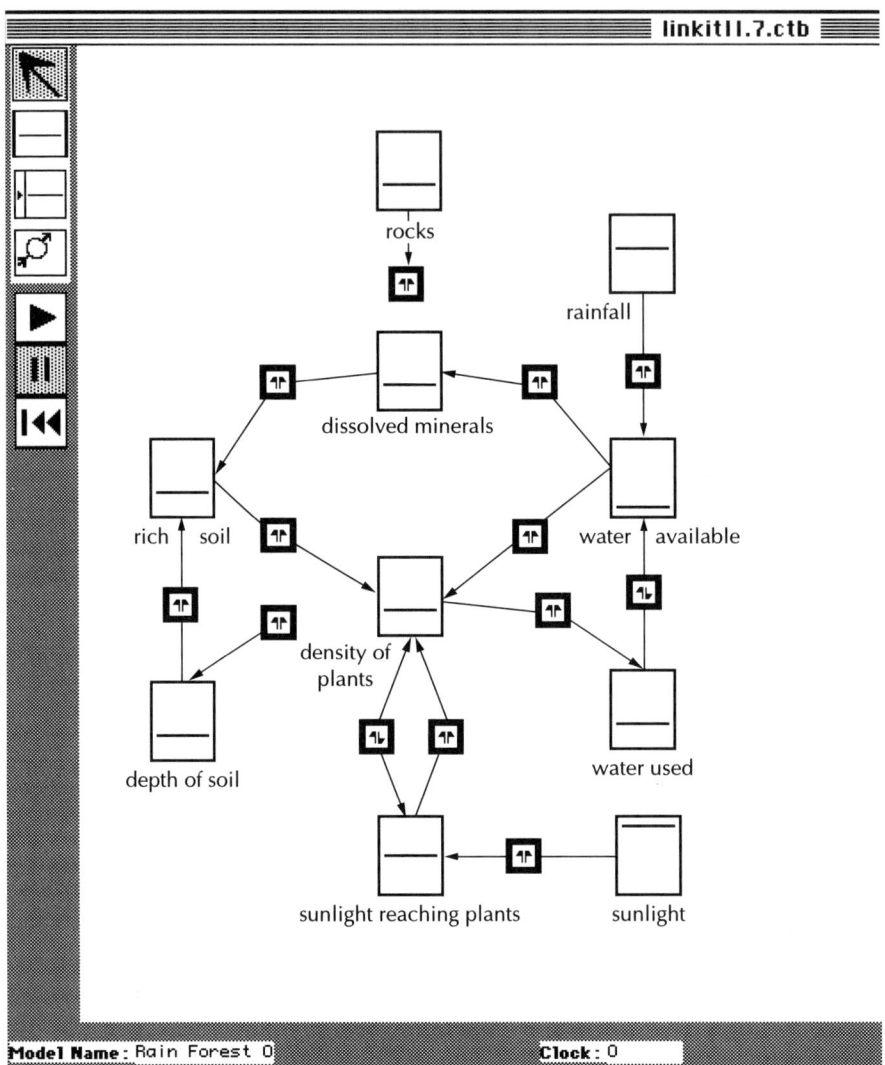

FIGURE 1.8 Rain Forest System of Figure 1.1 Translated into LinkIt.

other variables, and it is in fact set to be an average of the difference between the two. Another such "opposed" link reflects the fact that a high "density of plants" reduces the "sunlight reaching each plant," whereas a high value of "sunlight reaching each plant" makes for a high "density of plants." Where more than one variable acts as input to another, the net result (in this model but not always, as we shall see) is calculated by averaging them.

Unlike Figure 1.1, the system in Figure 1.8 can be run. In the Figure, the values of variables are represented visually by level markers—horizontal lines in the icon boxes whose heights are proportional to their values. Starting with high values for "rocks," "rainfall," and "sunlight," the model settles down to values much as shown in Figure 1.8 (after moving about a bit because of the way values are fed around loops). We can now see what happens if we increase or decrease the controlling variables (those that do not get input from any other)—"rainfall," "sunlight," and "rocks."

The example illustrates several aspects of LinkIt. The system being modeled is quite complex, yet as its effects unfold, it becomes amenable to understanding, because it can be run. The modeler *sees* the consequences of decisions about how to choose variables and link them together. This often leads to the *rethinking* of those decisions. The thinking required can be in the actual physical terms of the system being modeled: We think about "how dense the plants are" and what makes that variable great or small or what makes its value increase or decrease. We increase or decrease a value by dragging its level up or down. The computer system largely looks after the formalities, keeping the way variables are linked together consistent. All this helps us focus on the physical system, whereas unless we were very experienced at describing systems in terms of algebraic or differential equations, the formal aspects would demand a lot and perhaps all of our attention.

The values of variables displayed on the screen by their level markers are kept within the range of the boxes by "squashing" the values to fit within that range. Thus a value at the bottom of a box represents "nothing" and a value at the top represents "as large as you can imagine." How this is done is discussed below. The effect is that all variables are on a common scale, from nothing to very large. They are, as discussed previously, semi-quantitative in nature. They reflect values that are understood as larger or smaller, but not as absolute values with units.

In Figure 1.8 the rain forest is represented as a static system of variables,

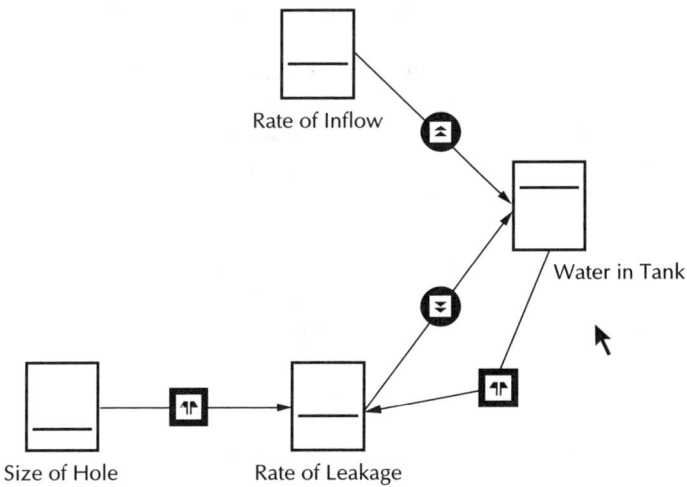

Rate of Inflow

Water in Tank

Size of Hole   Rate of Leakage

FIGURE 1.9 A Leaky Tank.

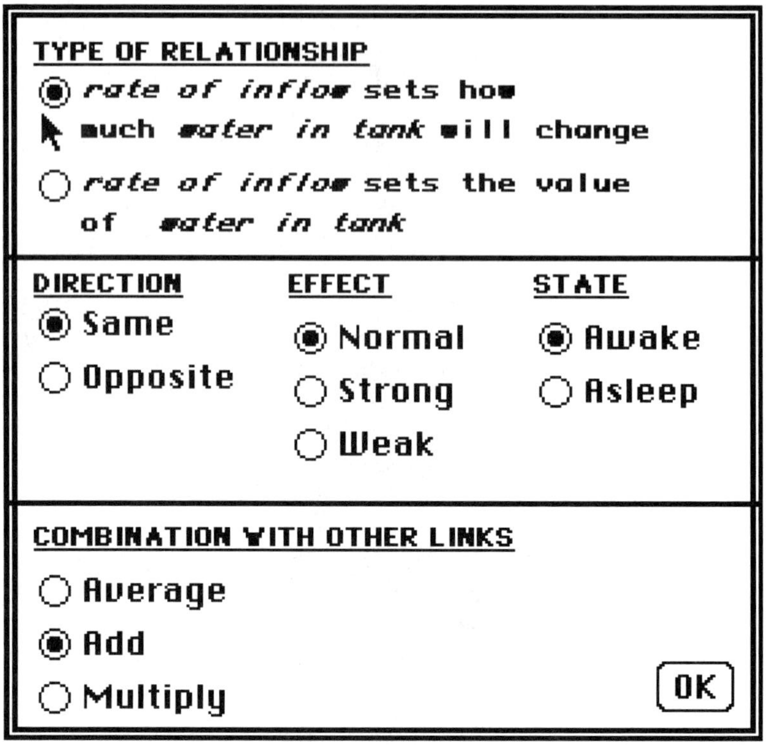

TYPE OF RELATIONSHIP

◉ *rate of inflow sets how*

▶ **much** *water in tank will* change

○ *rate of inflow sets the value*
   of *water in tank*

DIRECTION	EFFECT	STATE
◉ Same	◉ Normal	◉ Awake
○ Opposite	○ Strong	○ Asleep
	○ Weak	

COMBINATION WITH OTHER LINKS

○ Average

◉ Add

○ Multiply   ⸢ OK ⸣

FIGURE 1.10 Dialogue to Select Rate-of-Change Option.

combining to give a fixed result. If the job were done mathematically, we would have a set of algebraic relations to solve. This introduction to LinkIt illustrates only its most basic features, however. LinkIt can also be used to construct dynamic models in which certain variables determine the rate of change of others; thus it can solve differential equations. The effects of several variables on another can be combined in different ways, adding (and subtracting), and multiplying (and dividing) as well as being averaged as in Figure 1.8. And variables can be allowed to have values less than zero as well as greater than zero. Finally, a variable can be made Boolean in nature, triggering on (or off) when its input is above (or below) a level that the user can choose. All operations on a model are done by clicking or dragging; there are no equations to write. The next section illustrates these possibilities through a variety of models that are useful in science.

Examples Showing Possibilities of LinkIt

In Figure 1.8 it might have been better to represent the rain forest more dynamically. What sunlight, rich soil, and water actually do is to make plants

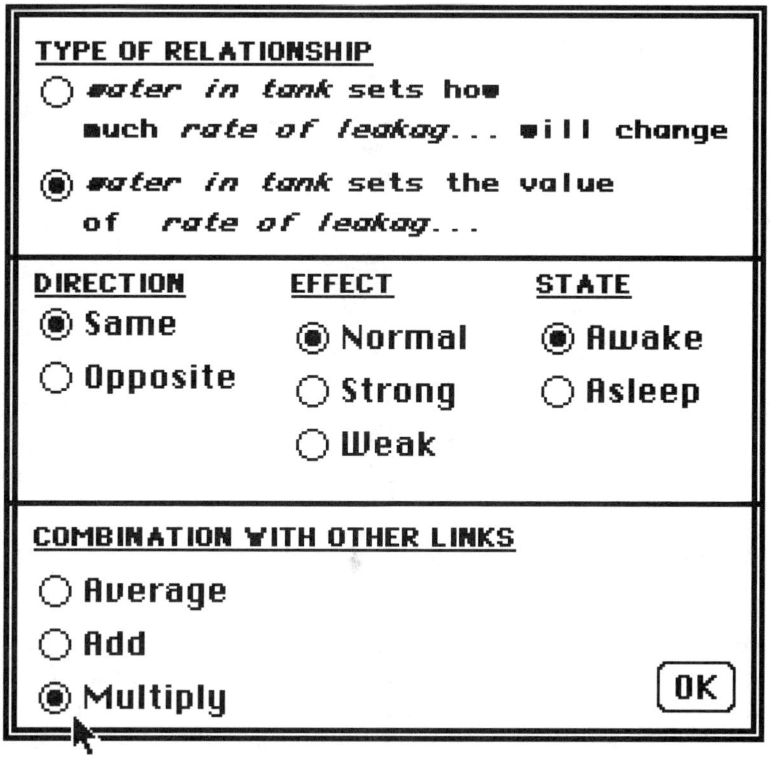

TYPE OF RELATIONSHIP

○ *water in tank* sets how much *rate of leakag...* will change

◉ *water in tank* sets the value of *rate of leakag...*

DIRECTION
◉ Same
○ Opposite

EFFECT
◉ Normal
○ Strong
○ Weak

STATE
◉ Awake
○ Asleep

COMBINATION WITH OTHER LINKS
○ Average
○ Add
◉ Multiply

OK

FIGURE 1.11 Dialogue to Select "Multiply" Option.

grow faster. High levels of these variables should make the variable "density of plants" *increase* rapidly.

A simpler example of rates of change is shown in Figure 1.9. The inflow to a water tank determines how rapidly the amount of water in the tank increases. But if the tank has a leak, water runs out as well, at a rate determined by the size of the hole(s). Accordingly, in Figure 1.9, the variables "rate of inflow" and "rate of leakage" are connected to the variable "water in tank" in a new way. As Figure 1.10 shows, the option *rate of inflow* sets how much *water in tank* will change" has been selected. This automatically sets the same type of option for all variables affecting "water in tank." The links to "water in tank" now have a different icon, small arrowheads pointing up or down (see Figure 1.9) according to whether the rate of change increases or decreases the variable it affects, and this is set by choosing the direction of the effect to be "same" or "opposite," as also seen in Figure 1.10.

There is one more feature to be set. If there is no water in the tank, none will leak out no matter how large the hole. Similarly, if the hole has zero size, no water will leak out no matter how much water is there. This means that the variables "water in tank" and "size of hole" must combine to determine

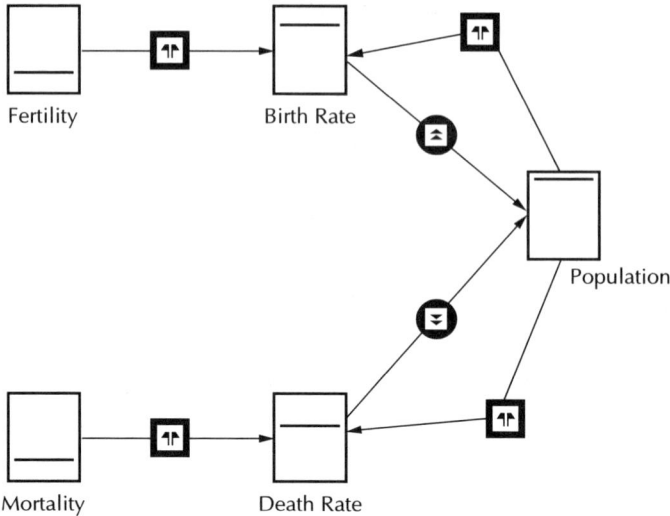

FIGURE 1.12 Unlimited Population Growth.

the rate of leakage in such a way that if either is zero, the result is zero—that is, they must multiply. This is how we often explain multiplication in teaching LinkIt: A good example is shopping for food. If the amount bought is zero, you spend nothing whatever the price; if by chance the goods are free, you spend nothing however much you take. Arguably, the rain forest model of Figure 1.8 would be better if the quantities "water available," "rich soil," and "sunlight reaching plants" were multiplied in determining the rate of growth of plants—all are conditions necessary for plant growth. Figure 1.11 shows the option "Multiply" being chosen for one link to "rate of leakage"; making this choice for one link selects it automatically for other links to the same variable. The solid black arrowheads on the links in Figure 1.9 show that this choice has been made—look for them in later models!

The result of all this is simple but important. When the model is set running, the level of water in the tank rises or falls, reaching a steady level after some time. How high that steady level is depends on the variables "rate of inflow" and "size of hole." The bigger the rate of inflow and the smaller the hole, the higher the final level—and, of course, vice versa. We have an example of a dynamic equilibrium, able to be approached from above or below: The level of water in the tank can be set by dragging it to any initial level one wants, and it can even be changed during a run (perhaps corresponding to throwing a bucketful of water into the tank).

Similar ideas apply to the growth of populations, as illustrated in Figure 1.12. A given population has a certain birth rate and a certain death rate, which together determine how rapidly it will increase or decrease. These variables are affected by the population itself: More people have more babies, and more people means that more die—both in a given time. The value of the

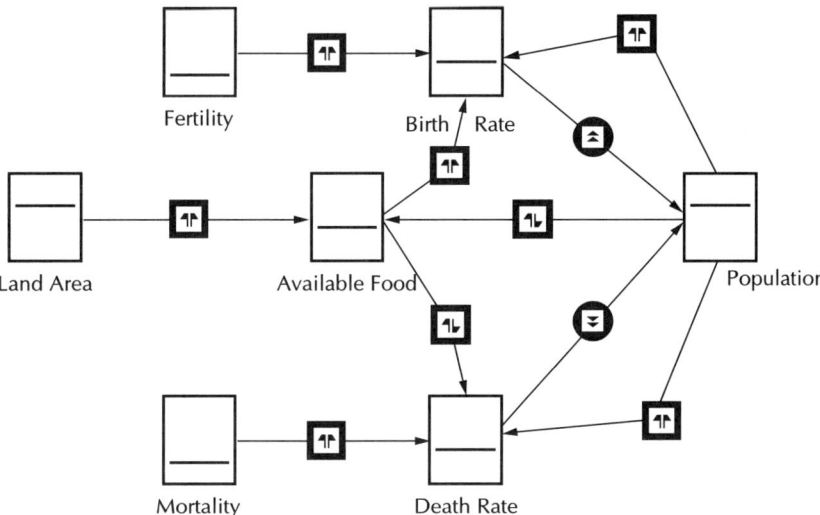

FIGURE 1.13 Limited Population Growth.

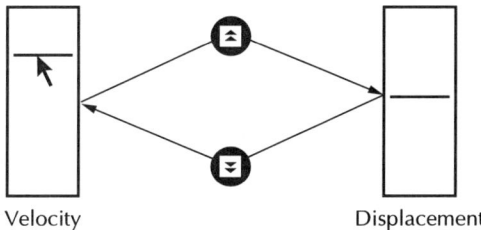

FIGURE 1.14 Stripped-Down Oscillator.

current population multiples with the inherent fertility or mortality to determine the birth and death rates. Unlike that of the water tank, this model has no simple equilibrium. The population either rises without limit or falls to zero; it illustrates the very important pattern of exponential growth or decay.

Real populations are limited by many factors, including space to live and food to eat. Figure 1.13 shows a modification to the model that takes account of limited food. The food supply affects the birth rate and the death rate, the latter inversely (low food supply tends to produce a high death rate). The food supply is improved by having a larger area of land available but is decreased by the existence of a large population. The model now has a stable equilibrium population (even with differing underlying fertility and mortality) at which the effect of the food supply is to make birth and death rates equal.

So far, all the variables used in models have had only positive values (though with negative values, we could have combined birth and death rates into one variable in Figures 1.12 and 1.13). A good example of a model with

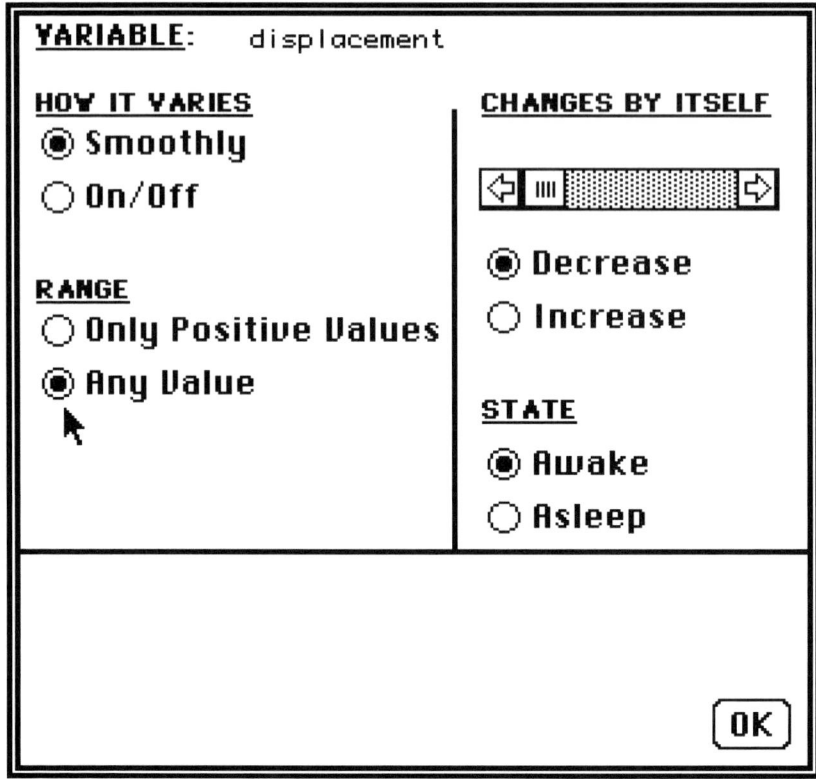

FIGURE 1.15 Variable Dialogue Box.

variables that may take on positive and negative values is the harmonic oscillator, shown in a stripped-down version in Figure 1.14 and in a more elaborate version in Figure 1.16.

The model of Figure 1.14 could hardly be simpler, though this does not mean it is easy to understand. It shows, in a form probably too abstract for most students at first acquaintance, the essentials of a harmonic oscillator. The velocity is the rate of change of the displacement. The displacement, acting through a spring, determines the negative rate of change of the velocity, because the force of the stretched spring acts to decelerate the oscillating mass. With just these two coupled variables, we get oscillations if we start the model off with a nonzero value for either variable. The model shows the two essentials of harmonic motion: the feedback from each variable back to itself via the other, and the fact that the relationship is of second order, each variable being the rate of change of a rate of change.

Both variables have to be able to be positive or negative in order to represent velocities and displacements both in one direction and in its opposite. This choice is made through another dialogue box obtained by clicking on a variable, as shown in Figure 1.15.

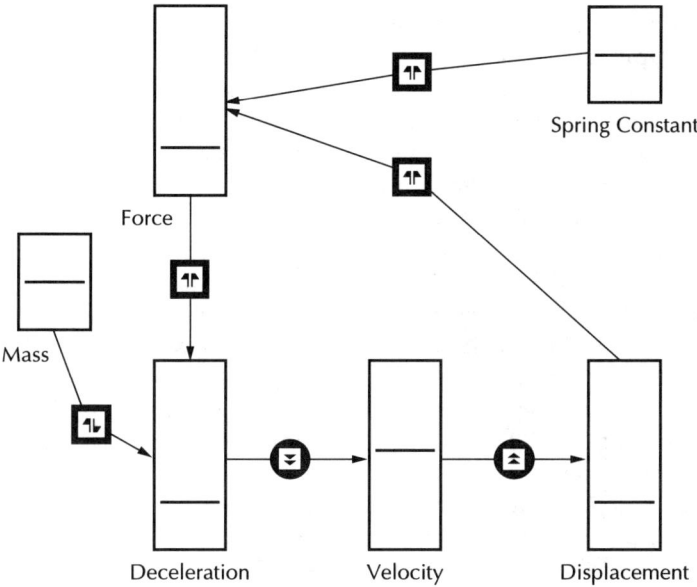

FIGURE 1.16 Harmonic Oscillator with Effects of Mass and Spring Constant.

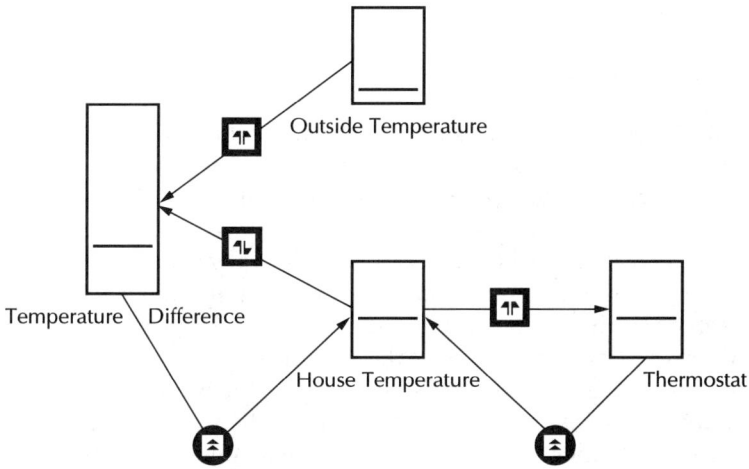

FIGURE 1.17 Home Heating System.

To show that a more comprehensive picture is possible, Figure 1.16 provides a more elaborate harmonic oscillator model, including explicit representations of the mass and the spring constant. Note that force and mass combine to determine the deceleration by multiplying by the force and dividing by the mass. Increasing the mass makes the oscillations go more slowly; increasing the spring constant increases their frequency.

So far, all the variables we have used have been continuous. Some situations (for example, a thermostat) involve something turning on or off according to the strength of its input. LinkIt provides just such a variable, as seen in Figure 1.17, which represents the heating system of a house. The net gain of energy to the house is determined by the net gain or loss from outside (which depends on the difference in temperature between inside and outside) plus any additional gain from the heating system when the temperature falls below a preset value.

In the state shown in Figure 1.17, the house is warmer than the surroundings, leading to a net loss of energy. But the house temperature is lower than that set on the thermostat, as indicated by the small arrowhead on the left-hand side of the variable "thermostat." The thermostat is therefore on, so the furnace is supplying energy to the house, which will warm up until its temperature reaches the preset value. At that point, the thermostat will go off and the house will cool down, in time triggering the thermostat to warm it up again, and so on. A further dialogue box for such "on/off" variables makes it possible to choose whether the variable will be on or off in terms of whether its input is above or below the threshold.

Such variables are useful in problems that contain some kind of potential crisis, such as a war or epidemic breaking out when conditions pass a certain level. They can also be used to represent networks of "on/off" circuit elements.

How LinkIt Works

LinkIt is designed so that behind the scenes, values of connected variables are calculated and displayed dynamically, in a series of iterations. First, values of independent variables are collected, together with the initial values of any variables whose inputs are set (so that their rate of change can be determined). These data are used to calculate the values of variables to which they provide input, and so on until all values are determined and can be displayed. After display, the next iteration begins.

The value of a variable shown on the screen is not the value calculated behind the scenes. It is, for the purpose of display, "squashed" into a fixed range (for example, to squash values into the range -1 to $+1$, we use the logistic transformation $[(\exp(x)-1)/(\exp(x)+1)]$. When collected for the next iteration, values are "unsquashed," via the appropriate inverse transformation, before being used.

The determination of the next value of a variable proceeds in two different ways, depending on whether its inputs determine its value or its rate of change. If the inputs determine its value, they are simply combined and used to set the new value. How they are combined depends on the "sign" of the link and on the method of combination. If inputs are set to add, they are added together, subtracting those that have "opposite" links. If inputs are set to multiply, they are multiplied, dividing by any that have "opposite" links. If they are set to average, that is what is done, taking account of sign.

Clearly, multiplication by a large value or division by a value near zero can lead to a very large result. A cut-off is imposed, such that the "squashed" value for display is very near the extreme end of the box representing the variable. Division by zero is avoided by replacing zero by a small number. The inputs to a variable are analyzed to see whether they could produce a negative result. If so, a message is displayed advising that the variable affected be changed to have "any value" if it presently allows only positive values. If this is not done, negative values are set to zero.

If the inputs to a variable determine its rate of change, the net input is calculated as above and is then used, after being multiplied by a suitable small fraction, to increment or decrement the current value.

Thus calculations take place in an essentially linear regime (ignoring the large-number cut-off), whereas the display is nonlinear and all displayed values are scaled to the same fixed range (0 to 1 or -1 to $+1$). This enables LinkIt to dispense with varying scales for quantities, and with units. Everything is on the same scale of nothing – small – big – very big.

Limitations of LinkIt

It is clear from the foregoing account of how LinkIt works that it provides an approximation to dynamic models involving differential equations, with equations of order higher than 1 represented as sequences of simple derivatives. The method of integration is essentially the simple one devised by Euler, which projects future values linearly from current values and which, as is well known, gives systematic errors such as tending to make the amplitude of an oscillator increase when it should stay constant. Systems of algebraic relations can be combined with relations involving derivatives.

Models that use no links representing rates of change are essentially systems of algebraic relations. These might be better represented as systems of constraints, so that in $pV = nRT$, for example, one could calculate any quantity given the others. This is not possible in LinkIt, which follows the direction of links in calculating quantities. Thus one could have (say) $p = nRT/V$ in LinkIt but not (with the same variables and links) any of the other possible relationships. Loops of links in systems of algebraic equations necessarily sometimes give trouble (in STELLA® they are automatically disallowed). LinkIt permits such loops, which may then sometimes give rise to unusual behavior when a variable switches its value back and forth from one iteration to the next. We let this happen because it is visually very obvious that something is wrong.

Because of the "squashing" function used in displaying variables, LinkIt will not (so far as we know) show chaotic behavior (for example, in such a case as $dx/dt = kx(1 - x)$). It is essentially suited to monotonic relations between variables, but it is not restricted to linear relations, though no provision is made for nonlinear relations other than those obtained by multiplication or division.

Students Using LinkIt

LinkIt and another simpler version called IQON have been used, in the United Kingdom and in Brazil, with students ranging from 13 to 18 years. We find that students of all these ages can easily learn the basics of the system and can construct models of their own. They quickly become able to build more or less reasonable models with around half a dozen variables and links. Something similar has also been found in the United Kingdom and in the United States, using STELLA®. Students find it easy to use STELLA® as a "drawing tool" for modeling, employing it as a sketch pad to help them think of variables of interest and link them more or less appropriately. And of course STELLA® was designed to facilitate just this kind of thinking. However, the point of using STELLA® is, after having sketched the outline of the model in this way, to go on and define relationships between variables in appropriate mathematical forms. Only then will the STELLA® model run. When it does run, observing differences between how the model works and what was expected often leads to further thought and development of the model.

What LinkIt and IQON do is short-circuit the process of defining the model in explicit mathematical forms—a step essential to STELLA®. A model drawn on the screen in LinkIt or IQON runs as soon as it is drawn. It runs even if it is only partially finished. Thus the user gets early and rapid feedback on whether the model is coming along well and can get plenty of surprises leading to critical thought.

A good proportion of students begin with what has been called "laundry list thinking," assembling a considerable set of variables that might affect the variable of interest. Asked what affects pollution, they might nominate size of factories, number of cars, effectiveness of cleanup, strength of wind, and so on. This leads them to build star-like models, with many variables giving input to one central variable (Bliss, 1994; Kurtz dos Santos, 1995). Such models are of limited interest. Models become much more interesting when there are chains of effects of one variable on another and when there are feedback loops.

Typically, students end up building models with four to six variables and a similar number of links, sometimes with feedback loops (Bliss, 1994; Sampaio, 1996). They have some trouble in LinkIt distinguishing an inverse link (a "dividing" link corresponding to $1/x$) from a subtractive link (corresponding to $-x$). It is not always clear to them why an "opposite direction" link from a variable that has a negative value has an effect opposite from what they may expect (*increasing* the value of the variable it goes to, for example). Indeed, much of their thinking about variables and the relationships among them is "positive thinking": They prefer "same direction" links to "opposite directions" links where possible, even if it means inverting the name and meaning of a variable.

We have compared what students do when constructing their own models (expressive use) and what they do when looking at and trying to improve models given to them (exploratory use) (Bliss, 1994; Kurtz dos Santos, 1995;

Sampaio, 1996). We consistently find, over a range of ages, that students' own models are initially simpler than those they can cope with when models are presented to them but that the improvements they make to their own models are much more interesting, and lead to more complex models, than the changes they make to models provided for them. Furthermore, when constructing their own models, older students often invent "devices" to achieve a given effect (such as a variable that receives feedback via another) and then use these devices in other models (Sampaio, 1996). Herein they show signs of beginning to think structurally—that is, mathematically.

This suggests a pedagogic strategy of using both exploratory and expressive modes. Exploratory use of the system can introduce systems of some complexity, whereas expressive use encourages further thinking and abstraction.

For us, the most valuable activity that LinkIt or IQON produces is not the models made or seen, but the talk and discussion that accompany their making and use. Because the models run as soon as they are drawn on the screen, they provoke immediate discussion as values rise or fall in ways not always expected. These discussions do have their failings. Students sometimes try arbitrary "fixes" to get the result they expect, without giving much thought to the meaning of the "fix." This of course arises in work with any computer program that makes changing things easy. Students also sometimes stop too soon, when the result seems to be what they want but the model does not in fact reflect their ideas. As we noted in connection with WorldMaker, students will from time to time project their knowledge of the world onto the model, "explaining" something the model does by invoking a feature of the world that is not in fact represented anywhere in the model.

Even so, much of the discussion is productive and in focus. LinkIt has given us a way to get students talking about variables and their possible effects on one another, for quite complex systems, long before they have anything like the mathematical ability to explore these relationships by using algebra or calculus.

References

Atkins, P. 1984. *The second law*. New York: W. H. Freeman Scientific American Library.

Bliss, J. 1994. Reasoning with a semi-quantitative tool. In Mellar, H., Bliss, J., Boohan, R., Ogborn, J., & Tompsett, C. (eds.), *Learning with artificial worlds*. London: Falmer Press.

Boohan, R. 1994. Creating worlds from objects and events. In Mellar, H., Bliss, J., Boohan, R., Ogborn, J., & Tompsett, C. (eds.), *Learning with artificial worlds*. London: Falmer Press.

de Kleer, J., & Brown, J. 1985. A qualitative physics based on confluences. In Hobbs, J., & Moore, R. (eds.), *Formal theories of the commonsense world*. Hilllsdale, NJ: Ablex.

Forbus, K. 1985. The role of qualitative dynamics in naive physics. In Hobbs, J., & Moore, R. (eds.), *Formal theories of the commonsense world*. Hillsdale, NJ: Ablex.

Gentner, D., & Stevens, A. 1983. *Mental models*. Hillsdale, NJ: Lawrence Erlbaum.

Hayes, J. 1985. The second naive physics manifesto. In Hobbs, J., & Moore, R. (eds.),

2

Training System Modelers: The NSF CC-STADUS and CC-SUSTAIN Projects

Ron Zaraza

Diana M. Fisher

Background

The CC-STADUS Project (Cross-Curricular Systems Thinking and Dynamics Using STELLA®) has an unusual origin and has experienced an even more unlikely evolution. Although some work had been done in computer modeling in K–12 classrooms prior to 1990, the number of teachers actually using system dynamics was very small. Almost no instructional materials were available, either commercially or in the public domain. The Creative Learning Exchange had been established to serve as a clearinghouse for K–12 materials, but by the early 1990s it had accumulated little.

One of the co-authors became aware of the STELLA® computer modeling software after attending a workshop and shared this information with the other co-author. Over a period of 2 years, we began to use STELLA® to teach our classes. The software opened up for us new instructional possibilities.

Mathematics and physics classes in the K–12 environment are among the most tradition-bound subjects. Despite publication of the National Council of Teachers of Mathematics Standards and numerous documents released by the National Science Teachers Association and the American Association of Physics Teachers, much of the instruction remains focused on sequences of topics and standard problems that have not changed appreciably for the last 40 years. Critics of math and science education note that instructors tend to teach as they were taught. Furthermore, teachers of mathematics and science *successfully* learned in the environment they experienced, so they often do not see a need to change. They learned mathematics by mastering arithmetic, algebra, geometry, trigonometry, and calculus, usually in a rigid, sequential curriculum. Physics teachers take these mathematical tools and learn to apply them, using well-defined rules, to a narrow range of limited (and unrealistic) problems.

These approaches exclude the use of mathematics and physics in exploring real and interesting problems for the students. Worse, these paths are not accessible to many—perhaps not even to a majority of students. It has been amply demonstrated that alternative approaches can work. The first transformational geometry text, written by Zalman Usiskin, introduced some new and powerful ideas about how mathematics can be done. "Conceptual Physics," a course first designed for college students, is now widely used in high schools as an alternative physics course. This text allows students to focus on ideas, not equations. Unfortunately, students do not solve the problems numerically, which is sometimes frustrating for those who want "the answer."

Computer modeling seemed to represent an even more important alternative approach to mathematics and physics. The STELLA® software uses four components common to all work in system thinking/dynamics. Stocks, or accumulators, store information or values. Flows change the values of stocks. Converters carry out logical or arithmetic operations. Connectors carry information from one model piece to another. The matching of these model pieces with the language and concepts of mathematics and physics is obvious.

Flows are often referred to by Jay Forrester, the founder of system dynamics, as rates. In mathematics, students use rates throughout computation-based courses (most often grades K–7) without any discussion of what rates do. Then, their first formal exposure to a rate is the concept of "slope" that they encounter in algebra classes. Any of these rates allow for the parallel introduction of flows in models. In calculus, the concept of the derivative is the ultimate example of a rate. *Stocks*, on the other hand, are the results of integration. Models allow students to explore these two key concepts of calculus while learning much more basic mathematics; thus their conceptual framework can be more solid before they formally encounter mathematical symbolism.

Physics, as a discipline, has sometimes been characterized as the study of rates. Position, velocity, acceleration, force, impulse, momentum, energy, current, and power are all concepts that either are rates or are controlled by rates. This makes them ideal subjects for representation by models. That such models can be easily modified to allow students to explore problems beyond their mathematical reach deepens their understanding of the physics content.

The stocks of mathematics or physics and the rates that change them are key to the study of these two subjects. Development of basic models to use in classes initially dominated our work with systems. As simple models were developed, our own interests and questions asked by students led us beyond merely using models in the classroom. Our broader need for information led us to extensive reading in systems thinking and system dynamics. This lead to a broader perspective. The focus moved from "modeling as an instructional tool providing an alternative path to knowledge" to "systems thinking as a structure for organizing information" and "system dynamics as a means of using/expanding that knowledge." This inevitably led to a different perspective on modeling as well. It became obvious that there were models common to both subjects and that only a few basic model structures were necessary to deal with most problems.

Perhaps more important was a shift in who used the models. Initially, in science, much of the work focused on a model used by a teacher to present and explore ideas with students. When students began to use models, or suggest changes in models, it quickly became apparent that more in-depth learning was possible if the students were actually using and (in some cases) building the models. In math, students built small models to represent the traditional functions taught in most algebra classes. This work led to the creation of a modeling class in one school and an independent-study modeling program at the other. Now, 5 years later, more than 120 students at the two schools are enrolled in modeling classes.

The power of modeling in the secondary classroom was becoming evident. As work continued, it became clear that system dynamics was a tool that could address a problem of education: learning disciplines in isolation, ignoring interconnectedness between subjects, ideas, and phenomenon. In 2 years it became obvious that the models used in our classes described patterns of behavior that were exhibited in other areas as well. Students working with models began to ask questions that transcended the boundaries of the subject for which they were created. It became clear that system dynamics models were inherently interdisciplinary and that they offered teachers an opportunity to address problems and ideas in a comprehensive manner not previously possible.

This realization ultimately led to the formation of the CC-STADUS Project. This National Science Foundation project was designed to create a critical mass of teachers to develop and use models and their accompanying curricula. An equally important goal was the development and release of a substantial number of models and curriculum materials, allowing others to carry the experiment forward without having to start from the very beginning.

CC-STADUS

A few key assumptions guided the initial design of the CC-STADUS Project. The central assumption was that for teachers to use models in the classroom, it was essential that they work with practitioners and be trained by practitioners. Thus the project has always been, and will always be, directed by working classroom teachers. Living up to the expectation that what is presented to teachers has already been successfully tried *by* teachers is necessary for creditability and effectiveness.

A second assumption was that systems work in classes would become viable on a large scale only if there was a broad range of models and curriculum materials available. Some teachers are innovators and creators. They generate new approaches, new ideas, and new tools for teaching. Most teachers, though not themselves innovators, are willing to use new, proven tools, techniques, and ideas if these are readily available. In 1992, the number of models available commercially or in the public domain was so small that there was little chance for system dynamics to catch on. To develop the models necessary to allow systems use to grow, the project would have to focus on training modelers.

The third major assumption was that modeling real-world problems is, by its nature, an interdisciplinary activity. Although much lip service is paid to the idea of interdisciplinary work in education, there has been very little progress beyond the most simplistic questions and approaches. The use of system models allows the interdisciplinary work to increase in complexity as students ask important questions. The model grows as student understanding grows. At the same time, even relatively simple models can pose problems for more than one discipline. Because modeling is a powerful way to generate interdisciplinary discussion, the project focused on training mathematics, science, and social science teachers, and it insisted on the development of cross-curricular models.

The choice of modeling software at the inception of the project was virtually predetermined. All of the principal investigators were Macintosh users. Most had experience on other platforms and had experience programming. However, it was assumed that most participants would not be experienced programmers. In fact, because one-third of the participants were expected to come from the social sciences, where computers had made little penetration in secondary schools, it was necessary to choose the easiest-to-learn software on the easiest-to-use operating system. This limited the choices to Macintosh software, which was fortunate: At that time, about 90% of all computers used in Pacific Northwest schools, outside of business departments, were Macintosh machines. The only modeling package available for the Macintosh—and the simplest modeling software available at the time—was STELLA II. This package was the one we were familiar with, and it used the language in which the few models already released were written.

At the present time, there are a number of other modeling packages available, including Powersim and Vensim. However, the project staff still feels that the STELLA® software is the easiest to learn and the most intuitive. There is also now a substantial "installed base" of software and models developed for STELLA®. As a result, our work will continue to be done with STELLA®.

Implementation

The CC-STADUS Project was designed to train 36 teachers each year in the use and development of dynamic models for their curricular areas. Preference was given to teams of three teachers from a single building if the team included a math teacher, a science teacher, and a social science teacher. Teachers were trained from the Portland metropolitan region the first year, from Oregon and southern Washington the second year, and from throughout the Pacific Northwest the third year. The project was given an extension and supplementary grant that allowed a fourth year of training to be offered. That training focused on the Portland area, but included a team of teachers from San Antonio, Texas, who planned to be a core group developing similar training in Texas.

The initial training was 3 weeks in length. The first week of training focused on familiarizing participants with basic systems concepts and instruction in the use of the STELLA® software. Participants began by playing the FishBanks simulation as an introduction to the idea of computer simulations. This was followed by training in STELLA® conducted in single-discipline groups. All instruction was done by the principal investigators and a core team. This core team consisted of two experienced modelers from each discipline. For most of the remainder of the first week, two core-team members worked with the 12 same-subject-area participants. This allowed the training experience to focus on simple models that were directly related to the participants' own teaching assignments. It also provided a familiar environment in which to ground the new systems concepts.

The second week of training focused on the use of computer modeling in business, research, and higher education. This instruction was carried out by practitioners who use modeling in their own work. Included was a short discussion on the mathematical basis of the STELLA® software and its limitations. Also included in this week, and carried over to the next, was discussion of some of the educational theory behind modeling and its classroom use.

The last week dealt with further background on systems and the actual project work. The participants, working in teams of three or more, developed cross-curricular models with curriculum materials for a minimum of two different disciplines.

This basic pattern was repeated for each of the next 3 years, with some changes that reflected our increasing understanding of how best to use and teach modeling. Changes were recommended by the core teams. Each summer, the best of the new participants were asked to work as the core-team instructors for the following year's workshop. New team members were debriefed with the old core team and the principal investigators.

After the first summer training, the new core team made a recommendation that has shaped all future workshops. Initially, each of the three disciplines received very different training, with topics and model structures chosen to fit the curriculum of each discipline. There was no real coordination of topics. As a result, each discipline group had different modeling experiences. When they came together to develop their cross-curricular model, the teachers did not have common experiences or a common vocabulary to discuss their ideas for models and curriculum. There were some common experiences, but dissimilarities outweighed similarities.

All the new core-team members strongly recommend that all groups work with similar modeling structures and that any other structures be added only after adequate work has been done on the "core structures." Some feel strongly that no other structures should even be presented. Long discussion led to identifying four basic structures. Since that time, all training has focused on models that feature linear relationships, exponential growth and decay, S-shaped growth, and quadratic equations. Even though the model structures are kept uniform, the subject matter of the models taught has continued to focus on the discipline, thus providing teachers with a core set of models easily transferable to their classrooms. Oscillatory models, as interesting and important as they are, were identified as too complex for initial training. With only 20 clock-hours for actual training in the STELLA® software, the strong recommendation that the focus be kept simple has proved correct. Each year, participants have achieved a greater "comfort level" and greater mastery of the basics of modeling than did the first group.

This recommendation also focuses the project on an important realization about the nature and future of modeling. The shared experience in learning STELLA® gives participants a common language for discussing models. This facilitates designing models for their final project. It also starts teachers thinking about how model archetypes translate across disciplines. This in turn starts a series of conversations on system dynamics as the language that allows ideas and concepts to cross the line between disciplines. The project has gradually moved into advocacy for system dynamics as the field that can bind the other disciplines into a coherent whole.

The teachers who were beginning to use systems gave us a list of suggestions about what we could have done to help them get started. Foremost among these suggestions was demonstration teaching during the training. Also needed was a discussion of where and how to introduce models into the curriculum. The result of all the recommendations was a summer institute that evolved into a 12-day training program with most of the educational and systems theory deleted and some demonstration teaching and discussion time added.

The Product: The Cross-Curricular Models

So far, only a few of the major cross-curricular models developed by participants in the CC-STADUS project have been released. The remainder are undergoing mathematical verification or editing (model or curriculum materials). Basic patterns have emerged that divide models into two types. The first type, and by far the style of model that teachers are most likely to attempt, consists of what we have identified as "content-rich" models. These models deal with a highly specific problem in great detail. The model tends to be somewhat complex, with many interrelationships and feedback relationships.

System dynamics is attractive to teachers because it allows discussion of the interconnectedness of real-world problems and phenomena. This leads teachers to adopt the "content-rich" model. These types of models are also attractive because they give real numerical results for complicated situations. Because of this apparent numerical precision, they are regarded as more real—and therefore more useful and important. Thus they are often the initial effort in the group modeling, and they frequently dominate the initial efforts of participants after the training program.

Two excellent examples of this type of model were developed in the first

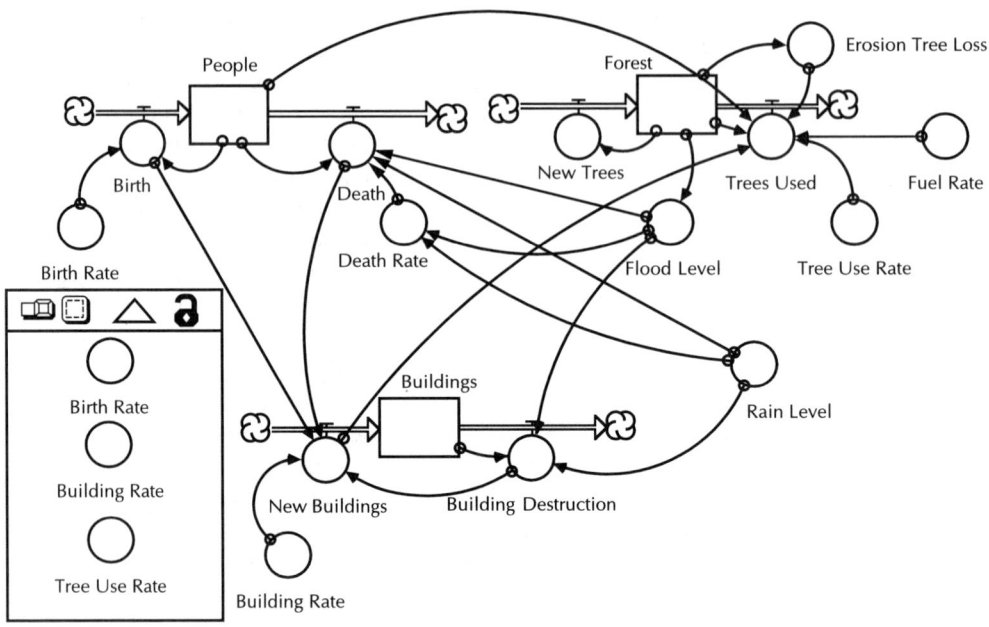

FIGURE 2.1 Mahenjo Daro Model.

2 years of the project. The first, Mahenjo Daro, deals with the growth and destruction of an ancient city (see Figure 2.1). The city was founded by a no-
madic tribe of about 300 who settled in a fertile flood plain below forested
hills and mountains. The richness of the soil, the availability of ample game
and of land for domestic animals, and an apparently boundless supply of
wood for building, heating, and cooking led to a city of 40,000 inhabitants en-
joying an extensive infrastructure in only a few hundred years. But 300 years
later the city was destroyed and abandoned. The model explores in great detail the reasons for its destruction and presents the students with a simple basic problem: Prevent the destruction of Mahenjo Daro.

Mahenjo Daro is an outstanding example of a detailed historical model that
explores the key factors that control events. It allows modification of key factors and exploration of their effects. In short, it is a well structured model that
accurately describes a historical reality and allows experimentation. Yet, despite its excellence, this model has been used by only a few participants. Its
downfall lies in its specificity. Unless a model fits exactly into a teacher's curriculum, it will not be used. Mahenjo Daro perfectly describes a very narrow
situation and is not easily modified or translated to any other. Its very com-
pleteness makes it unsuitable for adaptation to other problems. Fortunately,
it deals with a situation that fits into the global studies classes taught in many
high schools.

Another example of a content-rich model is the PERS model. This model
explores the effects on a school district of proposed changes in a state's pen-

sion plan. This model too is a marvel of detail, though simpler in basic structure. It also exhaustively describes a real situation and allows it to be modified. However, it is even more specific than Mahenjo Daro. Thus, the few times when it has been used, it has served as an example of how a model can be built to explore political decisions and their implications.

About half of all cross-curricular models fall into the content-rich category. They are excellent examples of how models can describe events and situations with great precision, and they have definite places in some curricula. However, the utility of these models will always be limited by their specificity.

Other models developed by project participants are referred to as "curriculum-rich." These models tend to be simpler and more generic, as do the curricula designed for them. Perhaps the best example of such a model is in the Rulers packet. This set of three models (see Figure 2.2) explores population growth, beginning with a basic population model. This model is then modified to look at the interaction between a population and a nonrenewable resource and, finally, that between a population and a renewable resource. This last model is far less complex than Mahenjo Daro.

The strength of the Rulers packet and of models like it is (1) their focus on basic ideas that are common to many problems and (2) the ability of the model to be expanded to address these problems in more detail. The basic model has been used to look at population growth in nations throughout the world and to examine all manner of growth of organisms. It has been used as a starting point for student research on the growth of *E. coli* in a stream system. In many classes, it is expanded to look at interactions of the base population with other factors that control births and deaths. The models can be expanded as needed and desired. New questions are developed and explored. The original package serves more as a catalyst than as a detailed, self-contained unit.

Such models open up the curriculum to exploration by students and teachers. Their simplicity brings out questions that lead to the creation of more realistic models. By provoking such questions, it encourages classes to explore the system that the basic model sketches out in simple form. Students ask better questions, probe more deeply, and even begin modifying and adding to the model themselves. These modifications are then tested for validity as students see whether they perform as they should. The models do not exhaustively cover a narrow topic. They present broad ideas that allow the curriculum to be expanded by the curiosity of the students. They pose many questions, allowing the teacher and students to choose which possibilities to pursue.

Verification/Validation and Documentation

Documentation has always had the potential to make computer software successful or cause it to fail. Well-documented programs can be readily learned

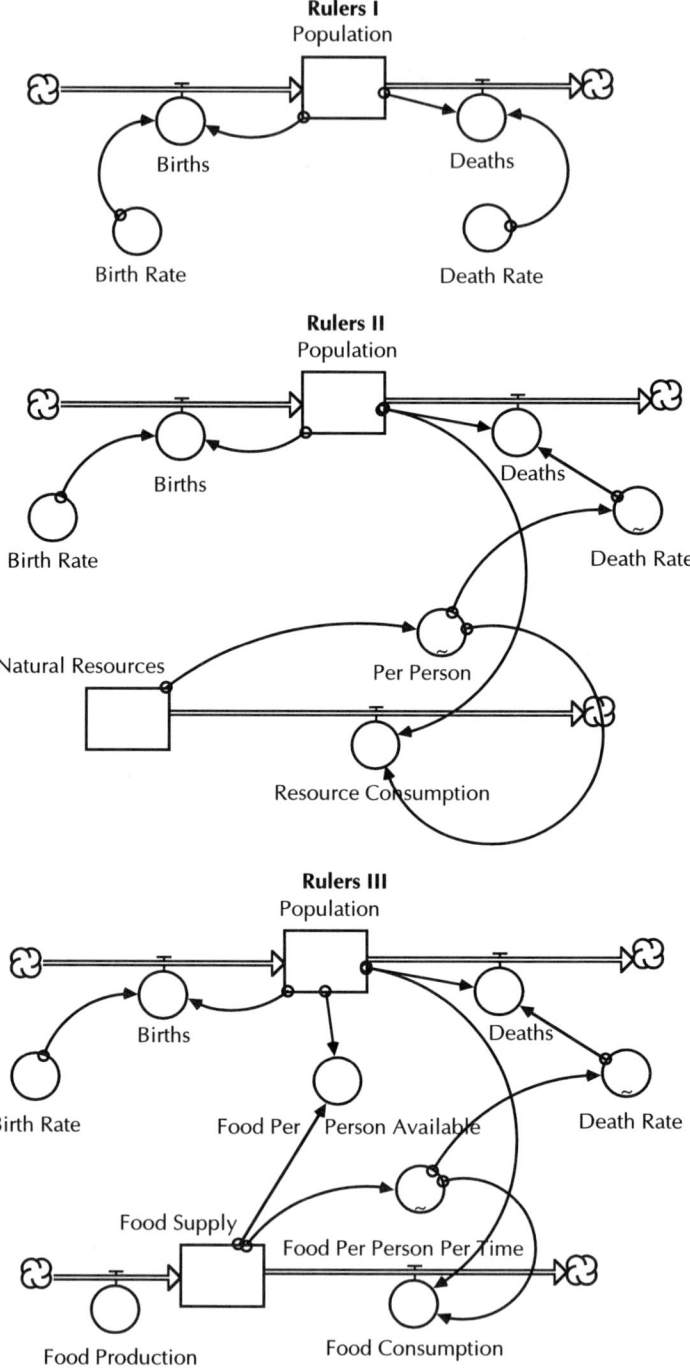

FIGURE 2.2 Rulers Models.

and problems quickly overcome. Poorly documented software may be discarded out of frustration. We emphasized this point repeatedly to participants in the summer institute. The models they develop are not just for their use. They are for distribution to the greater educational community. Therefore, their documentation has to be as thorough as possible, allowing new teachers to understand the model. Model dependencies have to be carefully specified. The dynamics of each component have to be clearly defined, and the set of initial parameters clearly specified.

As the participants develop their models, the internal documentation features of the software are used to provide extensive explanation of the model. This documentation is not limited to the "how"—that is, what a particular model piece does. This information explains the reasoning behind the model and the reason for the model's structure.

Inclusion of the "why" in the documentation provides justification for choices made and choices discarded. It places each part of the model in context. This makes the model easier to use, and it provides users with information that allows them to modify parts of the model. This explanation is also included and expanded in the curriculum packets developed for the models. The model itself provides a "snapshot" that ties the modeling topic and the modeler's interpretation together. Without a full explanation of the reasoning, the context of the model cannot easily be established, and its utility is diminished.

Participants are also asked to test their models carefully, examining the effects of changing the interval between calculations, DT (Delta Time). In the event that the model settles into a steady state, they are asked to see whether, when perturbed from that steady state, the model ultimately returns to it. These checks allowed the participants to do some verification of their models. Discussion of these tests is included in the curriculum packets.

Full documentation also captures the intent of the model. This can be vitally important in verification of the model. We became aware of the need for careful verification of models early in our work. Several models then in the public domain were discovered to have significant flaws. These flaws were related to the mechanism of the specific behavior being modeled. It became obvious that our participants would usually not be experts in the phenomena they were modeling, so there was the risk that the same sorts of flows would arise. Further, our own adventures with choice of DT, integration method, and quirks of the software made us aware that mathematical peculiarities could arise that did not represent the actual behavior of the system.

To minimize those possibilities, all cross-curricular models are reviewed twice. Initially, they are reviewed by Andrew Jonca, a Ph.D. mathematician specializing in numerical analysis. He checks model results with results produced by more traditional approaches. This step ensures mathematical validity. Then the models are reviewed by the project staff for content validity. Where there is uncertainty, source documents or outside experts are consulted. Finally, once the model has been validated, the curriculum materials are reviewed and edited. This entire process gives us a great deal of confidence in the models and materials.

After the Summer

The first year of training brought home the need for substantial revisions in the summer institute. It also made other needs very obvious. Participants, in addition to building a major cross-curricular model, are required to use models at least twice in classes during the ensuing school year, to develop at least one simple model specific to their curriculum area, and to make a presentation to co-faculty on system modeling. Most NSF summer institutes have similar expectations and get 50–65% implementation and compliance. Each year the CC-STADUS participants met or exceeded this figure. However, feedback—even from those who met all requirements—pointed out problems.

The most difficult requirement for teachers has been making presentations to other faculty. Many interpreted this requirement as necessitating a presentation to their entire faculty. This was often difficult to schedule, and many found it too intimidating to try. After the summer training, most participants were so drained that they did not use STELLA® again until the end of the summer or the beginning of the school year. They forgot things. Thus they lacked the confidence to make a presentation to a large group. They felt like a child taken on a long bike trip the day after the training wheels were removed. Those who did small-group presentations felt less stress and characterized their efforts as more successful. These presentations tended to be less formal and more personal. Many later applicants were people who had learned of the program from presentations done by past participants. Since that first year, participants have been encouraged to make small-group presentations rather than addressing the whole faculty.

Of the three requirements, the most important has always been use in the classroom. That is one area where a surprising number of problems emerged. Some teachers simply couldn't see how to fit one of the models into their classroom. Others were not sure exactly how to teach using models. Still others had difficulty getting access to computers.

The first year, attempts to deal with the three problems were not particularly well thought out or organized. It soon became clear that the problem of how to teach with models was the most serious. Core-team teachers and the principal investigators tried to deal with this difficulty one on one. Sometimes a few suggestions were sufficient. In other cases, offering to demo-teach the class or team-teach a unit took care of the problem. It was clear that we had to develop a systematic way to prevent difficulties in teaching with models. Appropriate changes were made in the summer training for the following years.

Similarly, the other problems were dealt with individually, as they arose. It was only when the end-of-the-year surveys were tabulated that it became apparent that some of the nonusers had the same concerns but never contacted the core team. It was clear that efforts had to be more proactive. The second and third years, more regular contacts were initiated by the core team. This reduced the number of nonusers but still left a significant minority who wanted to use the models but simply were not able to. The fourth year of the grant, we tried a new approach—one that will be carried into the new grant. The three summer days deleted from the training were added as work days on Saturdays. Those days were used to check participants' work and to deal with any problems. Although much of the time has been spent helping debug models, some has been devoted to making suggestions on classroom presentations. We have also placed participants in a position where they could easily arrange other help, as needed. This eliminated the "block" that so often seems to keep people from initiating a contact. The institutionalization of contact ensures that it actually happens in a timely manner.

Insufficient hardware was often cited as an impediment to implementation. This was curious, because access to hardware was a requirement for acceptance into the training program. Some teachers who found they were not going to complete their requirements may have used inadequate hardware as an excuse. In any case, the problem seemed to diminish as more schools naturally purchased additional computers. On the other hand, lack of sufficient copies of STELLA® arose as an issue repeatedly during the early years. It had been anticipated that the teachers would have their own copy with which to practice and that the school would purchase additional copies for student use. This was not to be. Consequently, starting in the second year, we scraped together enough money from the grant to purchase ten run-time versions of the STELLA® software for each participant, in addition to the full version, so they would have software to put on additional machines. At least the students would be able to run models and to create, within one class period, small models that did not need to be saved (as was generally the case for most math exercises). This was a reasonably satisfactory solution.

Building the subject-specific model was less difficult for most teachers than we anticipated. Most used one of the models built during training as a starting point and then made additions. The result was usually a fairly simple model that was easy to use in the classroom. In those cases where teachers encountered difficulty, a quick phone conversation usually sufficed. The Saturdays scheduled during the fourth year have resulted in a significant pattern change. The teachers have built more ambitious models, or sequences of models (such as the Rulers packet). The improved summer institute, coupled with better support through structured meetings, has resulted in dramatically improved modeling work. This reinforces the belief that support is essential for teachers to progress as modelers. The more assistance provided, the more rapid the progress. This support will also bring more teachers to the level where they can begin providing support to others.

Student Response

The real goal of all the work with modeling has been to provide students with an opportunity to use and build models in the classroom. Underlying all the work is the belief that using and building models allows students to explore problems in more depth, ask and answer better questions, and develop an understanding that their world is full of systems of varying complexity. Students come to realize that simple solutions rarely exist for complex problems. Simple solutions ignore the inherent complexity of human endeavor of an natural phenomena. In short, the whole point of bringing modeling into the classroom is to create an environment in which students become better thinkers.

The 4 years of the CC-STADUS project have provided ample anecdotal evidence that modeling does create such an environment. Moreover, the project directors have observed activities in more than 100 classrooms where systems were used this last year. And the authors have been using models in their math, science, and modeling classrooms since 1991. Some definitive conclusions about the use of system dynamics are possible.

First and foremost, the use of models has proved to be engaging for the students. Attention during the presentation of modeling ideas is substantially greater than in other classrooms. This is corroborated by the classroom teachers, who report that students tend to get more involved and are more attentive during both group and individual work with models.

Models tend to pique the student's curiosity. Secondary students have been the victims/witnesses of many educational trends and have been subjected to literally tens of thousands of commercials through TV. They are skeptical consumers, whether of new teaching approaches or of correct answers in classes. Models tend to focus and direct that skepticism in an intellectually positive way. For example, in many of the biology and social science classes, the first modeling work is done with population models—often those that represent growth patterns in Third World countries. The dramatic results of these models, showing 10- and 20-fold increases in population in the next century, are frequently met with student assertions that it won't happen the way the model shows. When pressed for reasons why, the students began to talk about variable birth rates, increased death rates due to malnutrition, and societal pressures to change land use.

In some cases, modeling never goes any further. Instead, discussion moves to pressures in the system that will either prevent or accelerate the predicted disaster. In these cases, although the model is not used to continue the learning, it provides the trigger that stimulates it. Often, the discussion makes reference to the model in an effort to evaluate what effect various contingencies would have on the outcome without ever actually modeling it.

Equally often, the model is revisited, usually at the direction of the teacher. One of the most effective ways of doing this is simply to ask students what changes they think need to be made to the model. The tendency is for students to want to incorporate a lot of changes all at once. However, when teachers work through changes one at a time, students get a clearer idea of how elements of the system interact. This approach builds a detailed understanding of very complex systems and allows students to generate their own knowledge. It also illustrates the fact that systems are often resistant to change because of their complexity and feedback structure. Understanding the concept of a leverage point in a system is a powerful idea that was developed time and again in the classes we observed.

Less common, but tremendously exciting, are the instances where students asked whether they could work with the model themselves. In only a few cases were these students who had had some formal training with STELLA®. In most cases, no previous exposure had occurred. These were simply students who found exploring an idea using a computer model exciting. These students usually needed help from the teacher (although the manual itself sometimes sufficed). The students often pushed the model well beyond what the teacher would ever have attempted with the entire class. Occasionally, the students used the model as a starting point for a new model.

Critiquing model results and asking questions based on models are excellent ways to facilitate student learning by modeling. Other approaches also work well. Some teachers have found it very useful to build a model in class, asking questions as they go. Student responses and suggestions shape the model structure. This has been particularly useful in two different situations. In a physics class, after discussing the equations on which the three common forms of heat flow (conduction, convection, and radiation) are based, the teacher and class built the three models. Comparing the structures of the three models was a good way to discuss similarities and differences among the processes. Actually running the models allowed the students to explore the relative importance of the processes. In a biology class, a discussion format led to a model of the environmental factors that affected bacterial growth in a local stream.

Whatever the approach, using models presents a means of addressing complex questions in class. This approach also brings home the idea that "right answers" don't really exist for many real-world problems. Instead, a range of solutions exist, and each spawns other problems and complications. This leads to excellent opportunities for students to write position papers, presenting arguments for and against potential solutions. Student work in this area has often been truly remarkable. Using a model to run trial solutions, students are able to develop written recommendations that rival the work of professionals.

Teacher Receptiveness

Assumptions we made at the beginning of the project about the openness of teachers to systems modeling in their classes have been almost completely wrong. As already noted, the two fields wherein traditional work translates most easily into systems models are physics and mathematics. Thus it seemed reasonable to assume that teachers in those areas would be the most open to the use of systems. In fact, the exact opposite has been true. Physics and mathematics teachers are the most likely to see no need for computer modeling. They concede that it may be useful in "those other fields" where problems are "ill-defined" and answers are not "precise and real." However, paths to solutions in their fields are well defined, are well understood, and do not need other approaches that "circumvent" the rigor of those disciplines. (*Note*: The quotations are not from any single individual but are, rather, a collection of statements made by a number of teachers over the last 5 years). Only when such teachers are shown a series of problems that their students (and in some cases, they themselves) cannot solve, but that yield readily to solution by models, do they begin to acknowledge that there is some usefulness in modeling. It is interesting that the most successful math and physics students, those taking advance-placement level courses, show a similar reluctance to use models. They also tend to moderate their views when confronted with problems outside their reach.

Other science teachers tend to be more open to modeling. They see it as a tool that allows experimentation with ideas and relationships, not merely thinking about how they should change. This same perception has been used as a compelling argument to go beyond systems thinking to system dynamics in discussing the appropriate use of systems concepts in education. Biology, chemistry, and earth science teachers all see numerous areas of their field that lend themselves to computer modeling. The main problem here is the tendency to make models too grand and all-encompassing. Modelers in these fields, particularly chemistry, find that it is hard to locate sufficient data upon which models may be based.

Next to science teachers, the group that has embraced system dynamics most enthusiastically has been the social science teachers. They seem to perceive systems as a way to represent visually the concepts with which they have always worked. The ability to work easily with numbers is seen as a plus, conferring some more traditional (mathematical) validity on their ideas and solutions. Our expectations initially were that this group would be most resistant, because fewer social science teachers had experience in computer programming. The first 2 years, this group showed more initial anxiety but still evinced the greatest enthusiasm for the potential of computer modeling. In the last 2 years, differences in familiarity and skill with computers have diminished, eliminating that initial concern for most teachers in this group.

Another factor that has led to increasing "comfort" for the social science group is the assurance that they will not be required to learn the mathematics necessary to make their models work. Because the participants are grouped into teams, optimal use is made of the particular expertise of each team member. The social science teacher provides many problem scenarios that are perfect for study with system dynamics techniques. The math teachers bring their understanding of functions and techniques to analyze data and define growth patterns via the appropriate systems structures. Actually, the math teachers were the most resistant to the group work until their role was defined. The roles each participant is to play in the team project toward the

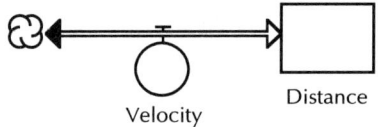

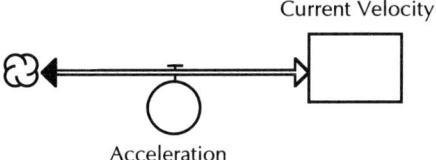

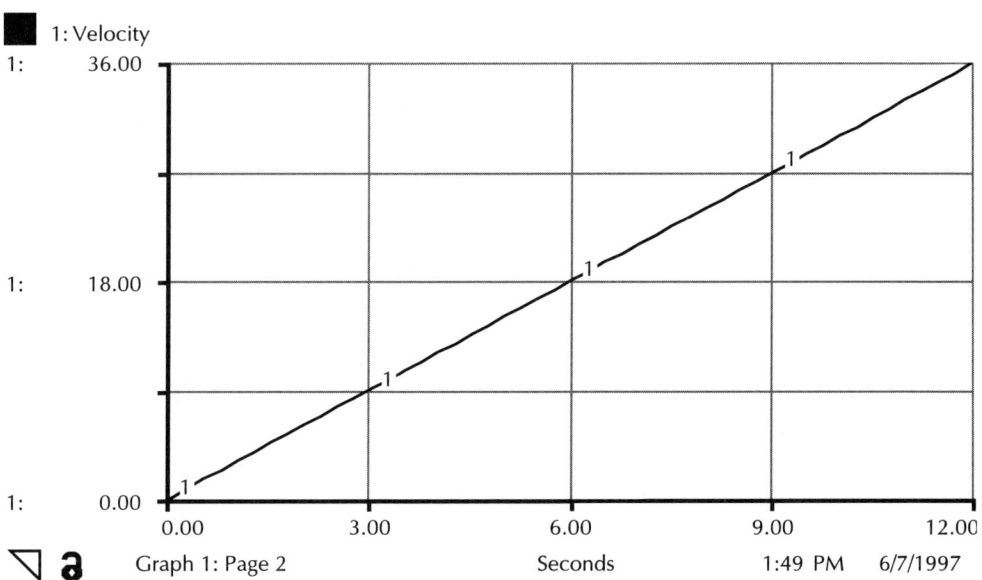

FIGURE 2.3 Constant-Velocity and Constant-Acceleration Models and Graph.

end of the workshop have to be explained early in the training so that expectations are clearly understood by each person.

Use of Models in Physics Classes

Because the language of system dynamics so closely matches the concepts of physics, initial work in developing models, both prior to the formal CC-STADUS project and during it, has focused on developing model analogs of the basic equations used in physics. Thus the initial models depicted nonac-

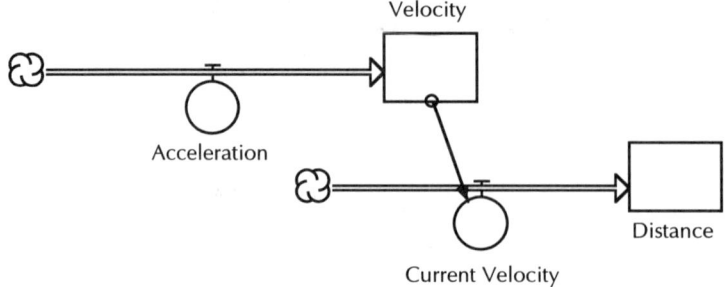

Velocity

Acceleration

Current Velocity

Distance

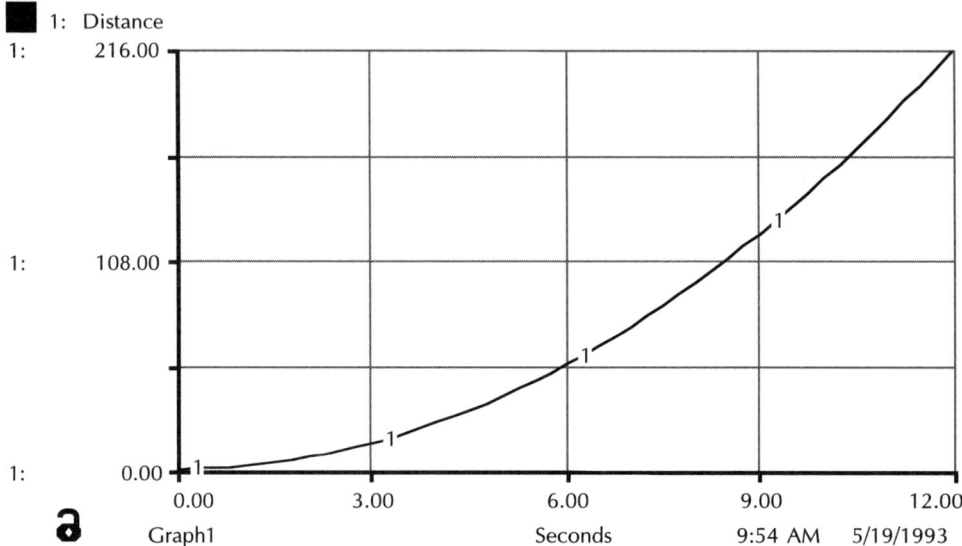

FIGURE 2.4 General Accelerated-Motion Model and Graphs.

celerated and accelerated motion (see Figure 2.3). These models are ex-amples of the linear and quadratic relationships taught to all CC-STADUS participants during the first week of their training. Using the language of stocks and flows to supplement the mathematics describing position, velocity, and acceleration has proved to be an effective way of introducing the concepts to students, as well a providing a means of checking their work as they ad-dress problems. The structure of the accelerated-motion model offers a unique advantage in explaining the differences among average, constant, and instantaneous velocity.

The constant-velocity model is the starting point for the accelerated-motion model. A second piece is added that changes the velocity as the ob-ject accelerates. Students in traditional physics courses learn that

$$\text{Change in position} = \text{velocity} \times \text{time}$$

for constant velocity only. However, when confronted with accelerated motion, they often fall back on the constant-velocity equation, which yields a wrong answer. The STELLA® model for accelerated motion provides a graphical clue that when acceleration is present, something very different is happening. The velocity is constantly being changed, so the rate of change in position (velocity) is going to change constantly as well. This makes the simple equation obviously wrong. Using graphs produced by the model, it is possible to look at the instantaneous and average velocities, emphasizing their relationship to distance (see Figure 2.4). Recently, some physics teachers associated with the project have begun to label the models differently, using "rate of change in position" and "rate of change in velocity" to emphasize the concept of rate when discussing velocity and acceleration. This added emphasis is expected both to reinforce the concept and to strengthen association of the model and concepts with the mathematical representation.

This use of STELLA® models to clarify and reinforce traditional methods of teaching concepts is typical of much of the use of models throughout physics. But the accelerated-motion model also makes available learning options that are unique to modeling. Consider: The traditional equations used to describe the motion of objects are all idealized. In particular, they ignore all consideration of friction, air resistance, and nonconstant accelerations or forces. Students develop a sense of frustration with this limitation. They find the concepts and equations difficult enough to learn and work with. Discovering that these do not fully describe any real problem calls into question the reasonableness of making the effort. The common teacher response, "You'll be able to answer that question when you learn partial differential equations," doesn't help.

Simple modifications to the accelerated-motion model make it possible for students to explore the real problems—even to get actual numerical results. One model developed for classes was the result of student responses to a classic physics problem. Students are presented with a situation in which a car at rest is passed by a moving truck. The car at rest then accelerates and passes the moving truck. The problem asks how long it will take for the car to pass the truck and how far it will go in so doing. Traditional solutions require students to write two equations of motion (one for the car and one for the truck), set them equal to one another, and then solve for time. Again, this problem usually puts unrealistic limitations on the situation, including constant truck speed and constant car acceleration. Some student inevitably observes, "I'd like to get my hands on that car, because *mine* sure doesn't accelerate uniformly to 60 mph!"

That comment can derail the entire process. It undermines the key idea that is the goal of the problem: recognizing that both vehicles travel the same distance in the same time interval, which makes a solution possible, if cumbersome. A simple modification to the accelerated-motion model (see Figure 2.5) allows the acceleration to change. The model, developed by students using acceleration data from *Road and Track* magazine, allowed learners to

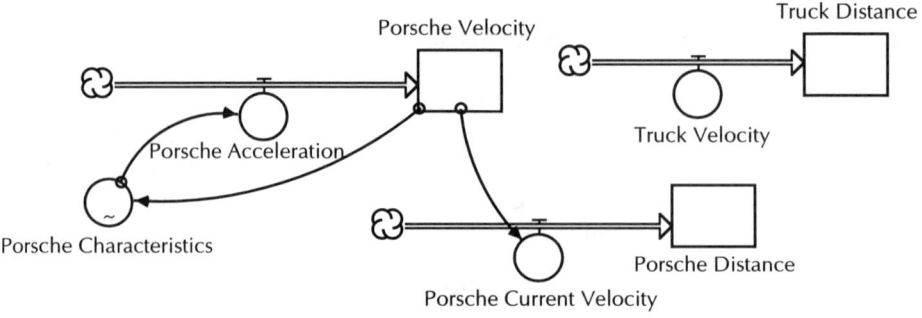

Graph for Porsche Accleration

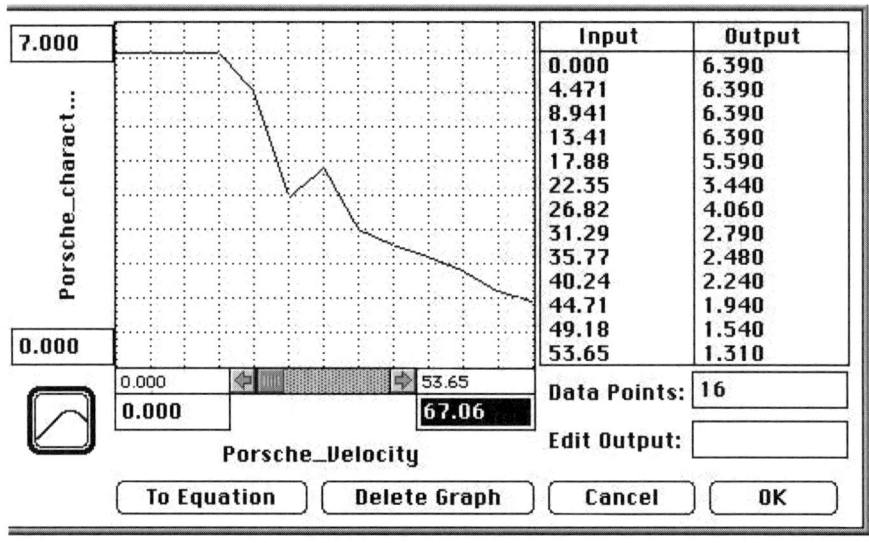

FIGURE 2.5 Car Model for Real Car and Graphs.

deal with a problem conceptually identical to the traditional problem, but more realistic. It also gave them an inkling of the ability of higher mathematics to take simple ideas and solve complex problems.

Students encounter similar situations throughout physics. *PSSC Physics*, regarded as the most rigorous secondary physics text, includes a short section on air resistance, including a problem dealing with a Ping-Pong ball (Figure 2.6). Most students are utterly frustrated by the problem. However, another simple modification to the accelerated-motion model enables them not only to solve that problem but also to explore solutions for other objects. It also, through the model structure, emphasizes the conceptual basis for air resistance and its dependence on velocity.

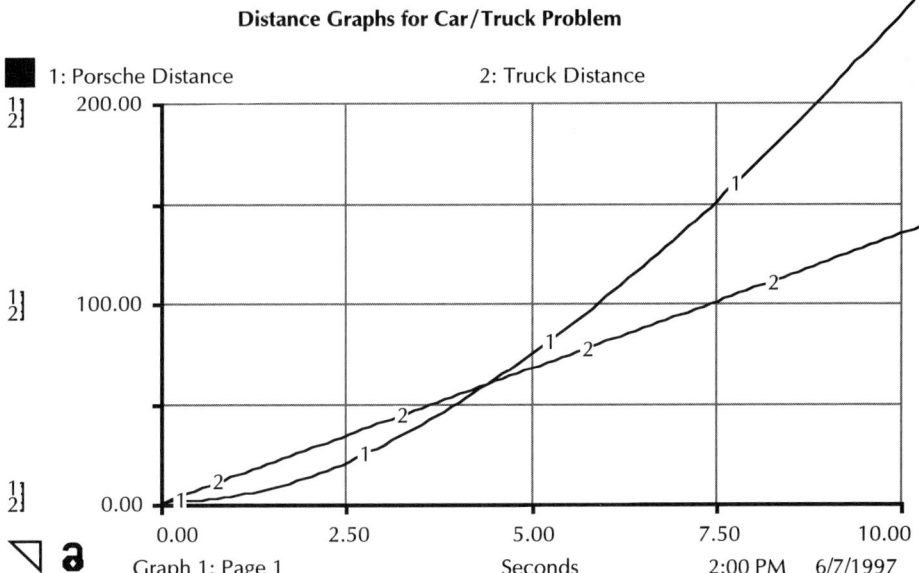

FIGURE 2.5 (Continued).

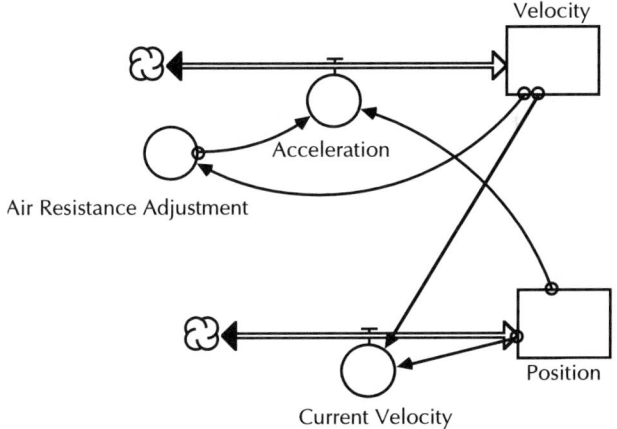

FIGURE 2.6 Ping-Pong Ball Falling Model.

Models have been developed and used to illustrate other commonly taught concepts in physics. Force models, dealing with changes in both position and momentum serve as frameworks for other models that deal with problems not generally addressed mathematically. A rocket model (Figure 2.7) uses the more basic force–momentum model to look at the effect a changing mass has on the impulse equation taught in all physics classes. This model has been used by some teachers to predict the performance of commercially available

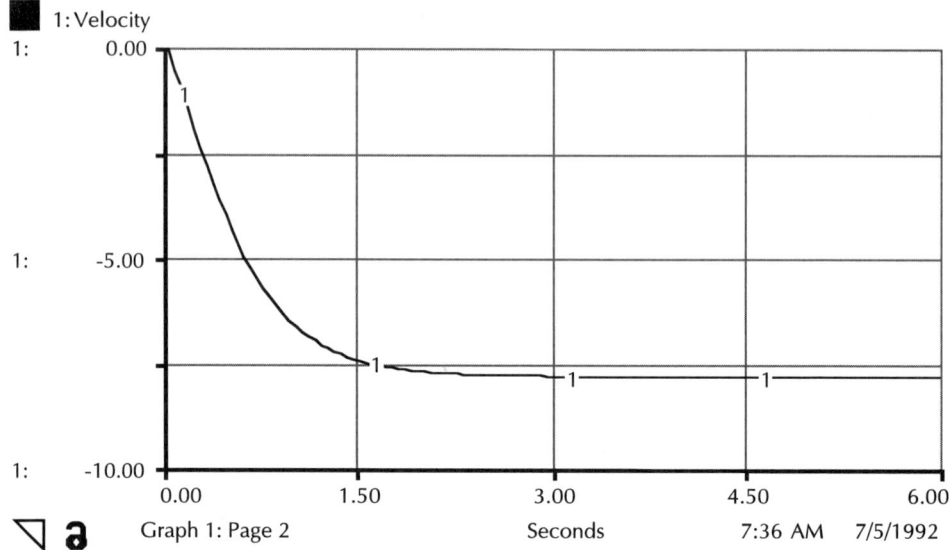

FIGURE 2.6 Ping-Pong Ball Falling Graph.

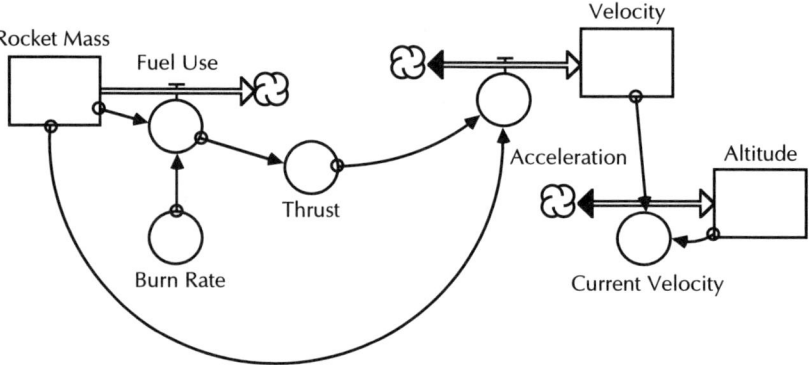

FIGURE 2.7 Rocket Model.

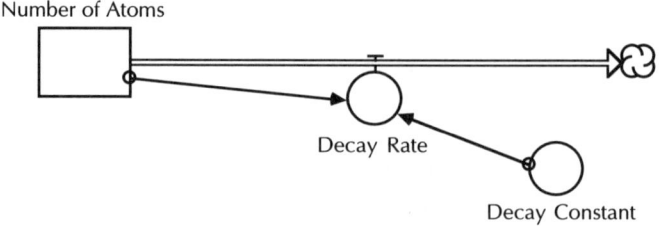

FIGURE 2.8 Generic Decay Model.

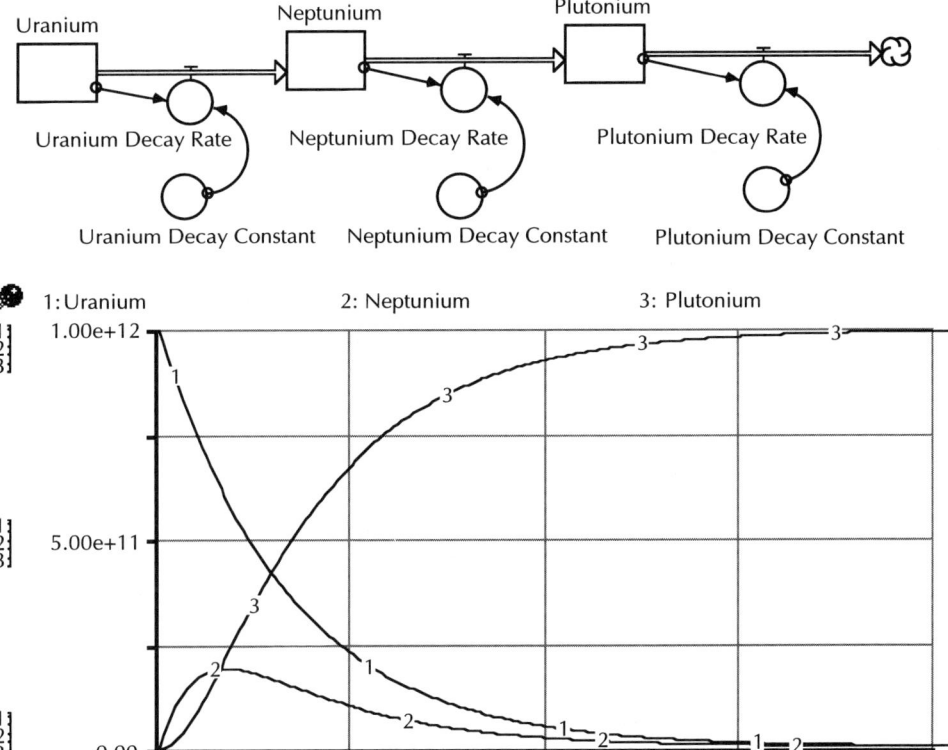

FIGURE 2.9 Decay Sequence Model and Graph.

model rockets, with good results. It has also been used to show how some major concepts, such as conservation of energy, can lead to erroneous assumptions if close attention is not paid to rates of energy transfer. This work involves the determination of optimal burn rates for rocket engines.

All physics models developed within the project or used by the project revolve around the simple model structures taught in the CC-STADUS summer institute: linear growth, quadratic growth, exponential growth/decay, and S-shaped growth. Although the bulk of the models focus on mechanics and kinematics, a number have been designed for use in thermodynamics, electricity and magnetism, and radioactivity. Many are presented as generic models for both whole-class situations and individual student use. Experience suggests that for developing new ideas, a whole-class environment, in which building or using the model is part of a discussion process, is most appropriate. That setting ensures a sort of "quality control," providing all students with the same experience as the concept is developed. Further exploration is usually done by students individually or in pairs.

A few models in use provide students with insights that are not normally developed in a secondary physics class. One is made possible by an extremely simple model referred to as the generic decay model (Figure 2.8). This simple exponential decay model is built (under careful direction) by students and used to explore radioactive decay and half-life, including radioactive dating. After completion of that activity, students move on to examining a three-stage decay sequence (Figure 2.9). Building this model allows them to understand what is actually happening in such a sequence (the stock–flow structure clearly shows the supply of one element flowing/changing to a supply of another), as well as the role half-lives play in determining the speed of progression of the sequence. Both are key ideas in understanding the problems associated with the disposal of radioactive wastes.

Work in Biology Classes

Although the material covered in biology classes is not normally presented in mathematical formulations that translate easily into systems models, the concepts dealt with are natural for modeling. Most biology teachers who have participated in the CC-STADUS Institute build and use variations on the simple population models while exploring exponential growth and decay. Some use these models without modification, but most regard them as a starting point from which they take a number of different paths.

Some teachers begin to explore links with other variables in the environment. Most commonly, they restrict these links to available food and living area. This investigation may be focused on humans or on other organisms. Often this idea is then carried further, using one of the cross-curricular models developed by the project (Mahenjo Daro or the Sahel) or expanding the models created. In one case, where students are conducting stream studies, students working with the teacher and an experienced modeler developed models that tied bacterial growth to some of the ten or more variables they had been following. A few biology teachers, after doing basic work with population models, explore interacting species using models such as the classic Lynx and Hare model developed by High Performance Systems, the publisher of STELLA®.

Although a few commercial models of cellular metabolism have been released, none are currently being used by biology teachers associated with the project. Several are planning to develop such models for their specific curriculum.

System Dynamics in the High School Mathematics Classroom

The past 10 years have seen some changes in two areas that have set the stage for the introduction of system dynamics in the mathematics classroom. Physics study expanded to include, in many high schools, a course in conceptual physics, so a foundation could be laid for more rigorous study in the future. The conceptual materials focused on developing an understanding of the key principles governing mechanics, sound, light, and electromagnetic phenomena. This course laid very important groundwork for future study with a much broader student audience than would have been reached by traditional physics, which is usually taught to juniors and seniors.

Also implemented at many universities and some high schools was the reform calculus, which broadened the vision of "doing calculus" to include more conceptual understanding of integration and differentiation. Students study functions from a numerical, symbolic, and graphical perspective and include written explanations wherever appropriate. (This has come to be referred to as the "Rule of Four.") These approaches have gained widespread acceptance and have opened to many students an avenue of understanding and study that is not provided by traditional courses in these areas. Developments in computer technology have been instrumental in supporting this transformation.

The study of system dynamics brings to high school mathematics those opportunities provided to students via the inclusion of conceptual physics and reform calculus. System dynamics provides an opportunity to study the traditional functions presented in most mathematics classes from a conceptual perspective. That context is the behavior of the function over time, emphasizing its rate of change. The teacher merely begins with the study of distance-versus-time situations using a motion detector. Students walk in front of the detector and produce a graph of their motion. Discussion revolves around the shape of the graph and how it is connected with the velocity of the person moving. Balls thrown into the air or oscillating springs can be analyzed in a similar fashion, embedding these concepts in the concrete experience of the students. More students understand this approach than the symbolic approach used to differentiate among functions. This does not exclude or obviate the need for the symbolic approach. It just increases the "Rule of Four" to a "Rule of Five," the fifth being a systems perspective on functions. The experiences with the motion detector can provide students with conceptual understanding of some of the characteristics of functions, generally a whole year before they are studied in the traditional, symbolic form. It is not necessary to make the symbolic connection at the same time the activity is introduced; one need only discuss the important "behavior-over-time" characteristics of the function/phenomenon being studied. This gives the student concrete examples on which to "hang" the abstract symbolic representation, once they study it. It provides the same preparation that conceptual physics classes do for mathematically based physics classes.

Having introduced functions in such a concrete manner leads directly to expressing each function type using STELLA® symbols. The focus of the STELLA® representations follows so directly from the rate-of-change study that it crystallizes the concepts introduced using the motion detector. The STELLA® diagrams show the actual "internal system behavior" of the functions. It is as though one were able to look at the internal workings of a black box that one had tried to study looking only at inputs and outputs. The view

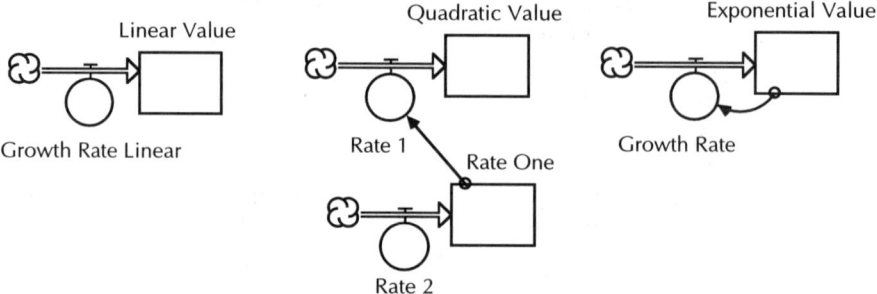

FIGURE 2.10 Linear, Quadratic, and Exponential Generic Models Diagram.

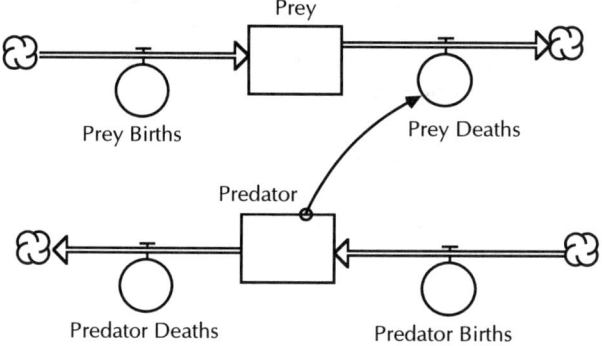

FIGURE 2.11 Predator/Prey Model.

is awesome, illuminating, and powerful. Nearly all of the standard function types studied in traditional algebra I, algebra II, precalculus, and calculus classes can be easily represented by STELLA® diagrams. Linear, quadratic, exponential, convergent, logistic, and sinusoidal functions are all expressed simply and easily using STELLA® (see Figure 2.10). And other, more complicated functional relationships can be introduced if desired.

This visual, conceptual view of function structures is very useful for students and gives them another way to understand what is already being presented in math classes. If this were all that could be accomplished, it would be worth the effort. But it is just the beginning. The introduction of system dynamics for representing functions opens a door that allows many students to make a quantum leap in numerical analysis of real-world applications. An altered view of the world—a system dynamics view using the STELLA® language—will change what we teach students at the high school level in the future. Because the STELLA® language enables students to represent problems using a visual-diagram structure, students understand better how to design and analyze problems. As one designs the diagram, dependencies become transparent. Problems can be studied that contain both increasing and decreasing rates of change. Consider that 98% of the problems currently studied in algebra and precalculus contain only a growth or a decay component, not both. But real-world problems nearly always contain both. It is possible to introduce more interesting problems at earlier levels by using STELLA® diagrams, because the visual design is conceptually easier to create. Problems such as predator–prey interactions that create oscillations (see Figure 2.11) can be introduced in algebra II, during the study of circular functions, expanding the typical swing/spring/Ferris wheel examples.

When problems are introduced conceptually, via the STELLA® interface, students have a better understanding of the symbolic representation they encounter in later courses, because they have something more concrete to hang it on.

It does not take a significant amount of time to implement these changes in mathematics classes, given an interested teacher who is willing to experiment building models herself or himself. Ten to twelve days in the school year are sufficient to include most of the motion detector and STELLA® activities we have discussed. The key, however, is to have the students involved in the action part of the motion detector exercises and, even more important, to have the students *always* build the STELLA® models for the functions they are studying. The value is reduced significantly if either of these activities is done merely as a demonstration. "If they build it, they will learn," has been the motto of our project and the teaching of system dynamics from its inception. The building can be done as a classroom exercise in which the teacher, working with one computer at the front and assisted by student input, develops the model. In math it works best to have the students in a lab, building their own solutions to problems after key concepts have been presented or discussed in a classroom setting.

Assessment is not difficult. Merely presenting a problem for which students must draw an appropriate STELLA® diagram, with identifying labels and numerical definitions, works well. Students can explain motion graphs or draw motion graphs, given a physical description. Students can be given problems as projects to be turned in as a part of reports, requiring that they design STELLA® models and/or use motion graphs.

Although much can be done in traditional mathematics classes to enhance learning via system dynamics, teachers are often constrained by a traditional syllabus that mandates spending significant time on required topics. Consequently, it is difficult to realize the full impact of the study of system dynamics in traditional mathematics courses. An independent course that focuses on system modeling, however, brings out the power of this method of study beyond question. At Franklin High School such a course has been in the curriculum for 5 years. The number of students taking this class as a math elective has grown from 11 to 60; students find this method of study interesting and useful. At Wilson High School, similar work is being done in a science class devoted to dynamic modeling. In both courses, exercises are provided during the first semester that include techniques of data analysis, learning to design a variety of models using the STELLA® software, and learning to vali-

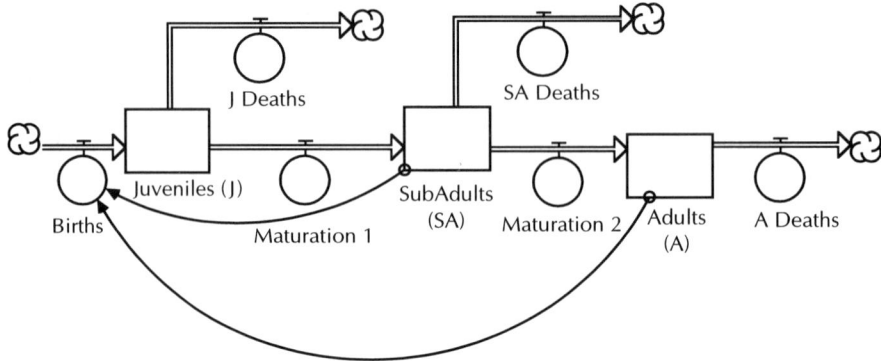

FIGURE 2.12 Pronghorn Population Model.

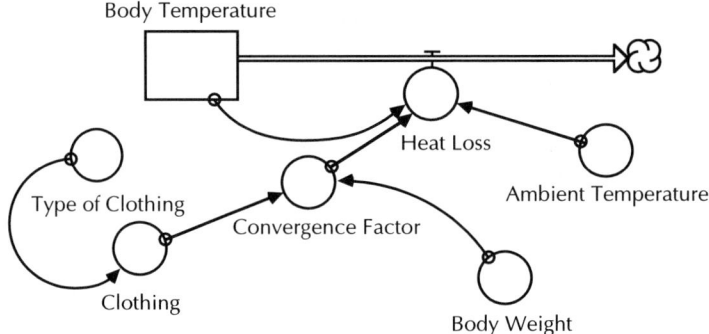

FIGURE 2.13 Time of Death Model.

date and explain the small models that students are expected to create. During the second semester, students work in teams on projects. They decide on a problem to study, contact a resource person or expert on the topic, collect data, design a working model, validate the model, and write a 10- to 20-page technical paper explaining their model.

One model created by two high school juniors studied the population dynamics of a herd of pronghorn antelope in central Oregon. A wildlife biologist in that area had recently modeled the population, and he faxed the students his spreadsheet results and the observed data. The students created a model similar to Figure 2.12 and were able to obtain results closer to the observed data than the results of the wildlife biologist.

Two seniors, one of whom had taken the forensic science class at Franklin High School, studied some of the factors that are considered in determining the time of death of a corpse. They spoke several times with a medical examiner in Portland, who generously lent them charts and articles to read on the subject. They created a model that took into account three primary factors: ambient temperature, weight of the body, and whether the body

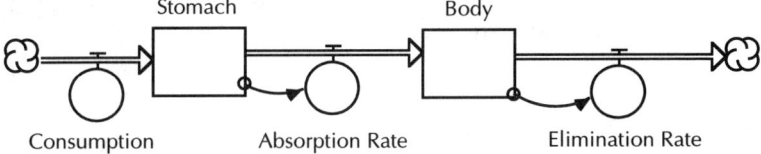

FIGURE 2.14 Alcohol Consumption Diagram.

wore dry clothing, wore wet clothing, or had no clothing (Figure 2.13). This model, although simple, won first place in the modeling competition (see SYM*BOWL in the following section).

A very interesting model relates the increase in blood alcohol level (BAC) to strength of the alcohol, gender of the drinker, size of the drinker, and length of time over which the alcohol was consumed was created by two high school seniors. This model set the stage for studying not only the variables included but also legal arguments involving driving under the influence of intoxicants. A similar model is shown in Figure 2.14.

Over the past 5 years, students have created over 80 models that study problems ranging from how one becomes addicted to cocaine to whether a current tax-cap ballot measure will undermine school funding, to expenses involved in owning and operating a bed and breakfast. The problems are interesting to the students, who create the models and write papers beyond what would ordinarily be considered high-school work. Indeed, university professors who have seen these models and papers have commented that they have some graduate students who could not produce work of the quality that these junior and senior high school students are producing. Next year the Franklin modeling course will have 20 high school freshmen among the 60 students who will take the course. At Wilson, 25 freshmen will be among the 57 students taking first- or second-year modeling. The prerequisite for the Franklin modeling course is completion of algebra I with a grade of A or B. This course is a perfect application of the mathematics concepts that are being taught in traditional courses.

An interesting characteristic of the modeling classes is that 50% of the class enrollment is female. No special efforts made were to this end. Of the models we have presented, the "Pronghorn" model was created by a female–male team, the "Time of Death" model was created by two females, and the "Alcohol Consumption" model was created by two females.

Bear in mind that Franklin High School is an inner-city high school that serves primarily a blue-collar population. Only about 20% of the school population attends 4-year institutions of higher learning after graduation. After next year, system concepts and the use of STELLA® will be expanded to pre-algebra courses to determine whether students who are less successful in the traditional track respond well to a systems approach. Wilson High School is in a considerably more affluent city environment. Many of the students enrolled in systems classes there, however, are not among the highest-achieving students. It has become a desirable course for creative students who are not excited by traditional math and science classes.

Plans for the Near Future

Both Wilson High School and Franklin High School will offer system (magnet) programs, which started in 1997. Wilson will focus on system study in science and social science, providing 2 years of system modeling and a course in science, society, and technology. Franklin's program will focus on system study in mathematics and science and will require the system students to create, each year, a model on a topic relevant to their current level of mathematics and science. In both programs, students will be required to take 4 years of math, 4 years of science, and 2 years of system dynamics modeling. Ultimately, the programs at Wilson and Franklin will be shared to expand the systems courses offered at each school. The goal is to train students to do systems modeling early in their high-school career so that they can use the theory and tools in most of their courses throughout high school.

Both Wilson and Franklin have system dynamics modeling classes. Three more high schools in the Portland area will add systems modeling courses to their curriculum in the 1997–1998 school year, and two other schools plan courses for 1998–1999. As more participants are trained in the NSF CC-SUS-TAIN project, it is expected that the number of schools that offer systems modeling courses will grow. A growing number of teachers in the middle schools are also giving their students system dynamics experience by providing simple models that students manipulate or using activities like the FishBanks simulation.

In 1996 a research pharmacologist who has used STELLA® in his drug research for about 10 years directed a high school modeling competition called SYM*BOWL. The competition was held at Oregon Health Sciences University in Portland, Oregon. The first year, 16 high school teams participated from four high schools in the Portland metro area. The second competition was held in 1997, and 20 teams participated, one of which flew in from Vermont. Students provide their written technical papers to the judges a week before the formal competition. At the competition, students bring posters and computers and spend the morning answering questions posed by judges who are assigned to cover each student group. From this group of 20, five or six top teams are selected. In the afternoon each team presents its model to the entire audience. From the top five, three teams place. Each of the students in the top five teams receives a copy of the STELLA® software. The students in the top three teams receive plaques and money prizes. The number of participants continues to grow, as does public interest in the competition. As the number of schools who have trained teachers grows, the number of teams competing will grow as well, so we expect there will have to be regional competitions, the winners of which will compete nationally.

The future of teacher training in systems is part of the design of the NSF

CC-SUSTAIN (Cross-Curricular Systems Using STELLA®: Training and Inservice) grant. The acronym represents the intent of the grant. If the preparation of teachers who learn systems modeling is to be sustained, it must be institutionalized. The project directors are currently working with faculty at both Portland State University and the University of Portland to include systems modeling as part of the teacher training programs. Additionally, the Math Learning Center, a not-for-profit organization that provides workshops in 40 states, will work with the directors to design workshops of four lengths to include in their offerings. Workshops will be half a day, 2 days, 5 days, or 12 days long (similar to the training provided each summer by the NSF grant). The directors will design the training materials, and the CC-STADUS and CC-SUSTAIN participants will become the instructors for these workshops. Finally, as part of the CC-SUSTAIN grant, a satellite group of nine teachers from one location in the United States will be trained each of the 3 years of the grant. These satellites will be supported by visits from the directors for 3 years so that, each year, they can provide training of increasing length in the summer. At the end of 3 years, each should provide a 12-day training program in its area similar to the one in Portland, using its own modelers as trainers, thus becoming a center for systems training.

It is necessary that published materials become available to support this effort. The directors will provide the Math Learning Center with training packets of materials for the workshops it will offer. Additionally, each director will produce booklets on the following topics: "The Curriculum for a Year-Long Course in System Dynamics Modeling at the High School Level," "Systems Models and Curriculum to Support the Study of Physics at the High School Level," and "Systems Modeling Curriculum to Support the Study of Mathematics at the High School Level."

It is quite apparent that current developments in both computer hardware and software require changes in how we teach and what we teach. The power of system dynamics theory, coupled with an intuitive software language like STELLA®, provides a vehicle for students to understand complex nonlinear problems that are significantly more advanced than can be accessed by any tools previously available to high school students. The fact that the tool provides a visual representation of the problems studied opens doors to students who have not found traditional methods helpful. Not only does system dynamics theory make it possible to reach more students, helping them understand the concepts we are currently trying to teach them; it also provides new areas of study for students who have found the traditional approach to a subject uninteresting because of its oversimplification or the lack of a variety of problems to study. We have found students much more receptive to using a tool such as STELLA® than are adults who have invested a lot of time learning to find solutions "the old way." Tools like STELLA® and techniques of analysis like system dynamics will become part of the regular methods of solution for students in the future. As young students have led the computer revolution, so will they lead the transition to system dynamics—using tools like STELLA® in analyzing problems. They are so receptive because this is such a reasonable way to think about problems.

Long-Term Vision

Our perspective on the future of systems may be somewhat surprising. First of all, we like to compare the spread of systems work to an infection. Initially there are very few cells present. However, because infections are classic examples of exponential growth, about the time the growth becomes noticeable, it quickly "takes off." The infection can be easily controlled only in the early stages, when numbers are very small. We believe the use of systems is in that early, fragile stage. Although it is exhibiting exponential growth, it is still below the "level of notice" because the number of teachers who are using systems is tiny compared to the total number of teachers. The next 5-7 years will be critical. If the current level of growth in use can be maintained, systems will soon become highly visible, even though still used by a small minority of teachers. If this is as powerful a tool for learning as we think, that growth will continue.

What, then, will the role of systems be in education 15-25 years down the road? By that time, systems will be relatively widely accepted as a tool—as a way of learning. All curricular areas, with the possible exception of foreign language, will be using systems as one of the primary instructional tools. Not in every class everywhere, but in a significant number of schools—a number that increases steadily. That does not mean that all instruction and learning will be through the use of systems. System thinking/dynamics is not a cure-all. There are some topics for which it works extraordinarily well. For other topics it may not make a significant difference. With still other topics, it may actually be detrimental. However, systems will be a broadly accepted tool in general use.

That does not necessarily mean that everyone will be modeling. Building a model is a higher-order skill than using a model, and not everyone needs to do it. Just as relatively few people today take programming courses in machine language or in languages such as C++ and Pascal, though many of us use software written in those languages, many more students and teachers will use models than will build them. That simply means that models and system dynamics will have become an accepted tool for the masses, as word processing and spreadsheet software have already.

A strong indicator of the *failure* of system dynamics would be the emergence of System Thinking/Dynamics departments in secondary schools. Segregating and separating systems would undermine its greatest potential: the capacity for linking ideas across traditional boundaries. The systems revolution will have succeeded when most students see models in every class, every year. That goal is probably 50 or more years distant, but when it is achieved, the objective of systems thinkers—educating everyone to understand the interconnectedness of learning and ideas—will have been reached.

References

Callahan, J., & Hoffman, K. 1995. *Calculus in context*. New York: W.H. Freeman.

Forrester, J. W. 1971. *Principles of systems*. Portland, OR: Productivity Press.

————. 1973. *World dynamics*. Portland, OR: Productivity Press.

Goodman, M. R. 1983. *Study notes in system dynamics*. Portland, OR: Productivity Press.

Haber-Schaim, U., Dodge, John, Walter, J. 1986. *PSSC physics*. Lexington, MA: D.C. Heath.

Hewitt, P.G. 1981. *Conceptual physics*. Boston, MA: Little, Brown.

Kauffman, D. L., Jr. 1980. *Systems 1: An introduction to systems thinking*. Cambridge, MA: Pegasus Communications.

Meadows, D. H.; Meadows, D. L., & Randers, J. 1992. *Beyond the limits*. Cambridge, MA: Pegasus Communications.

Meadows, D. H. 1991. *The global citizen*. Cambridge, MA: Pegasus Communications.

Richmond, B., & Peterson, S. 1996. STELLA®: An introduction to systems thinking. Hanover, NH: High Performance Systems.

————. 1996. STELLA® applications. Hanover, NH: High Performance Systems.

Roberts, N., Anderson, D., Deal, R., Garet, M., and Shaffer, W. 1983. *Introduction to computer simulation: A system dynamics modeling approach*. Portland, OR: Productivity Press.

Smith, D., & Moore, L. 1996. *Calculus modeling and application*. Lexington, MA: D.C. Heath.

Stroyan, K.D. 1993. *Calculus using Mathematica*. San Diego, CA: Academic Press. Harcourt.

————. 1993. *Calculus using Mathematica, scientific projects and mathematical background*. San Diego, CA: Academic Press.

3

Construction of Models to Promote Scientific Understanding

Michele Wisnudel Spitulnik

Joeseph Krajcik

Elliot Soloway

Students should use scientific knowledge to build explanations of phenomena and to make informed decisions (American Association for the Advancement of Science, 1993; National Research Council, 1996). We propose promoting the *use* of scientific knowledge by involving students in the construction of models. This chapter (1) describes a curricular unit designed to promote inquiry, collaboration, and the building of dynamic models; (2) presents one case study that examines students' understandings as they engage in this type of learning environment; and (3) reflects on the usefulness of a model-building environment with respect to science understanding.

Model-It

Model-It, developed by the Highly Interactive Computing group at the University of Michigan, is designed to facilitate model building by allowing students to define what they want to include in their model, build relationships between variables in their model, and test and evaluate their model (Jackson, Stratford, Krajcik and Soloway, 1995). The software supports students constructing models in several ways (Jackson *et al.*, 1996). First, it allows students to define the things, or objects, to be included in the model. These objects are represented visually, either with digitized images or with other graphics such as clip art or student-generated drawings. Examples of objects include people, cars, the atmosphere, and factories. Once the students define the objects they want to include, they define the factors, or the measurable characteristics of the objects. Examples of factors include amount of carbon dioxide emitted by cars, level of chlorofluorocarbons in the atmosphere, level of ozone in the atmosphere, and amount of pollutants emitted by factories. Next, students define relationships between factors. Model-It allows students to define these relationships in a number of ways. They can, for instance, construct sentence-like qualitative relationships or input real data; in both cases, the software "translates" the qualitative or data-based representation into a graphical representation. An example of a student-generated, qualitative, sentence-like relationship is "as the level of chlorofluorocarbons increase in the atmosphere the level of ozone in the atmosphere will decrease." Finally, Model-It allows students to test these relationships and their model by "running" the model. Students use meters to change values for the factors they want to test. They can also watch the status of factor values by viewing a graph. In this way, students can generate "what if" questions and predictions, and they can test different hypotheses within minutes.

The Global Climate Change Unit

Context

The global climate change unit was designed for the project-based science program called the Foundations of Science, or FOS. The project-based science class was developed with the support of a research project at the University of Michigan, funded by the National Science Foundation. The Foundations of Science class is designed as a 3-year interdisciplinary (earth science, biology, and chemistry) course, FOS I-III. The goals for this 3-year course are to integrate the separate science disciplines; to do real science, using projects as the driving force in the curriculum; and to create a classroom where the use of computational media is routine. Researchers at the University of Michigan have worked collaboratively with the Foundations of Science teachers for 4 years to develop curricular units that engage students in inquiry and dynamic model building.

Projects and Models

The Foundations of Science class uses the features of project-based learning (Blumenfeld et al., 1991) as a guide for designing curricula. Specifically, many of the projects include a driving question that serves as the focus for the whole unit of instruction; student-designed investigations; student-constructed models; students collaborating with other students, teachers, and outside resources to obtain information; and use of technological tools to support data gathering, analysis, telecommunications, and presentation. The students in the FOS I course engage in a couple of projects during the year. The first project, which lasts for one semester, focuses on the driving question "Is our local creek safe?" Students work collaboratively to collect and analyze stream water and, using Model-It, construct a model of the quality of the stream.

The second semester of the FOS I curriculum revolves around the question "Is our climate changing and does it matter?" This question is divided into three smaller questions: (1) "What is our weather like now?" (2) "What can we predict for our future climate and what impact does this have on our environment?" and (3) "What was our climate like in the past and what evidence

TABLE 3.1 Goals for the Global Climate Change Unit

Content Goals

General
- Describes the *purpose* of the content, how it might be used or how it relates to fundamental ideas
- Constructs *relationships* between concepts (and different symbolic representations) to develop in-depth explanations
- Explains phenomena and/or makes decisions using facts and concepts
- Constructs, interprets, and uses *many symbol systems* to explain phenomena

Specific
- Explains the natural factors that influence climate and climate change
- Describes the sources of human pollution
- Describes different types of pollutants
- Describes the impact of pollutants on climate and climate change
- Describes the impact of climate change on environment
- Describes any preventive measures or possible solutions

Nature of Science Goals

- Describes science, models, and theories as human constructs that are subject to change (tentative nature of science)
- Describes *purposes* of models, theories, and evidence and how they are used by different groups of people (scientists, politicians, business professionals)
- Evaluates (including limits and assumptions) models, theories, or evidence in terms of articulated attributes of a good model, theory, or set of evidence

Inquiry Goals

- Defines a problem area and constructs a systematic method to address the problem area
- Constructs and revises models (representations, explanations) to explain a phenomenon and/or make predictions
- Uses empirical evidence and/or models to construct or justify an *argument* for or against a stated position
- Evaluates arguments

do we have of paleoclimate changes?" The second question is the focus for this model-building unit.

Goals of the Global Climate Change Unit

The goals for this unit are divided among three main areas: Content, Nature of Science, and Inquiry (see Table 3.1 for summary). The Content goals include what students are expected to do with the scientific content while they construct their models. For example, students should be able to explain the phenomenon they are investigating and why this phenomenon is important. Specifically, the teachers want students to demonstrate an understanding of (1) the factors that influence climate change, (2) the sources and types of human pollution that contribute to potential climate change, (3) the impact of pollutants on climate, (4) the impact of climate change on the environment, and (5) preventive measures or possible solutions. The Nature of Science goals include what we would like students to understand about the scientific enterprise. For example, students should be able to describe the purposes of models, who builds models and why, and how one might evaluate the quality of a model. The Inquiry goals include what students should be able to do in researching a problem. For example, students are expected to define a problem area and design a systematic approach to address the problem; students build a model of their system to explain the problem; and finally, students should make some claims about their problem and justify the claims with empirical evidence.

The Global Model Planner

Along with Model-It, the Global Model Planner was designed as a way to support students' building of scientific understanding throughout the Global Climate Change Unit. Stratford highlights the importance of students planning first, before beginning to build models (Stratford, 1996). Perkins and others highlight the importance of students defining goals and purposes for the work they do (Perkins, 1996) . The intent of the Global Model Planner is to help students plan, reflect on, and evaluate their models. The planner prompts students to define goals for their project, define purposes for their model, and decide what objects and factors they would include in their models. It encourages students to reflect on their models by asking them what assumptions and limits they are building into their models. Finally, the planner encourages students to evaluate their models in terms of the goals they set and the use of evidence.

Translating Goals into Classroom Action: A Brief Account

The unit revolves around groups of students investigating the driving question "Is our climate changing and does it matter?" and building a model to represent some aspect of this phenomenon. Figure 3.1 is a schematic diagram that represents the main features of this project. With the teacher's guidance, students define a meaningful and reasonable problem and begin researching their issue. Students locate resources, refine their problem, and collect more information. As students continue to track down relevant resources, the teacher engages students in role-playing activities designed to highlight how and why people build and use scientific models and how different groups of people may respond to opposing models of the same phenomenon (Nature of Science goals). The teacher also structures discussions around how and where people are finding useful information. Lessons, demonstrations, and other laboratory activities are also incorporated to help students learn about some of the scientific concepts they encounter.

Once students have enough information, they begin to make sense of it by organizing it to give it structure (Content goals). Before students begin to build their models, they first plan their models with the Global Model Planner. The planner encourages students to identify the objects in their system (or the things out there in the world, such as people, cars, and plants) and the factors in their system (or the things about the objects that can be measured—for example, the number of cars people drive and the amount of carbon monoxide emitted from cars). The students also describe the relation-

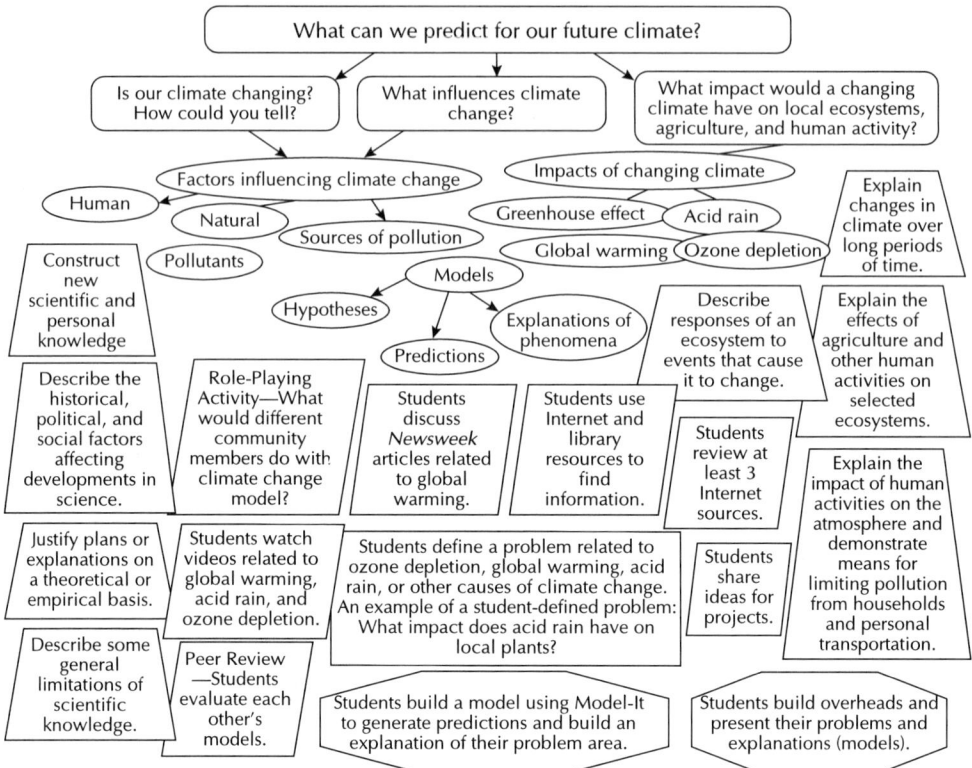

FIGURE 3.1 A Map Representing the Global Climate Change Unit.

ships between these factors. For example, as the number of cars increases, the amount of carbon monoxide increases (Content goals). Students present their plans to their classmates and teacher for comments.

Students begin to build their models, using Model-It, by incorporating the information they have found. As students continue to build their models, the relationships between factors become better defined and evolve from very descriptive, qualitative relationships to more quantitative, data-based relationships. Students explain each of the relationships (Content goals) within their models and then begin running and testing them. The teacher encourages students to make predictions about what they expect to happen and then to verify whether it actually happens. Students then evaluate their models and revise them so that they better reflect the problem they address (Nature of Science and Inquiry goals). A round of peer reviews is organized by the teacher so that every group of students evaluates another group's model. Students use an evaluation criteria sheet in the Global Model Planner to help give each group useful feedback. Finally, students present their work to their classmates, and the presentations are video-taped for Cable Access Television. During the presentations, students describe the problem they chose and the purpose of their model. They demonstrate and explain their model and justify any claims they make on the basis of scientific evidence.

"Ozone Depletion is a Real and Threatening Problem" A Case Study of Jamie, Lauren, and Rachel

This case study explores the understandings of one group of students as they engage in the Global Climate Unit. The case study presented here is one part of an ongoing research program within the Foundations of Science classrooms and is the culmination of 4 years of collaboration between the Foundations of Science teachers and the researchers at the University of Michigan. The students in this case had previously used Model-It to construct models of stream quality, so they were already familiar with the software.

The data used to support this case study come from several sources, including the model the students constructed, interviews with the students during and after construction of their model, copies of the students' Global Model Planners, and classroom video tapes. Each data source is analyzed and coded for categories within the areas of content, inquiry-related understanding, and nature of science. For example, the categories used to identify content understanding include use of purpose, construction of relationships and explanations, and use of symbol systems. Categories used to identify inquiry understanding include defining a problem, constructing a model, constructing an argument, and evaluating an argument. Categories used to identify nature of science include descriptions of the purpose of models and evaluation of models. The findings from all of the data sources are used to build a descriptive case for each group's content, inquiry, and nature of science understanding. The case of Jamie, Lauren, and Rachel is a representative example of average to above average students in the ninth-grade Foundations of Science classroom.

Building Content Understanding

The process of building a model allows Jamie, Lauren, and Rachel to construct robust content understanding. The students build this understanding by defining and revising a purpose for their model over time, by identifying the objects and factors to include in their model, and by building relationships and explanations of the phenomena within their model. The following paragraphs elaborate on how model building supports the construction of Jamie, Lauren, and Rachel's content understanding.

The Purpose

Jamie, Lauren, and Rachel decide early in the unit to focus their model building on ozone depletion. They explain that the purpose of their model is to identify "what causes ozone depletion, what is effected by it, how to prevent it, and how to deal with the existing problem (Global Model Planner)." Once they have a chance to work on their model, their purpose becomes more specific: "(We want to) show the process of ozone-depletion, show the effects on life on earth, show the sources of ozone-depleting substances, and talk about what's being done and what has been done for production (Classroom Video)." The students remain interested in "how to deal with the existing problem" and want to consider actions that will have an effect on the problem. In an interview at the end of the unit, the students stress this action component of their project. Rachel explains the purpose of their project:

Rachel: It's sort of like a community thing. Sort of like educating the community. What we wanted to do is, we just wanted to find out about it, and we really wanted to find out what's been done to stop it and maybe like a few alternatives that have been proposed.

Building Relationships and Explanations

Construction of the ozone depletion model gives Jamie, Lauren, and Rachel an opportunity to decide what objects and factors to include in their model and an opportunity to build relationships and explanations of the phenomena. Jamie, Lauren, and Rachel decide that they want four objects within their model. They describe these objects in their Global Model Planners: "atmosphere" will be included because "this is where ozone depletion takes place"; "chemicals that destroy" will be included because "these are what destroy the ozone layer"; "sources of chemicals" will be included because they "show the culprits of our problem. The chemicals come from places and we need to know where"; and the "earth" will be included because it will "show the effects of ozone depletion on life on earth."

These students also discuss some initial ideas for the factors associated with these objects. They describe four factors for the object "atmosphere." These are "stratosphere ozone layer, O_3, methane, chlorine nitrate" and they are included because "these chemicals are in the ozone layer and they play a role in ozone depletion." The object "chemicals that destroy" has several factors, including "CFC's, halons, methyl chloroform, carbon-tet, methyl bromide, HCFC." These factors are included because they are "the causes of ozone depletion." The object "sources of chemicals" has several factors, including "aerosol spray, foam blowers, solvents, cleansers, ingredients for other substances, agricultural uses, seaspray, substitutes for CFC—industry, cars, air conditioners." The students' reason for including all of these is "so we know what products not to use and know what uses of the chemicals there are." Finally, the object "earth" has three associated factors: plants, animals, and ecosystems. The students want to include these "to show what the effects of ozone depletion can do to us and our environment" (Global Model Planner).

As the students continue to build their model, they refine it by re-evaluating what should be included. They continue to have the objects "atmosphere, earth, and source," but the object "chemicals" becomes a factor of the object "atmosphere." Jamie explains in her Global Model Planner:

The issue of ozone depletion involves many parts of our world. The first is atmosphere, naturally because it is where ozone is and where ozone depletion takes place. The factors of atmosphere are the chemicals involved in ozone depletion. This makes

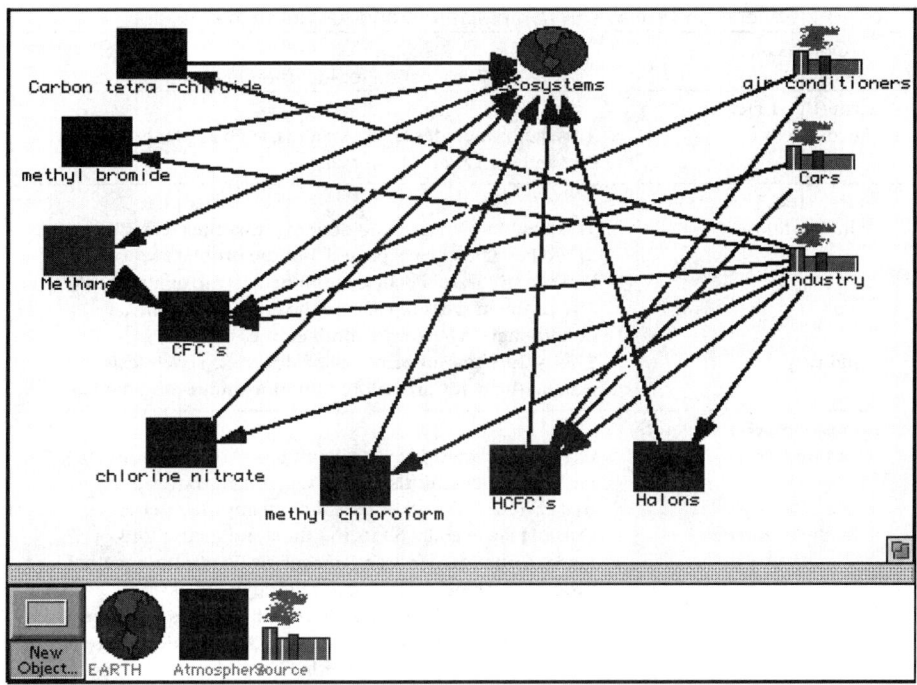

FIGURE 3.2 Jamie, Lauren, and Rachel Construct a Factor Map with Model-It. This is the final version of their model.

sense because the atmosphere is where they do their damage and where they are when we want to focus on them. Earth is an object and has ecosystems as the factor. Source is an object with all the sources of the chemicals as factors.

In their final model, the students use only the three objects: atmosphere, source and earth (Figure 3.2). The factors are similar to their original ideas, with just a few modifications. The object atmosphere has several factors, all of which are chemicals or pollutants that contribute to ozone depletion. These include carbon tetrachloride, methyl bromide, methane, chlorofluoro-carbons (CFC's), chlorine nitrate, methyl chloroform, hydrochlorofluorocar-bons (HCFC's) and halons. The students limit the number of factors the ob-ject "source" has by including only numbers of air conditioners, numbers of cars, and amount of industry (they omit specific sources such as aerosol cans and foam blowers in favor of more general categories). They also decide to limit the factors associated with the object "earth." They decide to group all of their original factors together to have one general factor, ecosystems. Table 3.2 lists their final factors and summarizes their explanations for in-cluding them.

The students build explanations as they construct their model. The expla-nations they give for the factors related to the "earth" and "source" objects are relatively simple. Their explanation for including air conditioners is "The

TABLE 3.2 Objects and Factors in Jamie, Lauren, and Rachel's Final Model

Objects and Related Factors	Explanation for Inclusion in Model
Earth (Has 1 factor)	
ecosystem	This factor includes all ecosystems, such as plants, marine life, and land animals / humans.
Source (Has 3 factors)	
air conditioners	The trouble with air conditioners is that they leak CFC's. There have been laws passed, but the problem remains that we still have old air conditioners in use emitting CFC's.
cars	The problem areas of cars are the exhaust and the air conditioners. These areas both emit CFC's.
industry	This is the cause of many ozone-destroying chemicals. Among the many are halons, chlorine nitrate, and methane.
Atmosphere (Has 8 factors)	
carbon tetrachloride	Carbon tetra-chloride is a chemical that is used in the making of other chemicals, like CFC's. It is restricted in industrial countries for other uses but not anywhere else.
methyl bromide	Oceanic algae emits 60 to 160 thousand metric tons of methyl bromide per year. Humans also contribute methyl bromide by biomass burning. This contributes 40 to 50 thousand tons. There are also agricultural uses which produce 20–60 thousand tons per year. All this added together is 120–270 metric tons of methyl bromide emitted per year.
methane	Methane decreases the amount of CFC's in the atmosphere. What happens is that methane replaces one of the hydrogen atoms in CFC's and makes the CFC molecule heavier, bringing it down to earth. Therefore, the level of CFC's in the atmosphere is decreased.
Chlorofluorocarbons (CFC's)	CFC's are the biggest destroyer of the ozone. They rise to the ozone layer, and because they are insoluble, they keep destroying over and over again for decades.
chlorine nitrate	This is a natural compound which forms over the polar areas during the winter season. From here it then drifts to the equator and little by little it destroys ozone over a widespread circle. This process is very gradual.
methyl chloroform	In 1988 there are 580,000 metric tons of methyl chloroform in the lower atmosphere. This is where methyl chloroform becomes chemically reactive. The objects that produce this chemical include cleaning products, precision parts, and dry cleaning.
hydrochlorofluorocarbons (HCFC's)	The difference between HCFC molecules and regular CFC's is that HCFC's are CFC's with extra hydrogen atoms. HCFC's were originally made to replace CFCs, but they do just as much damage.
Halons	This is made from a compound of CFC's and bromine. There are two types of Halons. The Halon, 1301, has a ODP of 16. While the other, 1211, has an ODP of only 4.

trouble with air conditioners is that they leak CFC's. There have been laws passed, but the problem remains that we still have old air conditioners in use emitting CFC's" (from final model). Their explanation for including cars is "The problem areas of cars are the exhaust and the air conditioners. These areas both emit CFC's" (from final model). Their explanations for including the atmospheric factors are much more detailed. These explanations involve more reasoning and are based on specific pieces of evidence the students gathered during their research. For example, their explanation for including methyl bromide in their model is, "Oceanic algae emits 60 to 160 thousand metric tons of methyl bromide per year. Humans also contribute methyl bromide by biomass burning. This contributes 40 to 50 thousand tons. There are also agricultural uses which produce 20–60 thousand tons per year. All this added together is 120–270 metric tons of methyl bromide emitted per year" (from final model).

Relationships Between Factors

In the early stages of their model building, the students describe generally what types of relationships they will construct: "The relationships are between which chemicals deplete or help, and where they come from, and then the effect the ozone depletion has on life" (Global Model Planner). They outline a few of these relationships: "Chemicals hurt earth and destroy ozone; industries cause ozone depletion and create chemicals; and the atmosphere protects the earth" (Global Model Planner). As the students continue to create their model, they better articulate these relationships. Specifically, they pinpoint the causes of ozone depletion by identifying the sources of each ozone-destroying chemical and by describing the different effects that each chemical/pollutant has on the ozone layer. As they build their model, they continue, however, to construct simplistic relationships between a depleted ozone layer and life on earth.

The students' final model contains 20 relationships between factors. These relationships are represented in their finished Model-It factor map (see Figure 3.2), and examples of some of these explanations are summarized in Table 3.3. The students define all of these relationships qualitatively. For example, they explain as a source of a chemical increases (sources are air conditioners, cars, and industry), the amount of chemical increases always *by about the same* (chemicals being chlorofluorocarbons, hydrochlorofluorocarbons, carbon tetrachloride, methyl bromide, chlorine nitrate, methyl chloroform, halons, and methane). The students do not differentiate among the different types of source and the amounts of chemical output. Likewise, the students explain that as most of the chemicals increase, the ecosystems decrease *by about the same*. The students include one interesting deviation. They describe methane as a different case with two relationships involving methane: As ecosystems increase, methane increases by about the same, and as methane increases, CFC's decrease by about the same.

TABLE 3.3 Examples of Relationships and Explanations in
Jamie, Lauren, and Rachel's Final Model

Relationships Between Factors	Type of Relationship	Explanation of Relationship
As air conditioners increase, CFC's increase	by about the same	This is due to the fact that old air conditioners (made before the ban on CFC's) keep emitting CFC's.
As cars increase, CFC's increase	by about the same	As countries recycle their old cars (with old air conditioners), the output of these cars (such as CFC's) keep producing the CFC's which have been banned.
As industry increases, chlorine nitrate increases	by about the same	Many factories in the U.S. (and other countries) emit chlorine nitrates among other ozone depleting chemicals. Therefore, as industries up-size, so does the amount of chlorine nitrates in the atmosphere.
As CFC's increase, ecosystems decrease	by about the same	CFC's (being the largest amount of ozone-depleting chemicals) also indirectly hurt all ecosystems. This is so because the less ozone there is, the more UV rays come in, and the more all the ecosystems suffer.
As chlorine nitrate increases, ecosystems decrease	by about the same	Because chlorine nitrate harms the ozone layer, it also harms all ecosystems in some way. This is so because as the ozone depletes, it lets in harmful UV rays which causes various hazards.
As ecosystems increases, methane increases	by about the same	A great deal of methane is caused by cows (believe it or not!). The relationship between the ecosystem of cows and methane is that as the number of cows increase, so does the level of methane.
As methane increases, CFC's decreases	by about the same	The methane molecule decreases CFC's by taking away one of the chlorine molecules and replacing it with a methane molecule. Because this different molecule is heavier, it drags the CFC molecule down into the ocean. This action has a result of an increase in ozone.

Explanations for Relationships

Jamie, Lauren, and Rachel build explanations for the model relationships while constructing their model. Examples of these explanations are given in Table 3.3. A typical explanation for the increase in a source of a chemical and, in turn, for the amounts of chemical emitted is "As countries keep recycling their old cars, chemicals, such as HCFC's (which have been banned) are once again released into the atmosphere" (from final model). A typical explanation for the negative impact on ecosystems due to increases in amounts of chemicals is "Because chlorine nitrate harms the ozone layer, it also harms all ecosystems in some way. This is so, because as the ozone depletes it lets in harmful UV rays which causes various hazards" (from final model). Finally, the students explain the unusual relationships with methane. They explain why methane increases as ecosystems increase: "A great deal of methane is caused by cows (believe it or not!). The relationship between the ecosystem of cows and methane, is that as the number of cows increases, so does the level of methane" (from final model). And they explain the relationship between methane and the level of CFC's: "The methane molecule decreases CFC's by taking away one of the chlorine molecules and replacing it with a methane molecule. Because this different molecule is heavier, it drags the CFC molecule down into the ocean. This action has a result of an increase in ozone" (from final model). The students expand on this explanation during an interview:

Interviewer: How does it (methane) decrease CFC's?

Lauren: I think you are talking about methane?

Jamie: Yeah, what happens is the CFC molecule, one of the chlorines, I think, I'm not positive, is taken off and a methane is put in its place. And that molecule is heavier than it was before, so it sinks. . . .

Rachel: To the earth.

Jamie: Yeah, into the ocean.

Lauren: Into the water cycle. And the water cycle pretty much gets rid of it.

Jamie: Yeah, but it's not in the atmosphere any more because it's too heavy. So it decreases the levels of CFC's, but not a lot. But it still does it.

The students explain how chemicals, like CFC's, destroy the ozone layer. The students do not include this explanation in their actual model but explain the process during a final class presentation of their model. Jamie uses transparencies with reactions and molecular drawings written on them (see Figure 3.3) and explains during the presentation:

This is the process of ozone depletion by CFC's. The process by other ozone-depleting chemicals is similar but it's not exactly the same. First, ultraviolet radiation strikes a CFC molecule and causes a chlorine atom to break away. The chlorine atom collides with an ozone molecule and steals an oxygen atom to form a chlorine monoxide and leaves a molecule of ordinary oxygen. When a free atom of oxygen collides with the chlorine monoxide, the two oxygen atoms form a molecule of oxygen and the chlorine atom is released and free to destroy more ozone. This creates a cycle where chlorine can destroy ozone over and over and over again. (Presentation Video)

Finally, during their class presentation, the students also explain some of the things that have been done to prevent ozone depletion. Rachel describes some of the conferences that have taken place to set limits on amounts of CFC's that can be produced and used:

I'm going to talk about the last part of our driving question, and that was what's been done to eliminate some of these chemicals or lower them, and so far there have been four big conferences. The Clean Air Act, that was just a U.S. conference, but there are also three other international conferences: the Montreal Protocol, the Copenhagen Amendment, and the London Amendment. The Montreal Protocol and London Amendment, and also the Copenhagen Amendment, were the three biggest conventions. In the Copenhagen Amendment, it was enacted November 1992, and all industrial countries were required to cut CFC's by 75% from 1986 levels, and by 1994 they were supposed to completely eliminate them. And, for all substances controlled by the Montreal Protocol and the London Amendment, developing countries were required to make the same cuts as the industrial countries, but they were allowed a 10-year delay. And the Clean Air Act cut class 1 substances, those were pretty much the eight substances that we had in our model. And there were a couple of other ones that were singled out (chemicals) and so in 1991 (Rachel puts up a data chart) they were cut by 85%, and they were cut by an additional 5% down to and later they went down to 15% and by the year 2000 they are supposed to be completely eliminated. There were two substances that were singled out, and they are carbon tetrachloride and methyl chloroform, and they were probably the most harmful class I substances. So, methyl chloroform has not been completely cut, its supposed to be cut by 2020. And the carbon tetrachloride is supposed to be cut by 1999, well, 2000 and so there have been cutbacks but there is still a lot that needs to be done. (Presentation Video)

During an interview, the students comment on the impact the conferences may have on ozone-depleting substances and on their model:

Interviewer: So if you included those plans (the conferences to cut down on specific chemicals) into your model, what predictions could you make?

Lauren: That the levels of CFC's and other chemicals would go down.

Interviewer: OK, so if the levels of the chemicals go down, then what else?

Lauren: The ecosystem would go up. The industries, well, the industries producing CFC's would probably go down, and I don't know about methane.

Rachel: Well they wouldn't necessarily go down, but what they were doing, they were looking for alternatives, so only the places that . . . only the industries that directly produced CFC's would go down. Like not the car industry or the air force industry.

Using Several Symbol Systems

In constructing their model, the students routinely use chemical names and chemical formulas in describing the pollutants that contribute to ozone depletion. They also use chemical reactions and molecular drawings to explain how chemicals that contain chlorine deplete ozone. Figure 3.3 is an example of a slide the students used during their class presentation to describe the process of ozone depletion.

Summary

Jamie, Lauren, and Rachel construct robust content understanding. They construct a specific purpose for their model. They spend time identifying the ob-

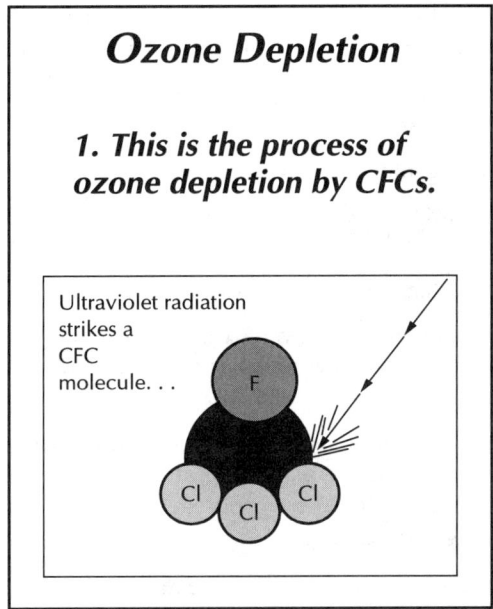

FIGURE 3.3 A slide that Jamie, Lauren, and Rachel use During a Class Presentation.

jects, factors, and relationships to include in their model, and they build detailed explanations for the phenomena they research. They construct several cause-and-effect relationships and explain them in some detail. Many of these explanations include data as evidence to support their explanations. Finally, the students use several symbol systems, including chemical formulas, reactions, and molecular-level drawings, to build explanations of the phenomena.

Building Inquiry Understanding

The process of building a model allows Jamie, Lauren, and Rachel to construct moderate inquiry understanding. The students construct this inquiry understanding by defining and refining their problem area over time, by assigning roles and responsibilities for the research required for model building, by building and revising a qualitative model, and by building an argument that includes the use of their model as justification.

Defining a Problem

Jamie, Lauren, and Rachel work to define the problem that becomes the basis for their model. Early in the model building unit, they decide they want to concentrate on ozone depletion, but they do not have a more specific focus. They do some initial research and use the research time to discuss and refine their problem. The students find quite a bit of information related to ozone depletion on the World Wide Web, and Rachel and Lauren want to clarify what kind of information to find and how to proceed with the search:

Rachel: Hey, Jamie, where are we going with this? Like what's our objective, like we need something central to it.

Jamie: We need as much information as possible.

Rachel: Yeah, but its hard to collect information when you don't know exactly where you are going with it.

Lauren: Well, you're going, what are you researching?

Jamie: We want to figure out what it is we can do about this problem, the position of the ozone. We want to find out as much as possible, that's where you are going, OK? (Class Video)

During this exchange, Jamie defines the focus of their model as "what it is we can do about this problem (ozone depletion)." As the students continue to research, they further refine their problem. Jamie talks with Lauren about the various roles they should have:

Lauren: I am covering the UV rays and stuff like that?

Jamie: Yeah, I will start with what happens with ozone depletion and what it gives off and you are going to say how it affects you.

Rachel, clarifying what they are doing: Jamie is doing "What is ozone depletion?" Lauren is doing "What is affected by ozone depletion?" and I'm doing "How to prevent it and what has been done already." (Class Video)

Later, after the students are finished with their model and presentation, they explain, during an interview, that they did not focus their model around one singular question. They explain that the problem for them was to educate themselves and the broader "community":

Interviewer: So what is your driving question? What's your purpose for this model?

Rachel: It's sort of like a community thing. Sort of like educating the community. What we wanted to do, is we just wanted to find out about it, and we really wanted to find out what's been done to stop it and maybe like a few alternatives that have been proposed.

Constructing a Model

Jamie, Lauren, and Rachel assign each other roles at the start of the model-building unit. They agree that each of them should be responsible for researching a different aspect of ozone depletion. Jamie is responsible for looking up information related to the chemicals that contribute to ozone depletion and how ozone depletion occurs. Lauren is responsible for looking up information related to the sources and/or causes of the chemicals that impact ozone depletion. And Rachel is responsible for researching some of the known solutions or preventions. The students use Internet resources for their research.

Jamie, Lauren, and Rachel build a qualitative model to address the questions they set out to research. They spend a great deal of time in the beginning discussing what factors they should include in their model and how to represent the connections between them. The following in-class exchange demonstrates some of their initial trials at building a model:

Jamie: (She writes on a piece of paper) So the environment is going to be the atmosphere and the earth? So, in here, we have the ozone. We have O_3 and we have methane.

Lauren: Are you drawing a model, or what?

Jamie: I'm just (drawing) words because I'm confused about what this is supposed to look like. How about you?

Lauren: Are you going to draw a concept map?

Jamie: Yeah, that's what this is (she points to the factor map). OK, so then we have the other chemicals, like the CFC's. (She writes the chemical names on the Global Model Planner). (Classroom Video)

Once the students get the hang of how to represent their factors in a concept map-like form, they construct the relationships among all of the factors they want to include in their model (Figure 3.4).

The main set of cause-and-effect relationships within the model is the following: As the number of air conditioners, cars, and industry increases, the amount of ozone-depleting chemicals increases, which leads to a decrease in ecosystems (this applies to all of the chemicals in their model except for

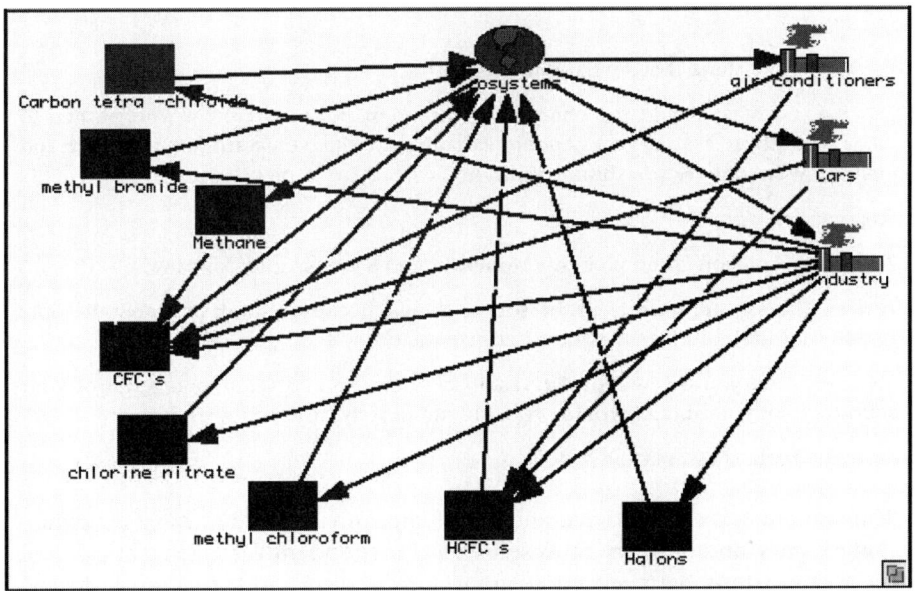

FIGURE 3.4 An Early Version of Jamie, Lauren, and Rachel's Model.

methane, which they show as decreasing the levels of CFC's and having a positive effect on ecosystems). In this early version of their model, the students also have a relationship between ecosystems and the numbers of air conditioners, cars, and industry. The students' reasoning for this is that the ecosystems (or the people in the ecosystems) are what produce the air conditioners, cars, and industry (as ecosystems increase, so do numbers of air conditioners, cars, and industry). This creates a cyclical model with no independent variables. During the class presentation, Jamie, Lauren, and Rachel show frustration with their model because they cannot get it to "work." Lauren presents the model and says,

The one in the middle (the meter), that's the earth's ecosystems, and so what we were planning to have happen, what we were trying to show, is if we up the air conditioners, and source cars and industry, that all of these chemicals that were affected in the factor map would go up and the earth ecosystem, the value of it would go down, because all of these are hurting the atmosphere more and more. And then if we brought these down (the amounts of air conditioners, industry, and cars) then all of the chemicals would go down and the ecosystems would go up. (Presentation Video)

A student in the class and the teacher help to point out the problem in their model during the presentation:

Andrew: Are there no independent factors that you can change?

Lauren: It's just the source, cars, and source, industry, and source, air conditioners.

Teacher: See when you don't have a meter on it (on a factor), that means that factor depends on something else, and you need an independent factor that you can move (a meter).

Lauren: Oh, I think that's what our problem is. . . .

Teacher: So what would you choose as your independent variable, if you wanted to take one of those as your independent variable, one that you want to manipulate and change? What one do you think would be the best one to pick?

Lauren: Only one?

Teacher: Well, you could pick other ones, but let's just say one for now.

Lauren: I'd say the source, all the source things, because I think probably the only reason they are not independent is because the ecosystems go to them and . . .

Teacher: So if you deleted a relationship between an ecosystem and some of your sources, then your model would probably run exactly how you want it to

Lauren: Yeah. (Presentation Video)

The students take this advice and revise their final model to have air conditioners, cars, and industry represented as independent variables (they delete the relationships between ecosystems, air conditioners, cars, and industry; see Figure 3.2).

Constructing and Evaluating an Argument

Throughout the construction of their model, Jamie, Lauren, and Rachel take the position that ozone depletion is a worldwide problem that needs serious attention. They state in their Global Model Planners, "Ozone depletion is a very real and threatening problem, and it affects everybody. It is imperative that the whole world comes to an agreement about this problem and comes up with alternatives to the things that are causing the break down" (Global Model Planner). They use their model as a way to justify this argument, and they elaborate on this in their Global Model Planners both at the beginning of the unit and later, during the unit.

Early in the unit, they describe in their Global Model Planners generally how models can be used to justify their argument:

Rachel: A model can show the effects of ozone depletion if nothing is done to delay or deal with it.

Jamie: The model can show ozone depletion and its effects on life on earth.

Lauren: The model can show what will happen to the earth if the ozone totally disappears, in other words, if we don't do anything to stop it.

Later in the unit, they describe in their Global Model Planners how their model in particular can be used to justify their argument:

Rachel: It [the model] showed that there were many ozone-depleting factors, with only a few that regenerated it. It showed the process of ozone depletion. It showed where most of the chemicals come from.

Jamie: It [the model] proves that increasing levels of chemicals hurt earth. It proves that chemicals increase ozone depletion. It showed the process of ozone depletion.

Lauren: It [the model] showed all the relationships, in other words, what chemicals do to the environment. It showed what factors causes ozone depletion directly and indirectly.

The students use their model to justify their argument, but they do not explicitly describe or use evidence to justify their argument.

Summary

Jamie, Lauren, and Rachel construct moderate inquiry understanding. They begin building their model with a few questions in mind and later focus and refine their problem area. The students assign roles and responsibilities to each other throughout the model building. They divide the questions they want to research among the three of them and pull most of the pieces together in a model; they incorporate into their model factors related to the sources and chemical causes of ozone depletion, but they leave factors related to prevention and solutions out of their model. The students build a qualitative model based primarily on Internet resources. The students originally create a model with one large feedback loop (no independent variables), but they revise this after presenting their model to the class. The students state a position and describe the use of their model (that includes data) as a justification for their argument, but they do not explicitly state the use of experimental evidence as a basis for evaluating their model and argument. These actions together demonstrate a moderate level of inquiry understanding: The students define a reasonable research problem for the amount of time they have to work; they construct a model to demonstrate the relationships among the factors in their system; and they revise their model to show the impact of increasing numbers of sources of ozone-depleting chemicals.

Building Nature of Science Understanding

The process of building a model gives Jamie, Lauren, and Rachel direct experience with defining and describing the purposes of models within scientific practice. The students build a high level of nature of science understanding by demonstrating a sophisticated understanding of the purposes of models and by articulating the limits and assumptions of their model. They describe the purpose of models as providing a means of testing ideas, making predictions, and educating a larger community. They evaluate their own model by describing the use of data and evidence to support the model and by describing some of the assumptions they made when constructing their model.

Purpose of Models

Early in the unit, Jamie, Lauren, and Rachel describe the purpose of models as a way to test hypotheses and make predictions related to future phenomena. They write about the general purposes of models in their Global Model Planners:

Lauren: Purpose of building models is to prove a hypothesis that you have come up with.

Rachel: To prove a point or make a prediction. They are also used to test a hypothesis.

Jamie: To predict or forecast what the outcome of a particular change might be. To make a smaller version of a large object—and to test out hypotheses of a larger version.

As the unit progresses, Jamie, Lauren, and Rachel continue to describe the purpose of models as a way to make projections, and they also suggest that models can be used to educate people. They explain this educational use of models during an interview after the completion of their own model. However, they say during the interview that they would not necessarily use their own model in educating people, because they feel it is too complicated:

Interviewer: How could a model be incorporated into educating a community?

Rachel: You could take a snapshot of a model and put it in . . . what we are doing here (showing the models on public access television). This is more in depth (their own model), but if you were educating the community, it would just be more basic stuff like don't use these types of aerosols or something . . .

Interviewer: So back to this idea of models being used to help educate people. I think Rachel said that this model is a little too complicated if you were going to show people?

Rachel: If you were (going to show this to) the community, I mean, if you research it to research it, they (the larger community) don't want it . . . I don't think they would want in-depth research, just basic information. I mean they (people) have other concerns, but I'm sure people would like to use products that aren't ozone-depleting, but they don't really know what they are, so I think this would be a little too confusing to show them that. They wouldn't understand this right off the bat.

The students also describe that different people or groups of people use models for different purposes. The students respond to the question "How do people use models?" in their Global Model Planner:

Lauren: Scientists use models as experiments. In other words, they use them to see what will happen to things under certain conditions. Politicians may use them to predict who will vote for them. Environmentalists use them to show what all the pollution we give off does to the world.

Rachel: Scientists use models maybe to play out some kind of scenario to see what might happen. Politicians use them maybe for their campaigning, and environmentalists use models to show the depletion of the environment so far and to predict its future. Environmentalists—they use them for the same reasons we did.

Jamie: Scientists might use them to test hypotheses or projects. Politicians to forecast polling results and in campaigns. Environmentalists use them to predict what the environment will be like in the future by what we do today.

Evaluation of Models

Jamie, Lauren, and Rachel describe the use of data in constructing their model and some of the assumptions they make when evaluating their model.

Global Model Planner: What processes do people go through to build models?

Lauren: 1. come up with a hypothesis to prove; 2. decide what kind of model they want.

Rachel: They collect data and research.

Jamie: 1. they observe; 2. they collect data; 3. apply their info; 4. make the model.

In evaluating their own model, the students base their evaluation on what they were able to "complete" and do not explicitly address the use of evidence to support their model. They explain in their Global Model Planners:

Jamie: We clearly stated what causes ozone depletion and what it is, we covered what is affected by it, we didn't cover how to prevent it besides the obvious, to reduce the chemicals. We covered what has been done in the past.

Lauren: We have found all of the factors that we are planning to put in the model. Such as how ozone depletion occurs and how we can prevent it.

The students are, however, able to discuss some of the assumptions that are built into their model. The students explain some of their assumptions in their Global Model Planners.

Jamie: That ozone depletion is bad.

Lauren: Even though there is no actual ozone [as an object, because it is invisible] it is still there.

Rachel: Ozone is a real problem and the measures being taken are working.

The students also describe their model as a "simulation" of phenomena, not the phenomena themselves:

Global Model Planner: How close does your model come to real-world phenomena?

Jamie: I think our details were close, but not all of the picture. We were missing a lot, but we covered a considerable amount.

Lauren: We tried to simulate all the factors as best we could, however it is impossible to create a situation that is exactly the same.

During an interview, the students recognize that they made a conscious choice about how to structure their model. They choose to emphasize the relationships between the sources of chemicals that influence ozone depletion and the chemicals themselves, rather than the impacts these chemicals have on elements within the ecosystem.

Rachel: Our factor map is kind of confusing because we had so many chemicals that related to the sources and the earth.

Jamie: Yeah, we did more detail rather than anything else.

And they write in their Global Model Planner,

Global Model Planner: Are there things you are realizing your model will not be able to show?

Jamie: It lacks specific problems with life on earth. It doesn't have more specific sources of industry. The numbers for the chemicals are hard to get. More details, if to be more exact. Details include effects and specific problems and info on chemicals.

Summary

Jamie, Lauren, and Rachel exhibit a high level of nature of science understanding by describing the purpose of model building as a way to test hypotheses, make predictions, and educate a community. They evaluate their model in terms of what they hoped to accomplish, but not necessarily with the eye toward interpretation and use of data. They do, though, describe some of the assumptions and limits of their model.

Conclusions of the Case

This case illustrates that model building provides Jamie, Lauren, and Rachel with an opportunity to build understanding of both the phenomenon of ozone depletion and the processes used to attain scientific understanding. The process of model building facilitates the students' identification of key concepts associated with ozone depletion, and the students begin to build explanations for phenomena, including explanations of which chemicals contribute to ozone depletion and what preventive measures are currently being used to control the problem. In building these explanations, the students use some data to justify their explanations and use symbolic expressions typical of scientists who study this phenomenon, including chemical reactions and molecular drawings. Construction of the model also provides the students with the opportunity to discuss the purposes of their model in particular, and the purposes of models in general. This allows the students to describe models as a way of testing hypotheses and making predictions. It also provides an impetus for discussing the limits of models and the assumptions that are made in the construction of any model. Finally, the process of model building engages these students in doing science. The students ask a question they are interested in and construct and revise a model to address the question. They use model building as a way to state a claim—that ozone is real and threatening problem—and construct a model (that is based on evidence) to support their claim. When defending their position, the students had difficulty in referring to the evidence they collected, and we see this as the next teaching and learning challenge.

Reflecting on the Purpose of Model Building

Engaging students in model building has the potential to transform what happens in our classrooms. It provides students with the opportunity to take ownership and responsibility for their learning. Students are able to ask and pursue answers to their own questions, while also being critical users of information. Students work with each other, much as they would in a real-world situation, sharing resources, debating ideas, and negotiating solutions to real, complex problems. The teachers in these dynamic classrooms are the quality control managers of ideas and products, rather than purveyors of information. They model the types of processes and the critical thinking required when students try to build explanations of complex problems.

Engaging students in model building has the potential to boost scientific understanding. The constructive process of model building requires that students sort out and build explanations of scientific phenomena, rather than merely memorizing facts and definitions. It requires that students define and revise problems over time. It requires that students search for information and data sources. If supported, the process of using data to build a model can give students a reason to determine the quality of different sources of information. Model building also involves testing ideas, making predictions, and revising previously held ideas if the model being constructed does not match up with empirical evidence. Model building provides a context for students to build scientific arguments, to state a position, and then to justify that position with evidence. It also provides a real context for students to think about the purposes of science and of the tools of science (such as models and theories). In this way, model building becomes a powerful activity for engaging students in doing and thinking about science. Science is no longer something that is read about in a book; rather, it becomes an activity through which phenomena are studied, manipulated, sometimes controlled, and perhaps even acted on.

The findings from this study suggest that given a collaborative, inquiry-based model-building environment, students construct scientific understanding. Model building provides a context for students to build and integrate content, inquiry, and epistemological understandings. The chapter demonstrates that model building engages students in defining problem areas, building explanatory mechanisms of scientific phenomena, and reflecting on the purposes of models. However, the findings also have many implications for further research.

One implication is that special emphasis must be placed on the use of evidence within models and arguments for students to build data-based justifications. Further research is needed to identify the supports that aid students in using evidence to justify models and scientific arguments. Several researchers describe this challenge, and it remains a critical key to engaging students in thoughtful inquiry (Hancock, Kaput & Goldsmith, 1992; Krajcik et al., 1996; Schauble, Glaser, Duschl, Schulze & John, 1995). Supports may take the form of teaching strategies (Spitulnik, 1998), or they may be curricular or technological supports; in either case, this essential element of understanding must be addressed to further students' understanding.

A second implication is that constructing understanding takes concerted effort and time. The students working in the Global Climate Change Unit did not squabble or complain (too much) about the cognitive demands of model building, but this may be due to the type of student who chooses to attend the alternative school or it may be due to the atmosphere in their school. Further research will help elucidate the conditions necessary for a productive work environment. These conditions may include explicit project goals, clearly defined authentic tasks, and expectations for collaboration (Crawford, 1996). Students also construct understanding over an extended period of time. Research could help identify the conditions under which it is advantageous to extend the time frame for a project or to determine appropriate endpoints. Research is also needed to track student understanding longitudinally. Students in this study engaged in only two model-building projects (a project on water quality and the Global Climate Change Unit). The potential for understanding over an extended time frame, such a full school year or longer, is tremendous.

Finally, a third implication is that model building would not be possible without the technological support of tools like Model-It. Further design, testing, and implementation of these types of tools will help identify the key features that support students in building models, in doing science (defining problems, making predictions, collecting data, analyzing data, building representations, and explaining phenomena), and in reflecting on the process. Tools are currently being designed and constructed (Loh *et al.*; 1997; Quintana *et al.*, 1998), but further research is necessary to examine their impact on flexible understanding.

The case presented in this chapter is just one example of how students can work collaboratively in a model-building environment to gain scientific understanding. The potential of this kind of learning environment is tremendous, but we need to continue developing the tools and teaching strategies that will better facilitate scientific understanding for all students.

References

American Association for the Advancement of Science. 1993. *Benchmarks for science literacy*. New York: Oxford University Press.

Blumenfeld, P., Soloway, E., Marx, R., Krajcik, J., Guzdial, M., & Palincsar, A. 1991. Motivating project-based learning: Sustaining the doing, supporting the learning. *Educational Psychologist, 26*; 369–398.

Crawford, B. 1996. *Examining the essential elements of a community of learners in a middle grade science classroom*. Unpublished doctoral dissertation, University of Michigan.

Hancock, C., Kaput, J., & Goldsmith, L. 1992. Authentic inquiry with data: Critical barriers to classroom implementation. *Educational Psychologist, 27*(3), 337–364.

Jackson, S., Stratford, S., Krajcik, J., & Soloway, E. 1996. Making dynamic modeling accessible to pre-college science students. *Interactive Learning Environments, 4*(3); 233–257.

Jackson, S., Stratford, S., Krajcik, J., & Soloway, E. 1995. Model-It: A case study of learner-centered software for supporting model building. Paper presented at the Working Conference on Technology Applications in the Science Classroom, Columbus, OH.

Krajcik, J., Blumenfeld, P., Marx, R., Bass, K., Fredricks, J., & Soloway, E. 1996. The development of middle school students' inquiry strategies in project-based science classrooms. Paper presented at the International Conference for the Learning Sciences, Evanston, IL.

Loh, B., Radinsky, J., Reiser, B. J., Gomez, L. M., Edelson, D. C., & Russell, E. 1997. The Progress Portfolio: Promoting reflective inquiry in complex investigation environments. In Hall, R., Miyake, N., & Enyedy, N. (eds.), *Proceedings of Computer-Supported Collaborative Learning '97*. Ontario, Canada

National Research Council. 1996. *National science education standards*. Washington, DC: National Academy Press.

Perkins, D. 1996. *Teaching for understanding*. Paper presented at the Annual American Educational Research Association Meeting, New York.

Quintana, C., Soloway, E., Krajcik, J., Carra, A., Houser, M., McDonald, M., Mouradian, M., Saarela, A., Vyas, N., and Spitulnik, M.W. 1998. Symphony: Exploring user interface representations for learner centered process scaffolding. In Karat, C.-M. and Lund, A. (eds.), *CHI summary sheet*. New York: ACM, pp. 311–312.

Schauble, L., Glaser, R., Duschl, R., Schulze, S., & John, J. 1995. Students' understanding of the objectives and procedures of experimentation in the science classroom. *The Journal of the Learning Sciences*, 4(2), 131–166.

Spitulnik, M.W. 1998. *Construction of technological artifacts and teaching strategies to promote flexible scientific understanding*. Unpublished doctoral dissertation, University of Michigan.

Stratford, S. 1996. *Investigating processes and products of secondary science students using dynamic modeling software*. Unpublished doctoral dissertation, University of Michigan.

4

A Visual Modeling Tool for Mathematics Experiment and Inquiry
Wallace Feurzeig

Introduction: The Function Machines Language

Programming languages are potentially powerful tools for helping students develop mathematical ways of thinking. However, helping students acquire fluency in developing mathematically rich programs, especially in a way that they find engaging and pleasurable, is a nontrivial task indeed. Even Logo, the most accessible functional language, poses significant conceptual barriers to the acquisition of the necessary knowledge and skill. The mechanisms for passing data and transferring control between procedures, particularly iterative and recursive control structures, are particularly difficult for beginning students. The Function Machines computer language was expressly designed to overcome these barriers through the use of visual representations that make control structures and program operation more transparent and accessible. Work with Function Machines enhances students' development of the notions and art of mathematical modeling and model-based inquiry.

Function Machines is a visual programming environment with the representational power of a universal programming language. It is based on a functional control structure and a data-flow model of program execution. Its key construct is the "function-machine," a visual isomorph of the function concept in mathematics. Machines are visual analogs of Logo procedures or Lisp functions. They communicate data to each other via "pipes" connecting the output of one to the input of another, in data-flow fashion. The sequence of execution between machines may be similarly directed by constructing "wires" connecting one machine to another. In the absence of such wiring, control flow among machines is unconstrained, and execution is essentially parallel.

The Function Machines language expresses program structures visually as two-dimensional graphical icons, in contrast with the symbolic textual expressions used in traditional (one-dimensional) languages. The system provides, as primitive constructs, machines that correspond to the standard mathematical, graphics, list processing, logic, and I/O operations. These machines are used as building blocks to construct more complex machines in an extensible fashion. Any collection of connected machines can be encapsulated under a single icon as a higher-order "composite" machine; proceeding in this way, it is possible to construct machines (programs) of arbitrary complexity.

Execution is essentially parallel—many machines can run concurrently. The program structures of Function Machines are modular and recursive. A composite machine may be used as a component of a more complex machine or of itself. The operation of recursion is made visually explicit by displaying a separate window for each instantiation of the procedure as it is created and erasing it when it terminates. The natural hierarchical organization of programs that is implicit in the notion of a composite machine fosters modular design and helps to organize and structure the process of program development. The explicit representation of data paths and control paths makes the semantics of functional operation transparent in Function Machines in a fashion readily accessible to beginning students. Function Machines thus provides a natural starting point for a constructive approach to the teaching of mathematics.

The underlyling idea in Function Machines (the "function machine" metaphor) is that a function, algorithm, or mathematical model can be thought of as a machine, represented visually in Figure 4.1. Machines can have one or more inputs. The output of a machine can be "piped" into the input of another machine, as with the addition and multiplication machines shown in Figure 4.2. When this two-machine network is activated, the addition machine will sum its inputs (3 and 2) and feed the output to the left-hand input of the multiplication machine, which will then run and output 35, the product of its inputs.

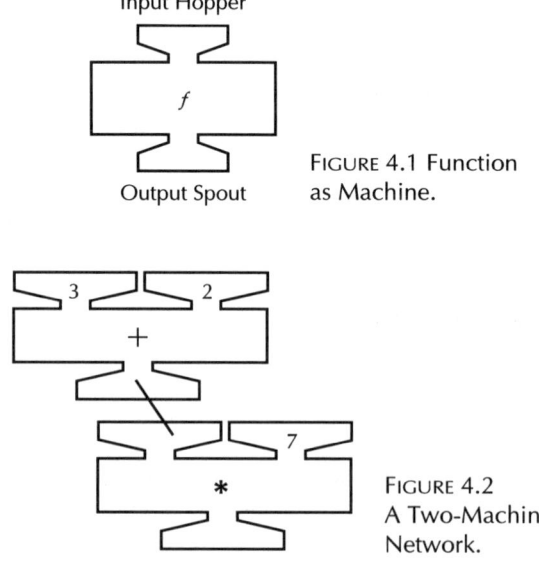

Input Hopper

f

Output Spout   FIGURE 4.1 Function
                    as Machine.

3   2

+

7

*   FIGURE 4.2
         A Two-Machine
         Network.

FIGURE 4.3
Backput Iteration.

FIGURE 4.4 After the
Count of 9.

FIGURE 4.5 Inside the
Counting Machine.

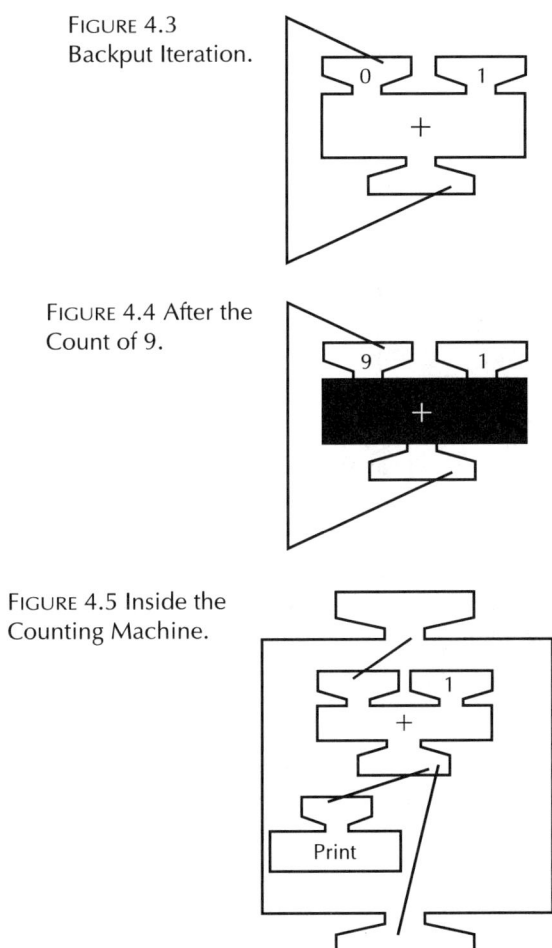

The output of a machine can be piped into its own input. This simple form of iteration, which is called "backput" iteration, is shown in Figure 4.3. Here the right-hand input is a constant, the number 1. Each time the addition machine runs, it adds 1 to the left-hand input and replaces the right-hand input by the resulting sum. Thus it successively generates the counting numbers.

The result after 9 iterations is shown in Figure 4.4. The current sum, 9, has been piped into the left-hand input and the machine is highlighted, signifying that it is ready to run again. The outputs of the + 1 addition machine can be printed to a display window by piping them to a Print machine, as shown in Figure 4.5. Further, this two-machine structure has been embedded within a higher-level "composite" machine, whose inner body is shown surrounding the two machines. When the composite machine runs, the addition machine receives its inputs from the input hopper of the composite machine, and it will pipe its outputs to the output spout of the composite (as well as to the Print machine.)

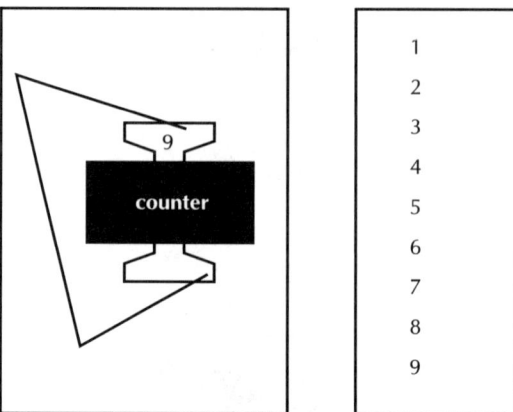

FIGURE 4.6 Running the Composite Counting Machine.

The outer body of this composite machine, which is labeled "counter," is shown in Figure 4.6. Counter is ready to run. Its current input is 9. When it runs, it will execute the machines in its inner body. Its output will become its next input. The figure shows the display window created by the Print machine, which has printed the sequence of outputs generated thus far by counter.

Mathematical models in Function Machines are developed through such an incremental process. Typically, models are composite machines constructed from simpler machines, starting with the primitive machines provided with the system.

Iteration and Recursion in Function Machines

Iterative and recursive processes are essential for building more powerful programs, but they are often very hard for beginning students to use. The dynamic visual data-flow representations of control structures in Function Machines facilitate students' understanding and development of mathematically rich and computationally powerful processes. The simplest form of iteration, called backput iteration, is introduced on the students' first day with Function Machines.

Figure 4.7 shows the use of a built-in iteration machine, the Repeat machine. Each time a Repeat machine runs, it decrements its first input (the number of iterations remaining) and carries out the computation called for in its inner body. In the example shown, the Repeat machine draws a turtle figure, a closed spiral, by repeating 100 times a left turn (initially 40 degrees, incremented each time by a constant of 30 degrees) and a constant forward step of 10 units.

Function Machines includes primitive machines for more complex forms of iteration than that provided by the Repeat machine, including a For-loop

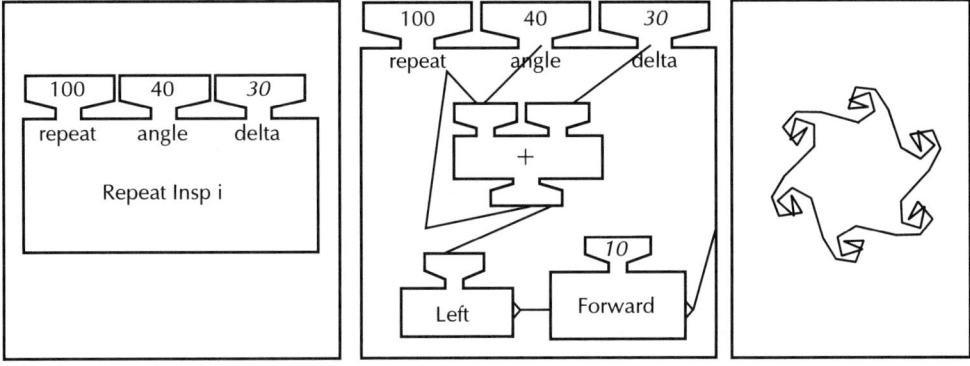

FIGURE 4.7 Using Iteration with the Repeat Machine to Generate a Closed Spiral.

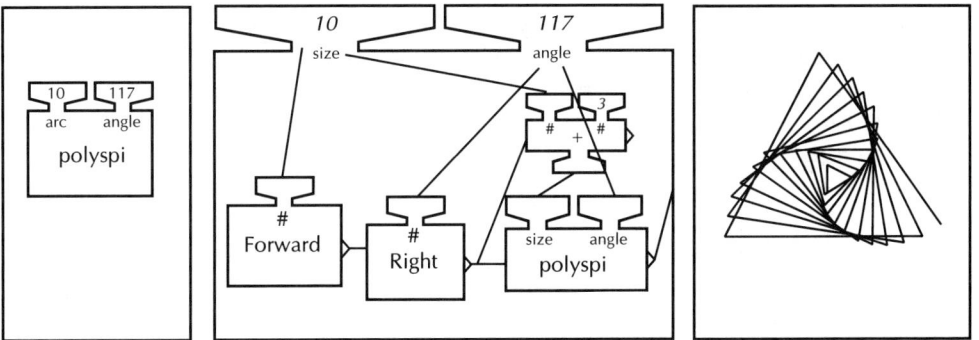

FIGURE 4.8 Recursive Computation of a Polygonal Spiral.

machine and a To-Each machine (which applies an iteration over a set of inputs.) Nested iterative operations are a great deal easier to follow in this iconic visual form than in the usual one-dimensional textual representation.

Even more striking in its clarity is the Function Machines representation of recursion. Recursion is much more confusing than iteration to beginning students. The idea of a procedure being defined in terms of itself seems cryptic, circular, even nonsensical. The sense and operation of recursion become marvelously clear in Function Machines, paving the way for students to experience and explore the extraordinary mathematical power of recursive processes. Figure 4.8 shows the use of recursion in a Function Machines program, the polyspi machine.

The inner body of polyspi is shown in the center. Each time polyspi runs, the turtle moves forward 10 units and turns right (117 degrees initially, incremented by 3 degrees on each new call of polyspi.) The inclusion of the polyspi machine icon inside its own inner body is accomplished in a straightforward way. First polyspi is saved in unfinished form (without the polyspi

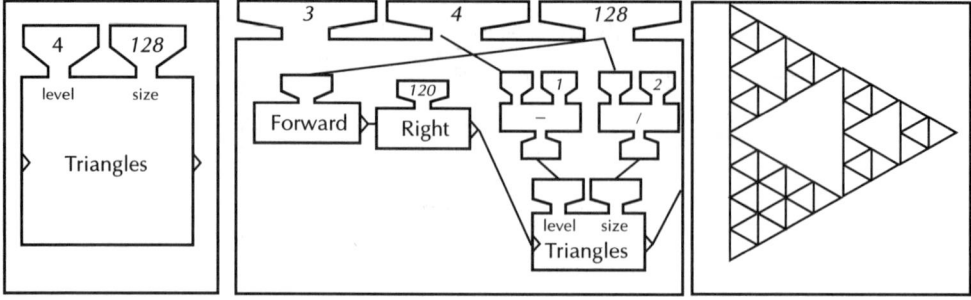

FIGURE 4.9 Combining Iteration and Recursion to Generate a Sierpinski Triangle.

machine icon in its body). This causes the icon to be included in the composite machine menu. Then polyspi is retrieved, and the polyspi icon is copied from the menu into the body of the machine.

Figure 4.9 shows the combined use of recursion and iteration. The inner body of the Triangles machine is an iterative Repeat machine (not shown) whose inner body is shown in the center. The repeat machine is run three times. Each time it runs, it calls the Triangles machine recursively. The resulting figure for a recursion level of depth 4 is shown on the right.

Classroom Investigations

Function Machines has been used extensively for mathematical explorations and investigations in elementary and secondary classrooms (Feurzeig and Richards, 1996; Feurzeig *et al.*, 1993; Wight *et al.*, 1988). Its visual representations of program structures significantly enhance students' understanding of key mathematical concepts such as algorithm, function, and recursion. By explicitly showing the passage of data objects into and out of machines, and by highlighting the data and control paths as machines are run, Function Machines renders computational processes as visual animations, and the program semantics becomes a great deal more transparent.

Function Machines is especially valuable for developing mathematical models. To understand a model, students need to see the model's inner workings as it runs. At the same time, they need to see the model's external behavior—the outputs generated by its operation. Function Machines supports both kinds of visualizations. The use of these dual-linked visualizations has unique and valuable learning benefits.

The following example illustrates the use of Function Machines for introducing mathematical modeling to middle-school students. The object is to help students develop, investigate, and reason about models of dynamic systems, with a focus on the effects of feedback. The example also illustrates the strategy of beginning with a simple model and moving to more complex and realistic model constructions through a sequence of successive developmental stages driven by inquiry issues. Students begin by building an apparently plausible model to address a real-world problem. They run the model to observe its behavior and its limitations in treating key aspects of the problem in a realistic fashion. They also improve the model by modifying or extending its structure. Proceeding incrementally in this way, they develop a sequence of models of increasing explanatory or predictive power. This example, which involves the modeling of population dynamics, begins with a story.

In a certain town, during the last several years, exactly 100 babies were born each year. In the town census this year (2020), the human population was found to be 5510. The town planners wish to have a model that will give an estimate of the future population each year for the next 20 years. What will such a model contain? To find the total number of people each year, we add the population at the start of the year to the births occurring during the year. Students build such a model in Function Machines by using a Repeat machine, a machine that repeats the computations within it a specified number of times. The inside of the machine is shown in Figure 4.10.

The left-hand input of the machine is set at 20, the number of times we wish to run the machine. Each time it runs, it will update the current population by 100 births and print the computed population for the corresponding year in a display window. The updated population is piped to the left-hand input of the addition machine as the starting population for the following year, via backput iteration. Then the process repeats. After it repeats its computation 20 times, the Repeat machine will exit.

This model will output a population of 5610 for year one, 5710 for year two, and so on up to 5510 + 100 * 20 = 7510 for the twentieth year, 2040. Ah, but isn't something wrong with the model? Thoughtful students realize

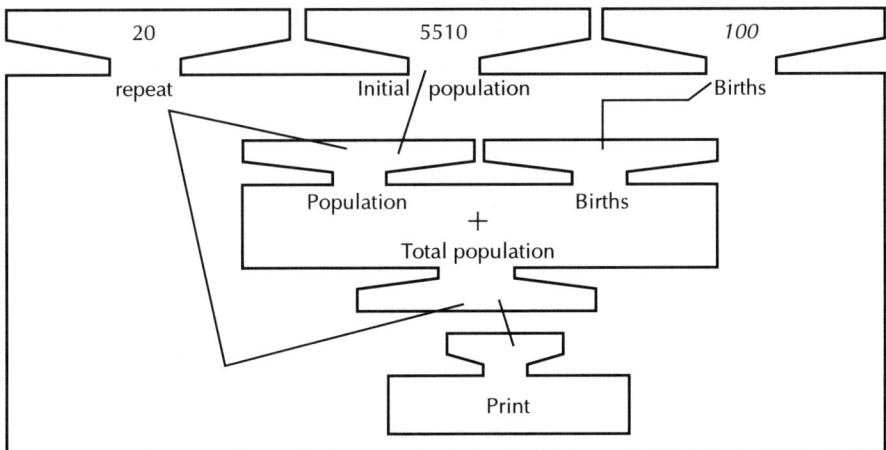

FIGURE 4.10 The Initial Population Model.

that the population will not continue to grow at exactly 100 births each year. As the population increases, the number of babies born each year will not be constant; it is more likely to grow in proportion to the current population size, by a certain percentage each year. Thus the notion of growth rate emerges. Students develop more realistic models, incorporating a nonconstant birth rate. Students engage in considerable discussion about how the computation should be done before coming to agreement on the mathematical formulation below. In the new formulation, the number of births affects the population, *and* the population also affects the number of births—students are thus introduced to the notion of a positive feedback loop. They then build and run their models, trying a variety of birth rates.

Figure 4.11 shows a typical new model, which extends the original model by computing births as a function of population size. The left-hand window shows the Repeat machine. Embedded in it is a Population Growth machine, whose inner body is shown in the right-hand window. The Population Growth machine computes the number of births in a given year by multiplying the population size that year by the birth rate. It then adds the number of newborns to the current population to compute the new population size at the end of the year. Both the new total population and the number of births are sent to a Print machine, which prints them in a display window.

Ah, but isn't something wrong with this model also? Even though the students now take into account a more realistic growth rate based proportionally on the size of the population, they realize that the model does not allow for death—in effect, it assumes that everyone lives forever. If this were true, the town would eventually run out of food and water, not to mention having

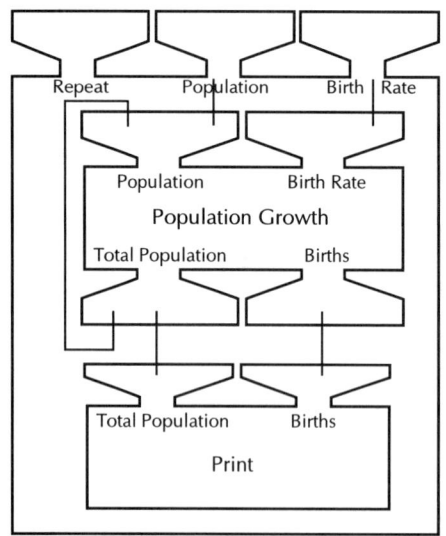
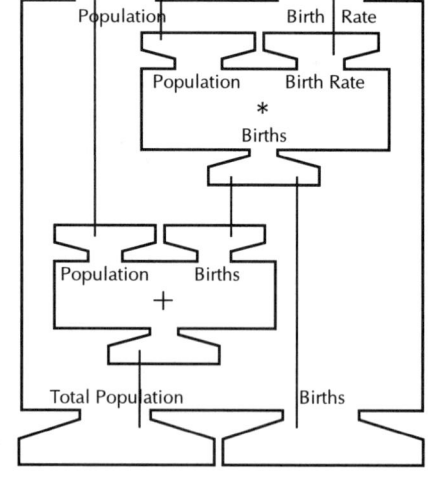

FIGURE 4.11 Incorporating Birth Rate.

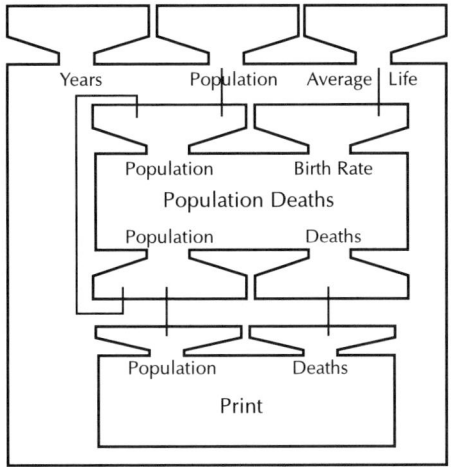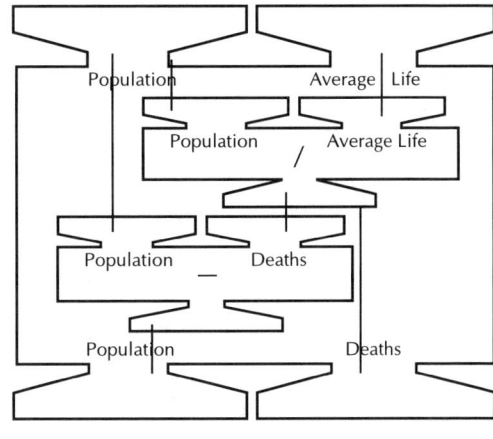

FIGURE 4.12 Incorporating Deaths.

other kinds of problems! They therefore decide that their models must incorporate deaths.

The town planners propose a goal of attaining a no-growth population so they can easily maintain a high quality of life for the townspeople. They think that living to 100 years will allow people a substantial life span and the joy of knowing several generations of their offspring. Students modify their models to include a computation of the effect of deaths on decreasing population size, given that the townspeople have an average lifetime of 100 years.

Figure 4.12 shows a simple mathematical model for computing the number of deaths per year and the corresponding reduction in population size. The left-hand window shows the Repeat machine. Embedded in it is a Population Deaths machine, whose inner body is shown in the right-hand window. The Population Deaths machine computes the number of deaths in a given year by dividing the population size that year by the average lifespan (100). It then subtracts the number of deaths from the current population, to compute the reduced population size at the end of the year, and outputs both results. These are sent to the Print machine, which prints them in a display window.

The students are then asked to build a more comprehensive feedback model to include both the birth rate and the death rate and to integrate their effects. A representative model that expresses the relationship among births, deaths, and total population is shown in Figure 4.13. The left-hand window shows the Repeat machine; the inner body of the embedded Population machine is shown in the right-hand window. Its operation is evident. Given the population at the start of each year, the births and deaths during the year are computed as before. Their difference, which is the net change in population, is added to the starting population to give the population at year's end.

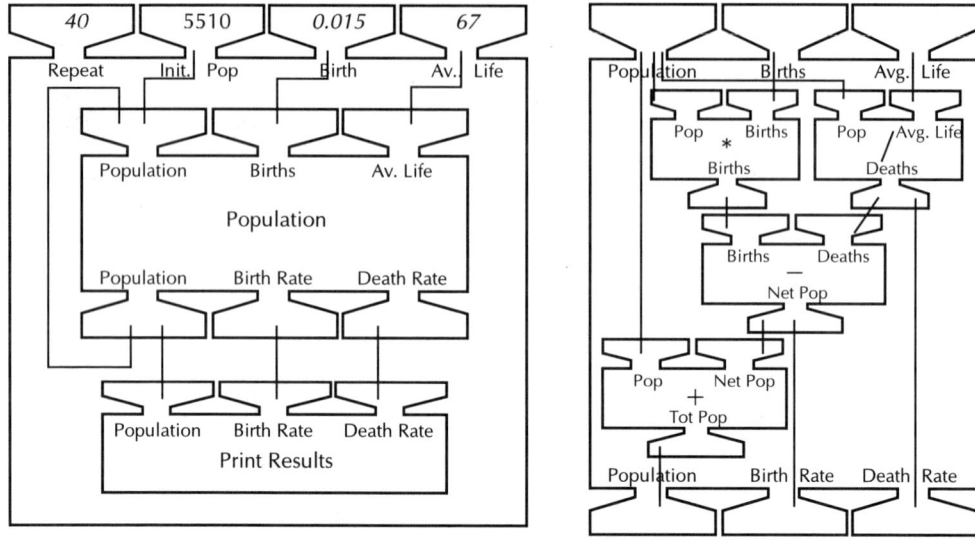

FIGURE 4.13 The Comprehensive Population Model.

Students try to determine what average life span would keep the population at a near-constant level over a 20-year period. They adjust the value of average life span until the model is close to equilibrium, maintaining a population level of around 5500. They then run the model for a 40-year period and try to adjust the value of the average life span until their model outputs are close to equilibrium again. Next, they investigate the effects of decrease in birth rate and increase in average life span on maintaining a constant population level. In the course of their work, students move from examining the processes that affect population size (birth and death) to considering the underlying rules that govern those processes, to investigating the behavior that the rules produce (dying out or explosion of the population), to studying modifications of the rules (either to produce a desired kind of behavior or to try to get model outputs that are more realistic.)

Modeling Complexity

Function Machines is beautifully suited for displaying the structure and behavior of mathematical models of dynamic systems. Simple physical systems, in certain regimes, display wild and erratic behavior that may cause arbitrarily close initial states to diverge exponentially, making it effectively impossible to predict the future behavior of the system. This phenomenon is intimately linked to the behavior of mathematical functions—often very simple ones—when they are iterated many times. Only one of three things can happen: Successive iterates of the function may approach a single fixed point; they may converge to a limiting orbit of points; or they may behave more er-

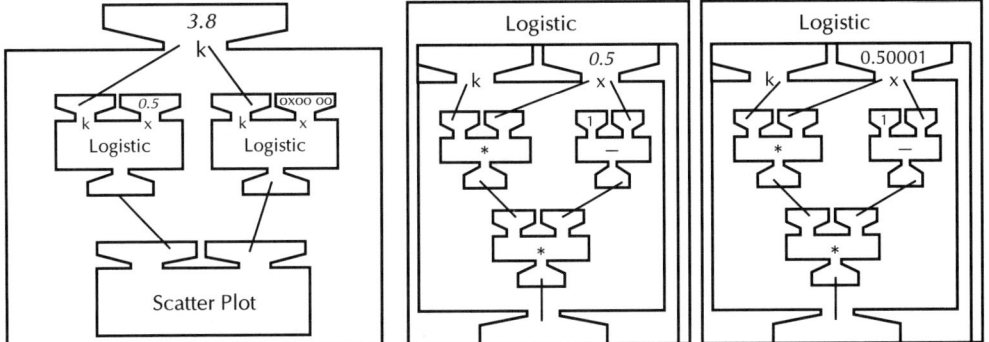

FIGURE 4.14 Iterating the Quadratic Function.

ratically, never quite returning to a value they have taken on before. In the last case, the iterated function sometimes displays an extremely sensitive dependence on initial conditions, so that neighboring starting points, when operated on repeatedly by the function, diverge very rapidly from one another, and all information about the starting point is lost. Behavior characterized by such an extreme sensitivity to initial conditions is termed chaotic.

Amazingly, a function as simple as the familiar quadratic, such as the function $k \cdot x \cdot (1 - x)$, exhibits chaotic behavior under iteration for some values of the parameter k. Figure 4.14 shows a Function Machines model for the quadratic function, called the Logistic machine. The left-hand window shows two Logistic machines, each of which is given a value of $k = 3.8$. Both machines use backput iteration on their second input, the input for x. The insides of the two machines are shown in the right-hand windows. The first has an initial value of $x = 0.5$. (This is in the middle of the range of allowable values $0 \leq x \leq 1$.) The second machine has an initial value of $x = 0.500001$, very close to that of the first machine. The corresponding outputs of the two machines are given as coordinates to a Scatter Plot machine, which prints the associated points in a display window.

The left-hand window of Figure 4.15 shows the Scatter Plot outputs from running the first several iterations of the twin Logistic machines. The outputs of the two machines are initially identical; thus the points fall on a diagonal line. However, as shown in the right-hand window of the figure, after a while the outputs of the two machines diverge wildly, and the resulting points in the Scatter Plot pepper the plane in an apparently random fashion. The behavior is chaotic.

As Figure 4.16 shows, the path to chaos is not, itself, chaotic. The left-hand window shows the result of running the twin Logistic machines with a value of the parameter $k = 3.65$. The ouput has a clearly defined structure— the values of all iterates are confined within the two symmetrically related squares. If one backs down a little further to a parameter value of $k = 3.5$, the regularity of the output is evident. The right-hand window displays the Scatter Plot output produced by hundreds of iterations. It clearly shows that

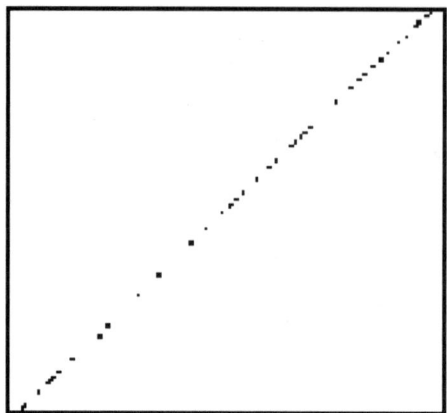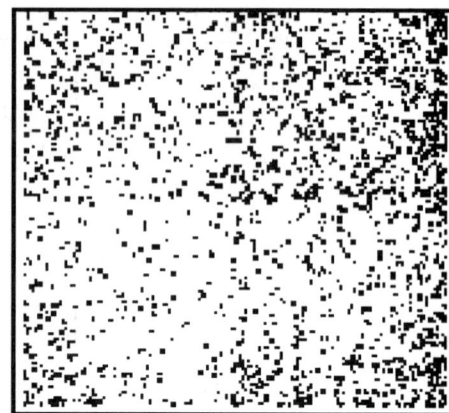

FIGURE 4.15 Scatter Plot Output for $k = 3.8$.

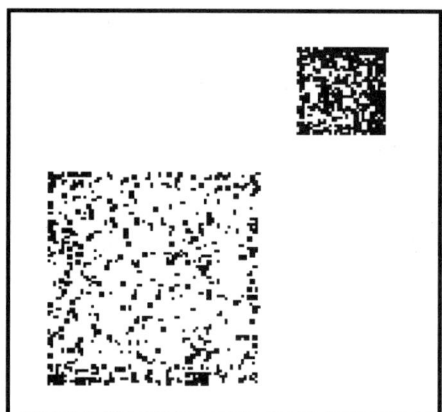

FIGURE 4.16 Scatter Plot Outputs for $k = 3.65$ and $k = 3.5$.

the iterates exhibit four-fold periodicity. Smaller values of k produce even simpler regularities. For example, the $k = 3.25$ iteration shows period two behavior, and the $k = 2.5$ iteration shows rapid convergence of the iterates to a fixed point. The result of the $k = 2$ iteration is left as a mental exercise for the reader.

Mathematical chaos is founded on a set of remarkable discoveries: (1) that nonlinear processes can give rise to very complex, unpredictable behaviors in a rich variety of systems—physical, chemical, and biological, (2) that these chaotic behaviors are nevertheless deterministic and can be modeled by simple mathematical equations (those with few variables or with a small number of degrees of freedom), (3) that the processes by which systems approach chaos are themselves orderly, and (4) that the underlying deep structure of chaotic behavior is very similar across diverse domains and systems, perhaps even universal. Function Machines is a useful tool for investigating nonlinear processes and for gaining insight into these wonderful discoveries.

Parallel Computation

The ability to run multiple processes simultaneously in a computer frees the machine from the "von Neumann bottleneck," where the length of a computation is largely the time it takes to move data between the processor and memory. Parallel processing avoids the situation where everything is waiting for the execution of a single operation. There are basically two software approaches to handling parallel constructs. The first divides the data of the program into independent units that can be operated on in parallel. This idea grew out of the need, in scientific programming, to operate on homogeneous data structures. In order to perform an operation on a vector of independent elements, for example, the same operation must be performed on every element in the vector. Because these are independent, the operations can be performed in parallel. The second approach divides the operations of the program into independent processes that can be computed in parallel. This is the approach that is implemented in the Function Machines mathematical modeling language. Function Machines simulates parallel execution on a machine with a single processor.

In building a model with embedded parallel processes, the designer confronts new problems. Typically, conflicts arise in attempting to coordinate and synchronize the various processes. We agree that "Concurrent programming languages (particularly those targeted at novice programmers) should make processes as 'concrete' as possible. That is, languages should make it easy for programmers to think about and identify with computational processes" (Resnick, 1990). The key difficulties in process synchronization and contention that characteristically arise in the development of models with concurrent operations are readily diagnosed in Function Machines modeling. The dynamic visual representation of processes is helpful in identifying and resolving these conflicts. Simple examples of some typical conflicts are illustrated next.

Race Conditions

A race condition occurs when more than one machine may supply a value for another machine. The values will therefore be determined by which of the previous operations finished first. In the example shown in Figure 4.17, two addition machines are feeding values simultaneously to the same hopper. In part (a) both machines are firing. In part (b) they have both arrived at their respective values (5 and 9), and in part (c) the multiplication machine will operate on 9. What is important here is that in another run, the 5 might be taken instead of the 9. This leads to nondeterminism in the code—and to inherent unpredictability.

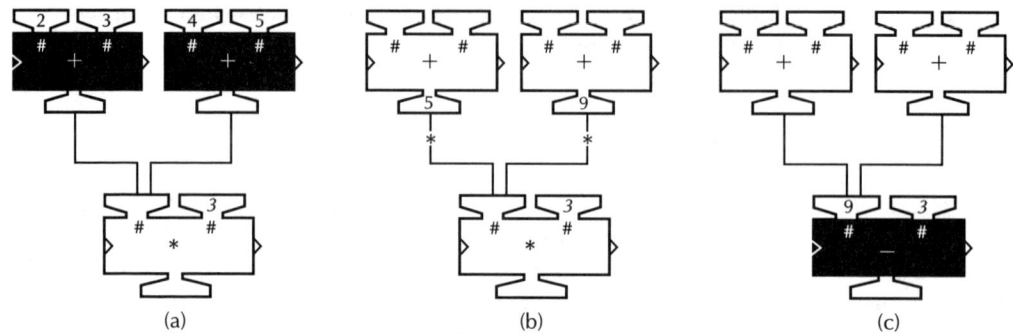

FIGURE 4.17 Race Condition.

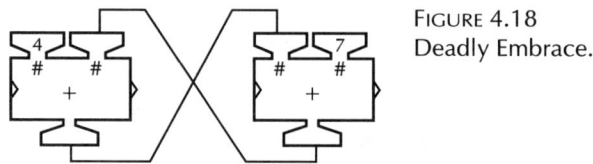

FIGURE 4.18
Deadly Embrace.

Deadly Embrace

In this situation, two machines are waiting for data from each other (see Figure 4.18). As a result, the program is unable to continue. If either machine were to run, then the process could resume.

Data Blocking

In Function Machines, a machine normally fires when its hoppers are filled. In Figure 4.19 we see a situation in which a machine is unable to fire, even though its hoppers are filled. In part (a) the addition machine fires, with inputs of 5 and 7. In part (b) the result, 12, is seen in the spout. It would normally be passed into the hopper of the multiplication machine. However, that hopper is already filled. Moreover, the multiplication machine cannot fire, because it is waiting for a value in its other hopper. Thus, when the addition hoppers are filled again (with 8 and 9) the machine tries to fire and cannot. It will continue trying to fire until the multiplication machine gets its second input. Then it will fire and empty its hopper.

In the Function Machines environment, a machine runs whenever its inputs are available. Because this can occur simultaneously for several machines, the system naturally supports concurrency and parallel computation. Thus, besides its unique and valuable potential as a starting computational language for beginning students, Function Machines offers opportunities for introducing, very early, the advanced subject of parallel algorithms and models.

We have piloted the use of Function Machines extensively in elementary and secondary classrooms. We have introduced the program to teachers as

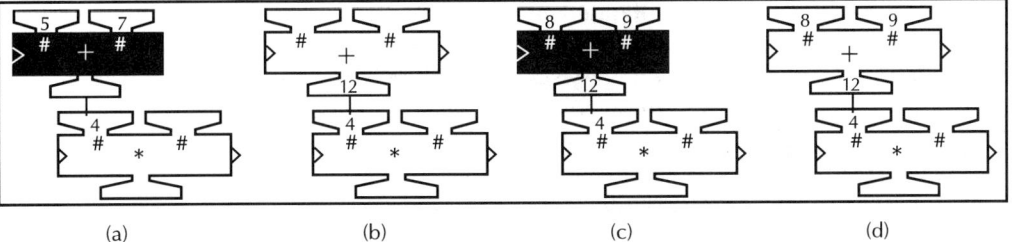

FIGURE 4.19 Data Blocking.

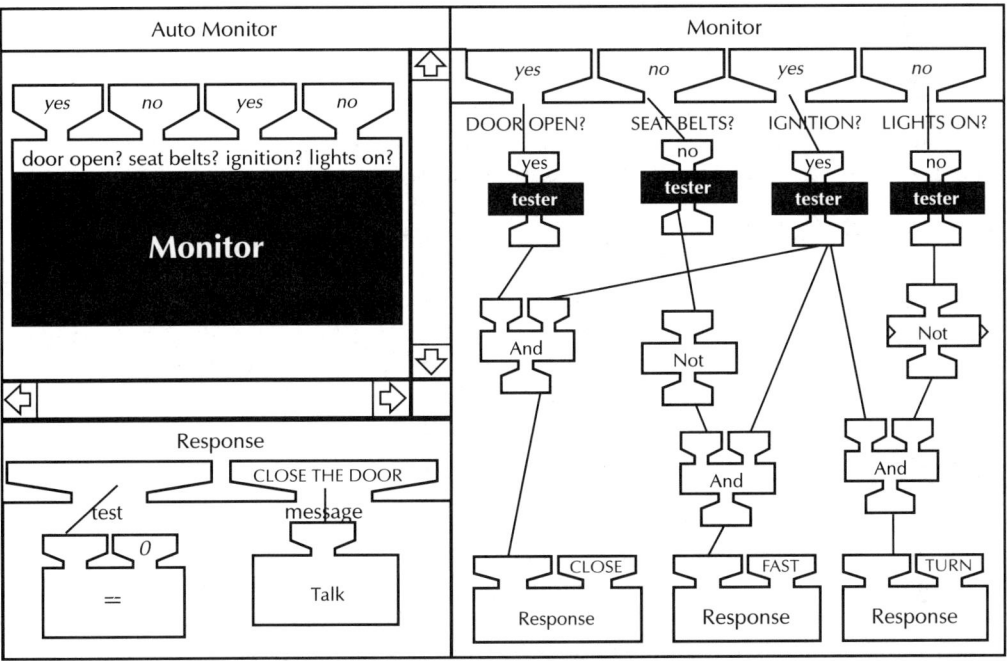

FIGURE 4.20 Function Machines Auto Monitor Model.

well as students through both undergraduate and graduate courses and in summer teacher institutes. Function Machines has been used in teacher workshops such as "Introduction to Computer Structures," given at Lesley College in Cambridge, Massachusetts. The teachers successfully create Function Machine models for mathematical machines—digital computing devices such as adders and multipliers. They also develop models of interactive games and computer-based devices. Some of these, such as the simple model described next, involve parallel computation.

Figure 4.20 shows a Function Machines model of an automobile system for testing the states of a set of sensors and making appropriate audible responses. The system monitors sensors that determine the on/off or open/ closed states of various automobile devices such as the ignition, brakes, fuel level, seat belts, and headlights and that respond with an audible signal or message if the state of some device is faulty. The program shown in the figure was developed by an elementary school teacher. The Auto Monitor machine includes sensors to determine whether a car door is open, whether a seat belt is buckled, whether the ignition is on, and whether the headlights are on. The top-level Monitor machine is shown at the top of the left-hand window. It takes the four corresponding yes or no inputs and passes them to its embedded machines, shown in the right-hand window. The four yes or no inputs are passed to "tester" machines that convert these yes or no values to 0 or 1 for input to the logic machines ("And" and "Not"), which have their usual functions. Note that this is a parallel model; all four tester machines are ready to fire "at the same time."

The "And" machines activate Response machines that invoke the speech output software to produce spoken utterances—in these instances, "close the door," "fasten your seat belt," and "turn on the lights." Depending on the inputs, the program may make one or more of these speech responses (or none), as appropriate. The figure shows the program at the point where it is about to run the four tester machines. It will run them all "at the same time," illustrating the capability for concurrent parallel processing inherent in Function Machines. The leftmost Response machine is shown at the bottom of the left-hand window. Its response message, "close the door," will be uttered by the Talk machine if the input piped to the left-hand hopper of the " = " machine is not 0; this corresponds to the condition (shown in the right-hand window) that the door is open and the ignition is on.

Tasks like the auto monitor pose significant logical challenges to generalist teachers. In pre-service and in-service workshops, we and others have shown that elementary school teachers who are not mathematically or computationally skilled can learn to use the visual modeling facilities of Function Machines to develop models of moderately sophisticated logical complexity, including stochastic as well as deterministic models.[1]

In the auto monitor model, the use of parallelism was incidental. The computation could have been done sequentially, completing the evaluation of the state of the seat belt sensors and the corresponding response before beginning the evaluation of whether the door was open or the lights on. In some models, however, parallelism is intrinsic to performing a correct computation. An example is the model for the classic "turtle tag" problem. The task is to describe the pattern generated by the tracks of four moving turtles, which are initially positioned at the vertices of a square. The turtles are traveling simultaneously, each one moving clockwise toward its nearest neighbor. Figure 4.21 shows a Function Machines simulation modeling this turtle tag dance.

The turtle display (in the right-hand window) shows the current positions of the turtles. As the program runs, each turtle first computes the heading of

---

[1] An extensive set of stochastic modeling projects has been developed for use in teacher workshops (Morrison and Feurzeig, 1993).

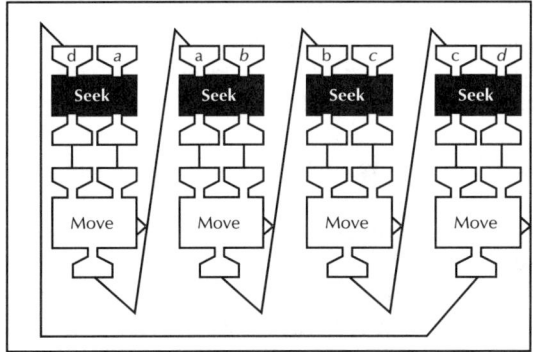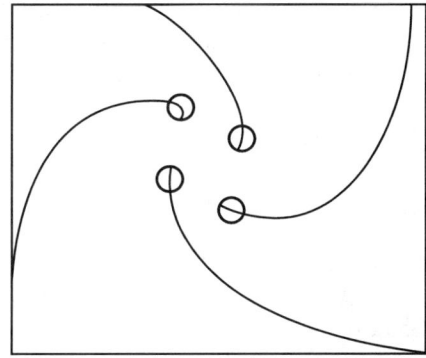

FIGURE 4.21 On the Way.

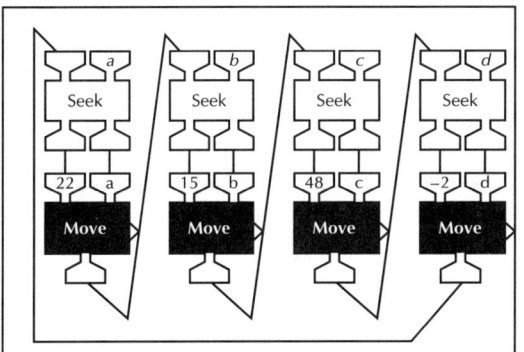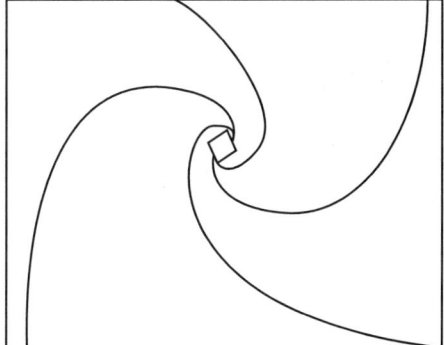

FIGURE 4.22 Rendezvous.

its nearest neighbor. Thus turtle a seeks turtle d, d seeks c, c seeks b, and b seeks a. Then each turtle moves a short distance along its new heading, and the process continues with further rounds of seeks and moves. The output of each Move machine passes the current position and heading of its turtle to the appropriate Seek machine to ready it for its next computation. The figure shows the Turtle Tag program in operation. As the left-hand window shows, the four Seek machines are ready to run. Note that all four have been activated at the same time so that they will run concurrently. The program has been in operation for some time. The right-hand window shows the tracks that have been generated thus far.

Figure 4.22 shows the turtles' final positions. It illustrates the use of simultaneous visualizations of the program structure and operation, as well as its output behavior. As the program runs, we can see the processes that are currently computing. At the same time, we also see what effects these processes have on the model's visual outputs. Moreover, we can study the relationship between the program description and the program output more intensively by running the program incrementally, one step at a time. Observing the dynamic visualization of the model *processes* can give students very direct insight into the mechanisms underlying the model's visual *outputs*. The benefits of working with both kinds of visualizations increase as models become more complex. Indeed, they have enormous potential for enhancing science research as well as science education.

The wide adoption of computational modeling experiences as a key and central component of precollege science and mathematics education has been hampered by the lack of accessible and informative visualization tools. Function Machines directly addresses this need.

The initial version of Function Machines was designed and implemented at BBN in 1987.[2] It currently runs on all Macintosh systems. The *Function Machines User Manual* (Morrison and Walters Associates), included on the accompanying CD, provides a complete reference to the language, along with sample sessions to familiarize users with the fundamentals of Function Machines programming.[3]

Acknowledgments

This research was supported in part by the National Science Foundation under NSF Grant MDR 8751519, "Intelligent Tools for Mathematics Inquiry," and in part by the Advanced Research Projects Agency under Contract N66001-95-D-8607, "An Organic Model for Learning."

References

Braunfeld, P., Dilley, C., & Rucker, W. 1967. A New UICSM approach to fractions for the junior high school. *The Mathematics Teacher*, March 1967, 215–221.

Feurzeig, W., Cuoco, A., Goldenberg, P., & Morrison, D. 1993. Special issue on Function Machines. *Intelligent Tutoring Media, 4 (3/4)*, 95–141.

Feurzeig, W., & Richards, J. 1996. Function Machines: A visual environment to support mathematical modeling. *Communications, Association for Computing Machines, 39(8)*, 88–90.

---

[2]The design team members were Wallace Feurzeig, Paul Horwitz, John Richards, Sterling Wight, and Richard Carter. Sterling Wight implemented the initial versions of the system. Notations similar to those in Function Machines were used in earlier mathematics education work—for example, in representation of fractions visually as machines called Stretchers and Shrinkers, in 1967 (Braunfeld *et al.*). The idea of developing a Logo-like visual computer language based on the metaphor of functions as machines was suggested by Paul Goldenberg in 1986, on the basis of these early noncomputer representations at the University of Illinois.

[3]We are currently completing work on a new version of Function Machines with major enhancements in functionality, interface, and performance. It will run on Windows machines as well as Macintosh systems. The scheduled release date is September 1999.

Morrison, D., & Feurzeig, W. 1993. Using Function Machines to model stochastic systems. *Intelligent Tutoring Media, 4 (3/4)*, 129-141.

Morrison, D., & Walters Associates. 1989. *Function Machines user manual*, Technical Report, Learning Systems and Technologies, Cambridge, MA: BBN Technologies, 1-164.

Resnick, M. 1990. MultiLogo: A study of children and concurrent programming. Interactive Learning Environments, 1 (3), 153-170.

Wight, S., Feurzeig, W., & Richards, J. 1988. Pluribus: A visual programming environment for Education and Research. *Proceedings, IEEE Workshop on Language for Automation*. College Park, MD: The Computer Society, The Institute of Electrical and Electronics Engineers, 122-128.

5

Decentralized Modeling and Decentralized Thinking
Mitchel Resnick

Introduction: The Era of Decentralization

It seems fair to say that we live in an era of decentralization. Almost every time you pick up a newspaper, you can see evidence of the growing interest in decentralized systems. On the front page, you might read an article about the transition of the former Communist states from centrally planned economies to market-based economies. Turn to the business page, and you might find an article about the shift in corporate organizations away from top-down hierarchies toward decentralized management structures. The science section might carry an article about new distributed models of the mind, and it might include a technology column about the role of the Internet in promoting distributed approaches to computing. And in the book review, you might discover how the latest literary theories are based on the idea that literary meaning itself is decentralized—always constructed by individual readers, not imposed by a centralized author.

But even as the influence of decentralized ideas grows within our culture, there is a deep-seated resistance to such ideas. At some deep level, people seem to have strong attachments to centralized ways of thinking. When people see patterns in the world, they often assume that there is some type of centralized control, even when it doesn't exist. For example, most people assume that birds in a flock play a game of "follow the leader": The bird at the front of the flock leads, and the others follow. But that's not so. In fact, most bird flocks don't have leaders at all. Rather, each bird follows a set of simple rules, reacting to the movements of the birds nearby it. Orderly flock patterns arise from these simple, local interactions. The bird in front is not a "leader" in any meaningful sense—it just happens to end up there. The flock is organized without an organizer, coordinated without a coordinator. Yet most people continue to assume the existence of a "leader bird."

This assumption of centralized control, a phenomenon I call the *centralized mindset*, is not just a misconception of the scientifically naive. It seems to affect the thinking of nearly everyone. Until recently, even scientists assumed that bird flocks must have leaders. It is only in recent years that scientists have revised their theories, asserting that bird flocks are leaderless and self-organized (Heppner and Grenander, 1990; Reynolds, 1987). A similar bias toward centralized theories can be seen throughout the history of science.

In this chapter, I discuss how computer-modeling activities can help people move beyond the centralized mindset and gain new insights into (and appreciation for) the workings of decentralized systems. In particular, I will discuss a programmable modeling environment, called StarLogo, that I developed to help precollege students model and explore decentralized systems. By telling "stories" of students' activities with StarLogo, I hope to shed light on the nature of the centralized mindset and on ways of moving beyond it. At the same time, I have a more general goal: to present (and defend) a set of principles to guide the uses of computer modeling in science education.

Learning Through Modeling

My "decentralized modeling" research project has been guided by five core principles. In my view, these principles apply not only to my own research project but to all applications of computer modeling in science education.

- *Principle 1: Encourage construction of models (not just manipulation of preexisting models)*. In many educational applications of computer modeling, students do little more than twiddle parameters on preconstructed models. For example, they are given a model of a spring with a mass on the end, along with sliders for controlling the spring constant and mass. That type of activity can have some value. But students are likely to make much deeper connections with the concepts underlying the model if they are given the opportunity to construct models on their own (Papert, 1991). Accordingly, I designed StarLogo as a *programmable* modeling environment, with which students can construct their own models.
- *Principle 2: Rethink what is learned (not just how it is learned)*. The activity of computer modeling provides a new opportunity for students to learn through exploration and experimentation. But often overlooked is the potential to use modeling to rethink not just the process but also the content of science education. Too often, educators use computer modeling as a new way to teach the same old things (such as the motion of springs). In my work, the emphasis has been on using modeling to help students explore ideas and concepts that were previously inaccessible. For example, ideas about decentralized systems and self-organizing systems have traditionally been taught at the graduate level, via advanced mathematics. StarLogo was designed to make these ideas accessible to pre-college students, without any advanced mathematics.
- *Principle 3: Support true computational models (not just computerization of traditional mathematical models)*. For several hundred years, mathematicians and scientists have used differential equations to model dynamic systems. Many computer-modeling tools reimplement this approach

on the computer, using the computer to solve differential equations numerically. These tools are certainly very useful, and some of them do a very good job of hiding the formal mathematics under graphical descriptions of the differential equations. But the most fundamental contributions of computer modeling are likely to come from tools that are based on totally new representations tailored explicitly for the computer. That is the case with StarLogo, which is based on hundreds of individual objects acting in parallel. This type of representation was not possible in the paper-and-pencil era, and it offers new ways for even young students to explore the workings of dynamic systems.

- *Principle 4: Facilitate personal connections (not just mathematical abstractions)*. In designing new types of learning tools, it is important to consider two types of connections (Resnick *et al.*, 1996). First, there are epistemological connections: How will the tool connect to important domains of knowledge and encourage new ways of thinking? But equally important are personal connections: How will the tool connect to users' interests, passions, and experiences? Many computer-modeling tools are "impersonal"; students must manipulate either mathematical abstractions or aggregate quantities. StarLogo aims to be more "personal," encouraging students to think about the actions and interactions of individual (and familiar) objects.
- *Principle 5: Focus on stimulation (not just simulation)*. Many computer models try to imitate some real-word system or process as accurately as possible. Computer simulations of nuclear reactors are used to predict when the reactors might fail; computer simulations of meteorological patterns are used to predict tomorrow's weather. In these cases, the more accurate the simulation, the better. But for educational applications of computer modeling, real-world fidelity should not have first priority. Instead, the real world should serve only as an inspiration—a departure point for thinking about some set of ideas or concepts. The goal is not to simulate particular systems and processes in the world; it is to probe, challenge, and disrupt the way people think about systems and processes in general. That is the goal of StarLogo: to stimulate people to develop new ways of thinking about decentralized systems.

The Centralized Mindset

Before exploring how computer modeling can help people move beyond the centralized mindset, it is worth examining the nature of that centralized mindset. In some ways, the pervasiveness of the centralized mindset might seem surprising. After all, aren't we living in an era of decentralization? Actually, however, it isn't so surprising if we look at the growing interest in decentralization from a different perspective: Why are people becoming more interested in decentralized ideas *now*? Why didn't it happen before? Why have people resisted decentralized approaches in the past? What underlies this resistance? What made people cling to centralized approaches so tightly for so long?

The centralized mindset can be seen throughout the history of science. Until the mid-nineteenth century, almost everyone embraced the idea that living systems were designed by some God-like entity. Even scientists were convinced by the so-called watchmaker argument (or the "argument from design") proposed by theologian William Paley in his book *Natural Theology* (Paley, 1802). Paley noted that watches are very complex and precise objects. If you found a watch on the ground, you could not possibly believe that such a complex object had been created by chance. Instead, you would naturally conclude that the watch must have had a maker. For Paley, the same logic applies to living systems: they, too, must have a maker.

It is not surprising that scientists accepted Paley's argument in the early nineteenth century, because there were no viable alternative explanations for the complexity of living systems. What *is* surprising is how strongly scientists held on to centralized beliefs even after Darwin provided a viable (and more decentralized) alternative. Science historian Ernst Mayr (1982) notes that biologists put up "enormous resistance" to Darwin's theories for a full 80 years after publication of *On the Origin of Species*, so persistent was the general preference for more centralized alternatives.

The history of research on slime mold cells, as told by Evelyn Fox Keller (1985), provides another example of centralized thinking. During their life cycle, slime mold cells sometimes gather together into clusters. For many years, scientists believed that the aggregation process was coordinated by specialized slime mold cells known as "founder" or "pacemaker" cells. According to this theory, each pacemaker cell sends out a chemical signal, telling other slime mold cells to gather around it, and this results in a cluster. In 1970, Keller and Segel (1970) proposed an alternative model, showing how slime mold cells can aggregate without any specialized cells. Nevertheless, for the following decade, other researchers continued to assume that special pacemaker cells were necessary to initiate the aggregation process. As Keller (1985) writes, with an air of disbelief: "The pacemaker view was embraced with a degree of enthusiasm that suggests that this question was in some sense foreclosed." By the early 1980s, researchers began to accept the idea of aggregation among homogeneous cells, without any pacemaker. But the decade-long resistance serves as some indication of the strength of the centralized mindset.

People also view the workings of the economy in centralized ways, assuming singular causes for complex phenomena. Children, in particular, seem to assume strong governmental control over the economy. Of course, governments *do* play a large role in most economies, but children assume that governments play an even larger role than they actually do. In interviews with Israeli children between 8 and 15 years old, psychologist David Leiser (1983) found that nearly half of the children assumed that the government sets all prices and pays all salaries. Even children who said that employers pay salaries often believed that the government provides the money for the salaries. A significant majority of the students assumed that the government pays the increased salaries after a strike. And many younger children had the seemingly contradictory belief that the government is also responsible for organizing strikes. As Leiser writes, "The child finds it easier to refer unexplained phenomena to the deliberate actions of a clearly defined entity, such as the government, than to impersonal 'market forces.'"

In some ways, it is not surprising that people have such strong commitments to centralized approaches. Many phenomena in the world *are*, in fact, organized by a central designer. These phenomena act to reinforce the centralized mindset. When people see neat rows of corn in a field, they assume (correctly) that the corn was planted by a farmer. When people watch a ballet, they assume (correctly) that the movements of the dancers were planned by a choreographer. When people see a watch, they assume (correctly) that it was designed by a watchmaker.

Moreover, most people participate in social systems (such as families and school classrooms) where power and authority are very centralized. These hierarchical systems serve as strong models. Many people are probably unaware that other types of organization are even possible. In an earlier research project, I developed a programming language (called MultiLogo) based on "agents" that communicated with one another. In using the language, children invariably put one of the agents "in charge" of the others. One student explicitly referred to the agent in charge as "the teacher." Another referred to it as "the mother" (Resnick, 1990).

Perhaps most important, our intuitions about systems in the world are deeply influenced by our conceptions of ourselves. The human mind is composed of thousands of interacting entities (see, for example, Minsky, 1987), but each of us experiences our own self as a singular entity. This is a very convenient, perhaps necessary, illusion for surviving in the world. When I do something, whether I'm painting a picture or organizing a party, I feel as though "I" am playing the role of the "central actor." It feels like there is one entity in charge: me. Thus it is quite natural that I should expect most systems to involve a central actor, or some entity that is in charge. The centralized mindset might be viewed as one aspect (and a lasting remnant) of the egocentrism that Piaget identified in early childhood.

Tools for Decentralized Thinking

In some ways, people already have a great deal of experience with decentralized systems. They observe decentralized systems in the natural world, and they participate in decentralized social systems in their lives. But, of course, observation and participation do not necessarily lead to strong intuitions or deep understanding. People observed bird flocks for thousands of years before anyone suggested that flocks are leaderless. Observation and participation are not enough. People need a richer sense of engagement with decentralized systems. One way to achieve that is to give people opportunities to *design* decentralized systems.

At first glance, this approach to the study of decentralized systems might seem like a contradiction. After all, how can you design decentralized phenomena? By definition, decentralized patterns are created without a centralized designer. But there are ways to use design in the study of decentralized systems. Imagine that you could design the behaviors of lots of individual components—and then observe the patterns that result from all of the interactions. This is a different sort of design: You control the actions of the parts, not of the whole. You are acting as a designer, but the resulting patterns are not designed.

Over the years, computer scientists have developed a variety of computational tools that can be used for this type of "decentralized design." Cellular automata represent one example (see Chapters 1 and 6). In cellular automata, a virtual world is divided into a grid of "cells." Each cell holds a certain amount of "state." Cellular automata have proved to be an extraordinarily rich framework for exploring self-organizing phenomena. Simple rules for each cell sometimes lead to complex and unexpected large-scale structures.

To engage students in thinking about decentralized systems, I wanted to provide an environment similar to cellular automata but more connected to students' interests and experiences. Although cellular automata are well suited for computer scientists and mathematicians, they seem ill suited for people who have less experience (or less interest) in manipulating abstract systems. The objects and operations in cellular automata are not familiar to most people. The idea of writing "transition rules" for "cells" is not an idea that most people can relate to.

Instead, I decided to create an environment based on the familiar ideas of "creatures" and "colonies." The goal was to enable students to investigate the ways in which colony-level behaviors (such as bird flocks and ant foraging trails) can arise from interactions among individual creatures.[1] Logo seemed like a good starting point for my computational system (Papert, 1980; Harvey, 1985). The traditional Logo "turtle" can be used to represent almost any type of object in the world: an ant in a colony, a car in a traffic jam, an antibody in the immune system, or a molecule in a gas. But traditional versions of the Logo language lack several key features that are needed for explorations of colony-type behaviors, so I developed a new version of Logo, called StarLogo, that extends Logo in three major ways (Resnick, 1991, 1994).

First, *StarLogo has many more turtles*. Whereas commercial versions of Logo typically have only a few turtles, StarLogo has *thousands* of turtles, and all of the turtles can perform their actions at the same time, in parallel.[2] For many colony-type explorations, having a large number of turtles is not just nice but necessary. In many cases, the behavior of a colony changes qualitatively when the number of creatures is increased. An ant colony with 10 ants might not be able to make a stable pheromone trail to a food source, whereas a colony with 100 ants (following exactly the same rules) might.

---

[1] I am using the terms *creature* and *colony* rather broadly. On a highway, each car can be considered a "creature," and a traffic jam can be considered the "colony."
[2] The initial version of StarLogo was implemented on a massively parallel computer, the Connection Machine. We have since implemented StarLogo on traditional sequential computers by simulating parallelism.

Second, *StarLogo turtles have better "senses."* The traditional Logo turtle was designed primarily as a "drawing turtle" for creating geometric shapes and exploring geometric ideas. But the StarLogo turtle is more of a "behavioral turtle." StarLogo turtles come equipped with "senses." They can detect (and distinguish) other turtles nearby, and they can "sniff" scents in the world. Such turtle–turtle and turtle–world interactions are essential for creating and experimenting with decentralized and self-organizing phenomena. Parallelism alone is not enough. If each turtle just acts on its own, without any interactions, interesting colony-level behaviors generally do not arise.

Third, *StarLogo reifies the turtles' world.* In traditional versions of Logo, the turtles' world does not have many distinguishing features. The world is simply a place where the turtles draw with their "pens." Each pixel of the world has a single piece of state information—its color. StarLogo accords a much higher status to the turtles' world. The world is divided into small square sections called *patches*. The patches have many of the same capabilities as turtles, but they cannot move. Each patch can hold an arbitrary variety of information. For example, if the turtles are programmed to release a "chemical" as they move, each patch can keep track of the amount of chemical that has been released within its borders. Patches can execute StarLogo commands, just as turtles do. For example, each patch could diffuse some of its "chemical" into neighboring patches, or it could grow "food" based on the level of chemical within its borders. Thus the environment has a status equal to that of the creatures inhabiting it.

StarLogo programs can be conceptualized as turtles moving on top of (and interacting with) a cellular automata grid. All types of interactions are possible: turtle–turtle, turtle–patch, and patch–patch interactions. StarLogo places special emphasis on *local* interactions—that is, interactions among turtles and patches that are spatially near one another. Thus StarLogo is well suited for explorations of self-organizing phenomena, in which large-scale patterns arise from local interactions. In addition, the massively parallel nature of StarLogo makes it well suited for explorations of probabilistic and statistical concepts—and for studies of people's thinking about these concepts (Wilensky, 1993).

Figure 5.1 shows a StarLogo simulation of slime mold cells aggregating into clusters. In this simulation, each cell emits a chemical pheromone, and it also moves in the direction where the pheromone is strongest (that is, it "follows the gradient" of the pheromone). At the same time, the patches cause the pheromone to diffuse and evaporate. With this simple strategy, the cells quickly aggregate into clusters, demonstrating as they do so that aggregation can arise from a decentralized mechanism.

In some ways, the ideas underlying StarLogo parallel the ideas underlying the early versions of Logo itself. In the late 1960s, Logo aimed to make then-new ideas from the computer science community (such as procedural abstraction and recursion) accessible to a larger number of users. Similarly, StarLogo aims to make 1990s ideas from computer science (such as massive parallelism) accessible to a larger audience. And whereas Logo introduced a

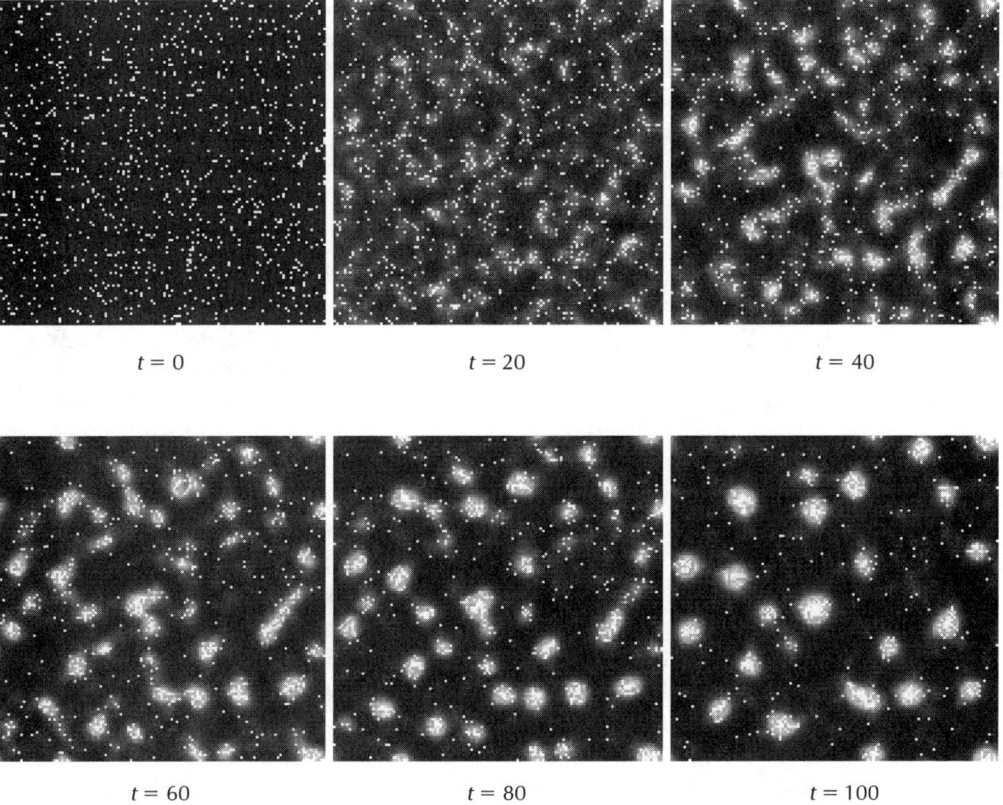

$t = 0$ $t = 20$ $t = 40$

$t = 60$ $t = 80$ $t = 100$

FIGURE 5.1 Slime Mold Cells Aggregating into Clusters.

new object (the turtle) to facilitate explorations of particular mathematical/scientific ideas, such as differential geometry (Abelson and diSessa, 1980), StarLogo introduces another new object (the patch) to facilitate explorations of other mathematical/scientific ideas (such as self-organization).

StarLogo Stories

This section presents stories of student projects with StarLogo, describing the models that students constructed and what they learned in doing so. The students typically came to M.I.T. for eight to ten sessions, each lasting 60 to 90 minutes. Most students worked together in pairs. I worked directly with the students, suggesting projects, asking questions, challenging assumptions, helping with programming, and encouraging students to reflect on their experiences as they worked with StarLogo. Computer interactions were saved in computer files, and all discussions were recorded on audio tape. In the early sessions, I typically showed students existing StarLogo programs. The students experimented with the programs, trying different parameters and making slight modifications of the programs. As the sessions progressed, I encouraged students to develop their own project ideas, and construct their own models, on the basis of personal interests.

Traffic Jams

Ari and Fadhil were students at a public high school in the Boston area. Both enjoyed working with computers, but neither had a very strong mathematical or scientific background. At the time Ari and Fadhil started working with StarLogo, they were also taking a driver's education class. Each had turned 16 years old a short time before, and they were excited about getting their driver's licenses. Much of their conversation focused on cars. When I gave Ari and Fadhil a collection of articles to read, a *Scientific American* article titled "Vehicular Traffic Flow" (Herman and Gardels, 1963) captured their attention.

Traffic flow is a rich domain for studying collective behavior. Interactions among cars in a traffic flow can lead to surprising group phenomena. Consider a long road with no cross streets or intersections. What if we added some traffic lights along the road? The traffic lights would seem to serve no constructive purpose. It would be natural to assume that the traffic lights would reduce the overall traffic throughput (number of cars per unit time). But in some situations, additional traffic lights actually *improve* overall traffic throughput. The New York City Port Authority, for example, found that it could increase traffic throughput in the Holland Tunnel by 6% by deliberately stopping some cars before they entered the tunnel (Herman and Gardels, 1963).

Traditional studies of traffic flow rely on sophisticated analytic techniques (from fields such as queuing theory). But many of the same traffic phenomena can be explored with simple StarLogo programs. To get started, Ari and Fadhil decided to create a one-lane highway. (Later, they experimented with multiple lanes.) Ari suggested adding a police radar trap somewhere along the road, to catch cars going above the speed limit. But he also wanted each car to have its own radar detector, so that cars would be alerted to slow down when they approached the radar trap.

After some discussion, Ari and Fadhil decided that each StarLogo turtle/car should follow three basic rules:

• If there is a car close ahead of you, slow down.
• If there are not any cars close ahead of you, speed up (unless you are already moving at the speed limit).
• If you detect a radar trap, slow down.

Ari and Fadhil implemented these rules in StarLogo. They expected that a traffic jam would form behind the radar trap, and indeed it did (Figure 5.2). After a few dozen iterations of the StarLogo program, a line of cars started to

FIGURE 5.2   Traffic Jams Caused by Radar Trap (shaded area). (Cars move left to right.)

form to the left of the radar trap. The cars moved slowly through the trap and then sped away as soon as they passed it. Ari explained: "First one car slows down for the radar trap, then the one behind it slows down, then the one behind that one, and then you've got a traffic jam."

I asked Ari and Fadhil what would happen if only *some* of the cars had radar detectors. Ari predicted that only some of the cars would slow down for the radar trap. Fadhil had a different idea: "The ones that have radar detectors will slow down, which will cause the other ones to slow down." Fadhil was right. The students modified the StarLogo program so that only 25% of the cars had radar detectors. The result: The traffic flow looked exactly the same as when all of the cars had radar detectors.

What if *none* of the cars had radar detectors—or, equivalently, what if the radar trap were removed entirely? With no radar trap, the cars would be controlled by just two simple rules: If you see another car close ahead, slow down; if not, speed up. The rules could not be much simpler. At first, Fadhil predicted that the traffic flow would become uniform; cars would be evenly spaced, traveling at a constant speed. Without the radar trap, he reasoned, what could cause a jam? But when the students ran the program, a traffic jam formed (Figure 5.3). Along parts of the road, the cars were tightly packed and

FIGURE 5.3 Traffic Jam Without Radar Trap. (Cars move left to right, but jam moves right to left.)

moving slowly. Elsewhere, they were spread out and moving at the speed limit.

Ari and Fadhil were surprised. And when I showed Ari and Fadhil's program to other high school students, they too were surprised. In general, the students expected the cars to end up evenly spaced along the highway, separated by equal distances. Several of them talked about the cars reaching an "equilibrium" characterized by equal spacing. No one expected a traffic jam to form. Some of their predictions:

Emily: [The cars will] just speed along, just keep going along . . . they will end up staggered, in intervals.

Frank: Nothing will be wrong with it. Cars will just go. . . . There's no obstacles. The cars will just keep going, and that's it.

Ramesh: They will probably adjust themselves to a uniform distance from each other.

When I ran the simulation, and traffic jams began to form, the students were shocked. In their comments, most students revealed a strong commitment to the idea that some type of "seed" (such as an accident or a broken bridge) is needed to start a traffic jam. Perhaps Frank expressed it best: "I didn't think there would be any problem, since there was nothing there." If there is *nothing there*—if there is no seed—there should not be a traffic jam. Traffic jams do not just happen; they must have localizable causes. And the cause must come from outside the system (not from the cars themselves). Some researchers who study systems talk about *exogenous* (external) and *endogenous* (internal) factors affecting the behavior of a system. In the minds of the students, patterns (such as traffic jams) can be formed only by exogenous factors.

Fadhil suggested that the jams were caused by differences in the initial speeds of the cars. Accordingly, the students changed the StarLogo program, starting all of the cars at the exact same speed. But the jams still formed. Fadhil quickly understood. At the beginning of the program, the cars were placed at random positions on the road. Random positioning led to uneven spacing between the cars, and uneven spacing could also provide the "seed" from which a traffic jam could form. Fadhil explained: "Some of the cars start closer to other cars. Like, four spaces between two of them, and two spaces between others. A car that's only two spaces behind another car slows down, then the one behind it slows down."

Next they changed the program so that the cars were evenly spaced. Sure enough, no traffic jams formed. All of the cars uniformly accelerated up to the speed limit. But Ari and Fadhil recognized that such a situation would be difficult to set up in the real world. The distances between the cars had to be just right, and the cars had to start at exactly the same time—like a platoon of soldiers starting to march in unison.

Termites and Wood Chips

Termites are among the master architects of the animal world. On the plains of Africa, termites construct giant mound-like nests rising more than 10 feet tall, thousands of times taller than the termites themselves. Inside the mounds are intricate networks of tunnels and chambers. Each termite colony has a queen. But, as in ant colonies, the termite queen does not "tell" the termite workers what to do. On the termite construction site, there is no construction supervisor, no one in charge of the master plan. Rather, each termite carries out a relatively simple task. Termites are practically blind, so they must interact with each other (and with the world around them) primarily through their senses of touch and smell. From local interactions among thousands of termites, impressive structures emerge.

The global-from-local nature of termite constructions makes them well suited for StarLogo explorations. Callie, one of the high school students, worked on a simple form of termite construction: She programmed a set of termites to collect wood chips and put them into piles. At the start of the program, wood chips were scattered randomly throughout the termites' world. The challenge was to make the termites organize the wood chips into a few, orderly piles.

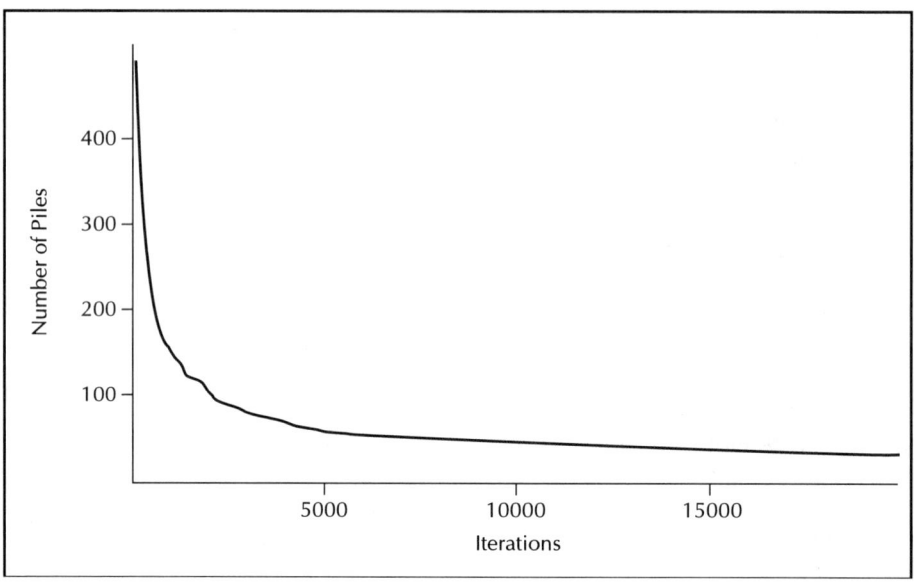

FIGURE 5.4 The Number of Piles Decreases Monotonically.

Callie and I worked together on the project. We started with a very simple strategy, programming each individual termite to obey the following rules:

• If you are not carrying anything and you bump into a wood chip, pick it up.
• If you are carrying a wood chip and you bump into another wood chip, put down the wood chip you're carrying.

At first, Callie and I were both skeptical that this simple strategy would work. There was no mechanism for preventing termites from taking wood chips away from existing piles. Thus, while termites were putting new wood chips on a pile, other termites might be taking wood chips away from it. It seemed like a good prescription for getting nowhere. But we pushed ahead and implemented the strategy in a StarLogo program, with 1000 termites and 2000 wood chips scattered in a 128 ×128 grid.

We tried the program, and (much to our surprise) it worked quite well. At first, the termites gathered the wood chips into hundreds of small piles. But gradually, the number of piles declined, and the number of wood chips in each pile increased (see Figure 5.4). After 2000 iterations, there were about 100 piles, with an average of 15 wood chips in each pile. After 10,000 iterations, there were fewer than 50 piles left, with an average of 30 wood chips in each pile. After 20,000 iterations, only 34 piles remained, with an average of 44 wood chips in each pile. The process was rather slow, and it was frustrating to watch, because termites often carried wood chips away from well-established piles. But, all in all, the program worked quite well.

Why did it work? As we watched the program, it suddenly seemed obvious. Imagine what happens when the termites (by chance) remove all of the wood chips from a particular pile. Because all of the wood chips are gone from that spot, termites will never again drop wood chips there. Hence the pile has no way of restarting.

As long as a pile exists, its size is a two-way street: It can either grow or shrink. But the *existence* of a pile is a one-way street: once it is gone, it is gone forever. Thus a pile is somewhat analogous to a species of creatures in the real world. As long as the species exists, the number of individuals in the species can go up or down. But once all of the individuals are gone, the species is extinct, gone forever. In these cases, zero is a "trapped state": Once the number of creatures in a species (or the number of wood chips in a pile) goes to zero, it can never rebound.

Of course, the analogy between species and piles breaks down in some ways. New species are sometimes created, as offshoots of existing species. But in the termite program, there is no way to create a new pile. The program starts with roughly 2000 wood chips. These wood chips can be viewed as 2000 "piles," each with a single wood chip. As the program runs, some piles disappear and no new piles are created, so the total number of piles decreases monotonically.

Rabbits and Grass

The great baseball manager Casey Stengel once said, "If you don't know where you're going, you might end up somewhere else." My experiences with computer-based modeling activities have taught me a corollary: "Even if you think you know where you're going, you'll probably end up somewhere else."

That's what happened to Benjamin, a high school student, when he set out to create an StarLogo program that would simulate evolution by natural selection. I had given Benjamin a *Scientific American* article (Dewdney, 1989) about a computer program called *Simulated Evolution* (Palmiter, 1989). Benjamin, who had just finished his junior year in high school, decided that he wanted to create a StarLogo program similar to the commercial program described in the article. His goal was to devise a set of computer "creatures" that would interact and evolve.

At the core of Benjamin's simulation were turtles and food. His basic idea was simple: Turtles that eat a lot of food reproduce, and turtles that don't eat enough food die. Eventually, he planned to add "genes" to his turtles. Different genes could provide turtles with different levels of "fitness" (perhaps different capabilities for finding food). But Benjamin never got around to the genes. Rather, on the road to evolution, Benjamin got sidetracked into an interesting exploration of ecological systems (in particular, predator–prey systems).

Benjamin began by making food grow randomly throughout the StarLogo world. (During each time step, each StarLogo patch had a random chance of growing some food.) Then he created some turtles. The turtles had very meager sensory capabilities. They could not "see" or "smell" food at a distance. They could sense food only when they bumped directly into it. Hence the turtles followed a very simple strategy: Wander around randomly, eating whatever food you bump into.

Benjamin gave each turtle an "energy" variable. Every time a turtle took a step, its energy decreased a bit. Every time it ate some food, its energy increased. Then Benjamin added one more rule: If a turtle's energy dipped to zero, the turtle died. With this program, the turtles do not reproduce. Life is a one-way street: Turtles die, but no new turtles are born. Still, even with this simple-minded program, Benjamin found some surprising and interesting behaviors.

Benjamin ran the program with 300 turtles. But the environment could not support that many turtles. There wasn't enough food, so some turtles began to die. The turtle population fell rapidly at first and then leveled out at about 150 turtles. The system seemed to reach a steady state with 150 turtles: The number of turtles and the density of food both remained roughly constant.

Then Benjamin tried the same program with 1000 turtles. If there wasn't enough food for 300 turtles, there certainly wouldn't be enough for 1000 turtles, so Benjamin wasn't surprised when the turtle population began to fall. But he *was* surprised at how *far* the population fell. After a while, only 28 turtles remained. Benjamin was puzzled: "We started with more. Why should we end up with less?" After some discussion, he realized what had happened. With so many turtles, the food shortage was even more critical than before. The result: mass starvation. Benjamin still found the behavior a bit strange: "The turtles have less (initial energy as a group), and less usually isn't more."

Next, Benjamin decided to add reproduction to his model. His plan: Whenever a turtle's energy increases above a certain threshold, the turtle should "clone" itself and split its energy with its new twin. That can be accomplished by adding another procedure to the program.

Benjamin assumed that the rule for cloning would somehow "balance" the rule for dying, leading to some sort of equilibrium. He explained: "Hopefully, it will balance itself out somehow. I mean it will. It will have to. But I don't know what number it will balance out at." After a little more thought, Benjamin suggested that the food supply might fall at first but that it would then rise back and become steady: "The food will go down, a lot of them will die, the food will go up, and it will balance out."

Benjamin started the program running. As Benjamin expected, the food supply initially went down and then went up. But it didn't "balance out" as Benjamin had predicted: It went down and up again and again and again. Meanwhile, the turtle population also oscillated, but out of phase with the food.

On each cycle, the turtles "overgrazed" the food supply, leading to a scarcity of food, and many of the turtles died. But then, with fewer turtles left to eat the food, the food became more dense. The few surviving turtles thus found a plentiful food supply, and each of them rapidly increased its energy. When a turtle's energy surpassed a certain threshold, it cloned, increasing the turtle population. But as the population grew too high, food again became scarce, and the cycle started again.

Visually, the oscillations were striking. Red objects (turtles) and green objects (food) were always intermixed, but the density of each continually changed. Initially, the screen was dominated by red turtles, with a sparse scattering of green food. As the density of red objects declined, the green objects proliferated, and the screen was soon overwhelmingly green. Then the process reversed: The density of red increased, and the density of green declined.

Depending on the particular parameters, the oscillations took on different forms. In Benjamin's initial program, the oscillations were damped: With each cycle, the peaks were a little less high, the troughs a little less deep. In the first cycle, the turtle population dwindled to just 26 turtles and then rose to 303 turtles. In the next cycle, the population shrank to 47 turtles and then rose to 244 turtles. Eventually, the turtle population stabilized between 130 and 160 turtles.

Benjamin recognized that this result depended critically on the parameters in his StarLogo program. He wondered what would happen if the food grew just half as quickly. He figured that this new world would support fewer turtles, but how many fewer? In the original version of his StarLogo program, each patch had 1 chance in 1000 of growing food. Benjamin changed it to 1 in 2000.

When Benjamin ran the program, he was in for another surprise: All of the turtles died. But Benjamin, who had just finished graphing the oscillations from the previous experiment, quickly realized what had happened. "The oscillation must be between some number and negative something," he said. That is, the trough of the oscillation must drop below zero. And once the population drops below zero, it can never recover. There is no peak after a negative trough. Extinction is forever; it is a "trapped state."

The problem lay in the initial conditions. Benjamin had started the simulation with 1000 turtles. If there were fewer initial turtles, the first trough wouldn't sink so deep. Benjamin came up with an ingenious solution. "I'll start with just one (turtle)," he explained. "It will definitely survive. I'll put money on it." Benjamin started the program again, this time with a single turtle. For a while, the single turtle roamed the world by itself. Benjamin cheered it on: "Come on. Hang on there. Come on. Get some food." Finally, the turtle cloned, and then there were two. "He's going to live," exclaimed Benjamin.

The turtle population rose to about 130 turtles, leveled off, and then fell. As before, the turtle population went up and down in a damped oscillation. Eventually, the population stabilized at about 75 turtles. Thus, with food growing at half the rate at which it grew before, the turtle population stabilized at about half the number at which it leveled off before. The "equilibrium population" seemed to be proportional to the rate of food growth.

Before running the program, Benjamin had predicted that the equilibrium population would be more drastically affected by the reduction in food growth. He expected the population to stabilize with considerably fewer than 75 turtles. But after watching the program run, he developed an explanation for the proportional relationship. Looking at the dots of food on the screen, he noted that the "food density" at equilibrium looked about the same as in the previous experiment, despite the change in the rate of food growth. That made sense to him: A certain food density is needed to keep the turtles just on the brink between death and reproduction. To reach a relatively steady state, the system needed to maintain that special food density. Given that the food was growing just half as quickly as before, it made sense that the system could support only half as many turtles.

Benjamin's reasoning is an example of what Hut and Sussman (1987) dubbed "analysis by synthesis." Traditionally, synthesis and analysis have been seen as opposed to one another—two alternative ways of solving problems. But with computer-based explorations, the two approaches get mixed and blurred. It is very unlikely that Benjamin could have developed his explanation without actually viewing (and manipulating) the simulation. Only by building and creating (synthesis) was Benjamin able to develop a well-reasoned explanation for the behavior of the turtles (analysis).

The oscillating behavior in Benjamin's project is characteristic of ecological systems with predators (in this case, turtles) and prey (in this case, food). Traditionally, scientific (and educational) explorations of predator–prey systems are based on sets of differential equations known as the Lotka–Volterra equations (Lotka, 1925; Volterra, 1926). For example, the changes in the population density of the prey (n_1) and the population density of the predator (n_2) can be described with the following differential equations:

$$dn_1/dt = n_1(b - k_1n_2)$$
$$dn_2/dt = n_2(k_2n_1 - d)$$

where b is the birth rate of the prey, d is the death rate of the predators, and k_1 and k_2 are constants. It is straightforward to write a computer program based on the Lotka–Volterra equations, computing how the population densities of the predator and prey vary with time (see, for example, Roberts et al., 1983).

This differential equation approach is typical of the way scientists have traditionally modeled and studied the behaviors of all types of systems (physical, biological, and social). Scientists typically write down sets of differential equations and then attempt to solve them either analytically or numerically. These approaches require advanced mathematical training; usually, they are studied only at the university level.

The StarLogo approach to modeling systems (exemplified by Benjamin's project) is sharply different. StarLogo makes systems-related ideas more accessible to younger students by providing them with a stronger personal connection to the underlying models. Traditional differential equation approaches are "impersonal" in two ways. The first is obvious: They rely on manipulation of abstract symbols (accessible only to students with advanced mathematical training). The second is more subtle: Differential equations deal in aggregate quantities. In the Lotka–Volterra system, for example, the differential equations describe how the overall *populations* (not the individual creatures) evolve over time. There are now some very good computer-modeling tools—such as STELLA® (Richmond and Peterson, 1990) and Model-It (Jackson *et al.*, 1996)—based on differential equations. These tools eliminate the need to manipulate symbols, focusing on more qualitative and graphical descriptions. But they still rely on aggregate quantities.

In StarLogo, by contrast, students think about the actions and interactions of individual objects or creatures. StarLogo programs describe how individual creatures (not overall populations) behave. Thinking in terms of individual creatures seems far more intuitive, particularly for the mathematically uninitiated. Students imagine themselves as individual turtles/creatures and think about what they might do. In this way, StarLogo enables learners to "dive into" the model (Ackermann, 1996) and make use of what Papert (1980) calls "syntonic" knowledge about their bodies. By observing the dynamics at the level of the individual creatures, rather than at the aggregate level of population densities, students can more easily think about and understand the population oscillations that arise. Future versions of StarLogo will enable users to zoom in and out, making it easier for them to shift back and forth in perspective from the individual level to the group level.

I refer to StarLogo models as "true computational models" (Resnick, 1997), because StarLogo uses new computational media in a more fundamental way than most computer-based modeling tools. Whereas most tools simply implement traditional mathematical models on a computer (numerically solving traditional differential equation representations, for instance), StarLogo provides new representations that are tailored explicitly for the computer. Of course, differential equation models are still very useful, and they are superior to StarLogo-style models in some contexts. But too often, scientists and educators see traditional differential equation models as the *only* approach to modeling. As a result, many students (particularly students alienated by traditional classroom mathematics) view modeling as a difficult or uninteresting activity. What is needed is a more pluralistic approach, recognizing that there are many different approaches to modeling, each with its own strengths and weaknesses. A major challenge is to develop a better understanding of when to use which approach, and why.

Decentralized Thinking

As students began working with StarLogo, they nearly always assumed centralized causes in the patterns they observed, and they nearly always imposed centralized control when they wanted to create patterns. But as students continued to work on StarLogo projects, most of them began to develop new ways of thinking about decentralization. In almost all cases, they developed an appreciation for and a fascination with decentralized systems. At one point, while we were struggling to get our termite program working, I asked Callie whether we should give up on our decentralized approach and program the termites to take their wood chips to predesignated spots. She quickly dismissed this suggestion.

Mitchel: We could write the program so that the termites know where the piles are. As soon as a termite picks up a wood chip, it could just go to the pile and put it down.

Callie: Oh, that's boring!

Mitchel: Why do you think that's boring?

Callie: Cause you're telling them what to do.

Mitchel: Is this more like the way it would be in the real world?

Callie: Yeah. You would almost know what to expect if you tell them to go to a particular spot and put it down. You know that there will be three piles. Whereas here, you don't know how many mounds there are going to be. Or if the number of mounds will increase or decrease. Or things like that. . . . This way, they [the termites] made the piles by themselves. It wasn't like they [the piles] were artificially put in.

For Callie, preprogrammed behavior, even if effective, was "boring." Callie preferred the decentralized approach because it made the termites seem more independent ("they made the piles by themselves") and less predictable ("you don't know how many mounds there are going to be").

Over time, other students shared Callie's fascination with decentralization, though they often struggled in their efforts to use decentralized strategies in analyzing and constructing new systems. As I worked with students, I assembled a list of "guiding heuristics" that students used as they began to develop richer models of decentralized phenomena. These heuristics are not very "strong." They are not "rules" for making sense of decentralized systems. Rather, they are loose collections of ideas associated with decentralized thinking. Pedagogically, they serve as good discussion points for provoking people to think about decentralization. They also serve as a type of measuring stick for conceptual change: As students worked on StarLogo projects, they gradually began to integrate these heuristics into their own thinking and discourse. In this section, I discuss five of these guiding heuristics.

Positive Feedback Isn't Always Negative

When people think about the scientific idea of positive feedback, they typically think of the screeching sound that results when a microphone is placed near a speaker. Positive feedback is viewed as a destructive force that makes things spiral out of control. By contrast, negative feedback is viewed as very useful—as keeping things under control. Negative feedback is symbolized by the thermostat that keeps room temperature at a desired level by turning the heater on and off as needed.

When I asked high school students about positive feedback, most were not familiar with the term, but they were certainly familiar with the concept. When I explained what I meant by positive feedback, students quickly generated examples that involved something getting out of control, often with destructive consequences. One student talked about scratching a mosquito bite, which made the bite itch even more, so she scratched it some more, which made it itch even more, and so on. Another student talked about stock market crashes: A few people start selling, which makes more people start selling, which makes even more people start selling, and so on.

Despite these negative images, positive feedback often plays a crucial role in decentralized phenomena. Economist Brian Arthur (1990) points to the geographic distribution of cities and industries as an example of a self-organizing process driven by positive feedback. Once a small nucleus of high-technology electronics companies started in Santa Clara County south of San Francisco, an infrastructure developed to serve the needs of those companies. That infrastructure encouraged even more electronics companies to locate in Santa Clara County, which encouraged the development of an even more robust infrastructure. Thus Silicon Valley was born.

For some students who used StarLogo, the idea of positive feedback provided a new way of looking at their world. One day, a student came to me excitedly. He had been in downtown Boston at lunch time, and he had had a vision. He imagined two people walking into a deli to buy lunch.

Once they get their food, they don't eat it there. They bring it back with them. Other people on the street smell the sandwiches and see the deli bag, and they say, 'Hey, maybe I'll go to the deli for lunch today!' They were just walking down the street, minding their own business, and all of the sudden they want to go to the deli. As more people go to the deli, there's even more smell and more bags. So more people go to the deli. But then the deli runs out of food. There's no more smell on the street from the sandwiches. So no one else goes to the deli.

Randomness Can Help Create Order

Like positive feedback, randomness has a bad name. Most people see randomness as annoying at best, destructive at worst. They view randomness as opposed to order. Randomness undoes order; it makes things disorderly.

In fact, however, randomness plays an important role in creating order in many self-organizing systems. People often assume that "seeds" are needed to initiate patterns and structures. In general, this is a useful intuition. The problem is that most people have too narrow a conception of "seeds." They think only of preexisting inhomogeneities in the environment, such as a broken bridge on the highway or a piece of food in an ant's world.

This narrow view of seeds causes misintuitions. In self-organizing systems, seeds are neither preexisting nor externally imposed. Rather, self-organizing systems often *create their own* seeds. It is here that randomness plays a crucial role. Random fluctuations act as the "seeds" from which patterns and structures grow. Randomness creates the initial seeds, and then positive feedback makes the seeds grow. For example, the differing velocities of cars on a highway create the seeds from which traffic jams can grow.

A Flock Isn't a Big Bird

In trying to make sense of decentralized systems and self-organizing phenomena, the idea of *levels* is critically important. Interactions among objects at one level give rise to new types of objects at another level. Interactions among slime mold cells give rise to slime mold clusters. Interactions among cars give rise to traffic jams. Interactions among birds give rise to flocks.

In many cases, the objects on one level behave very differently than objects on another level behave. For some high school students, these differences in behavior were very surprising (at least initially). For example, the students working on the StarLogo traffic project were shocked by the behavior of the traffic jams: The jams moved backward even though all of the cars within the jams were moving forward.

Confusion of levels is not restricted to scientifically naive high school students. I showed the StarLogo traffic program to two visiting computer scientists. They were not at all surprised that the traffic jams were moving backward. They were well aware of that phenomenon. But then one of the researchers said, "You know, I've heard that's why there are so many accidents on the freeways in Los Angeles. The traffic jams are moving backward and the cars are rushing forward, so there are lots of accidents." The other researcher thought for a moment and then replied, "Wait a minute. Cars crash into other cars, not into traffic jams." In short, he believed that the first researcher had confused levels, mixing cars and jams inappropriately. The two researchers then spent half an hour trying to sort out the problem. It is an indication of the underdeveloped state of decentralized thinking in our culture that two sophisticated computer scientists needed to spend half an hour trying to understand the behavior of a ten-line decentralized computer program written by a high school student.

A Traffic Jam Isn't Just a Collection of Cars

For most everyday objects, it is fair to think of the object as a collection of particular parts: A chair has four particular legs, a particular seat, and so on. But not so with objects such as traffic jams. Thinking of a traffic jam as a collection of particular parts leads to confusion. The cars that compose a traffic jam are always changing, as some cars leave the front of the jam and others join from behind. Even when all of the cars in the jam have been replaced with new cars, it is still the same traffic jam. A traffic jam can be thought of as an "emergent object"—it emerges from the interactions among lower-level objects (in this case, cars).

As students worked on StarLogo projects, they encountered many emergent objects. In the termite example, the piles of wood chips can be viewed as emergent objects. The precise composition of the piles is always changing, as termites take away some wood chips and add other wood chips. After a while, none of the original wood chips remains, but the pile is still there.

The Hills are Alive

In *Sciences of the Artificial* (1969), Herbert Simon describes a scene in which an ant is walking on a beach. Simon notes that the ant's path might be quite complex. But the complexity of the path, says Simon, is not necessarily a reflection of the complexity of the ant. Rather, it might reflect the complexity of the beach. Simon's point: Don't underestimate the role of the environment in influencing and constraining behavior. People often think of the environment as something to be *acted upon*, not something to be *interacted with*. People tend to focus on the behaviors of individual objects, ignoring the environment that surrounds (and interacts with) the objects.

Adopting a richer view of the environment is important in thinking about decentralized and self-organizing systems. In designing StarLogo, I explicitly tried to highlight the environment. Most creature-oriented programming environments treat the environment as a passive entity manipulated by the creatures that move within it. In StarLogo, by contrast, the "patches" of the world are equal in status to the creatures that move in the world. By reifying the environment, I hoped to encourage people to think about the environment in new ways.

Initially, some students resisted the idea of an active environment. When I explained a StarLogo ant-foraging program to one student, he was worried that pheromone trails would continue to attract ants even after the food sources at the ends of the trails had been fully depleted. He developed an elaborate scheme in which the ants, after collecting all of the food, deposited a second pheromone to neutralize the first pheromone. It never occurred to him simply to let the first pheromone evaporate. In his mind, the ants had to take some positive action to get rid of the first pheromone. They could not rely on the environment to make it go away.

New Media, New Mindsets

There is an old saying that goes something like this: "Give a person a hammer, and the whole world looks like a nail." Indeed, the ways in which we see the world are deeply influenced by the tools and media at our disposal. If we are given new tools and media, not only can we accomplish new tasks, but we also begin to view the world in new ways.

Often, we hardly recognize how our tools and media are influencing our ways of viewing the world. For several centuries now, scientists have described the world in terms of differential equations. Is that because differential equations are the best way to represent and describe the world? Or is it because the common media of the era (paper and pencil) are well suited to manipulations of differential equations? Could we say, "Give a scientist paper and pencil, and the whole world looks like differential equations"?

New computational media now hold the promise for radically reshaping how people model (and think about) the world. But this shift won't happen automatically. Computer modeling will bring profound change to the classroom only if modeling tools take full advantage of new computational representations. Just as sculptors need to understand the qualities of clay (or whatever material they are using), designers of computer-modeling tools need to understand their chosen medium. StarLogo, for example, leverages two new computational paradigms—massive parallelism and object-oriented programming. These new paradigms offer new design possibilities: new ways to create decentralized models. But even more important, these new paradigms offer new epistemological possibilities: a new decentralized framework for making sense of many phenomena in the world.

Adding new tools to the carpenter's toolkit changes the way the carpenter looks at the world. So, too, with computational ideas and paradigms. That is the central challenge for computer-modeling activities in education: not only to help students create models in new ways but also to help students develop fundamentally new ways of thinking about the systems and phenomena they are modeling.

Acknowledgments

Brian Silverman, Andy Begel, Hal Abelson, Seymour Papert, Uri Wilensky, and Larry Latour provided inspiration and ideas for the StarLogo project. The LEGO Group and the National Science Foundation (Grants 9153719-MDR, 9358519-RED, 9553474-RED) provided generous financial support. Portions of this chapter were previously published elsewhere (Resnick, 1994, 1996, 1997) and are reprinted with permission.

References

Abelson, H., & diSessa, A. 1980. *Turtle geometry: The computer as a medium for exploring mathematics*. Cambridge, MA: M.I.T. Press.

Ackermann, E. 1996. Perspective-taking and object construction: Two keys to learning. In Kafai, Y. & Resnick, M. (eds.), *Constructionism in practice*. Mahwah, NJ: Lawrence Erlbaum, pp. 25–35.

Arthur, W.B. 1990. Positive feedbacks in the economy. *Scientific American*, *262(2)*, 92–99.

Dewdney, A.K. 1989. Simulated evolution: Wherein bugs learn to hunt bacteria. *Scientific American*, *260(5)*, 138–141.

Harvey, B. 1985. *Computer science Logo style*. Cambridge, MA: M.I.T. Press.

Heppner, F., and Grenander, U. 1990. A stochastic nonlinear model for coordinated bird flocks. In Krasner, S. (ed.), *The ubiquity of chaos*. Washington, DC: AAAS Publications.

Herman, R., & Gardels, K. 1963. Vehicular traffic flow. *Scientific American*, *209*(*6*), 35-43.

Hut, P., and Sussman, G.J. 1987. Advanced computing for science. *Scientific American*, *255*(*10*).

Jackson, S., Stratford, S., Krajcik, J., & Soloway, E. 1996. A learner-centered tool for students building models. *Communications of the ACM*, *39*(*4*), 48-49.

Keller, E.F. 1985. *Reflections on gender and science*. New Haven: Yale University Press.

Keller, E.F., and Segel, L. 1970. "Initiation of slime mold aggregation viewed as an instability." *Journal of Theoretical Biology*, *26*, 399-415.

Leiser, D. 1983. Children's conceptions of economics—The constitution of a cognitive domain. *Journal of Economic Psychology*, *4*, 297-317.

Lotka, A.J. 1925. *Elements of physical biology*. New York: Dover Publications (reprinted 1956).

Mayr, E. 1982. *The growth of biological thought*. Cambridge, MA: Harvard University Press.

Minsky, M. 1987. *The society of mind*. New York: Simon & Schuster.

Paley, W. 1802. *Natural theology—or evidences of the existence and attributes of the deity collected from the appearances of nature*. Oxford, England: J. Vincent.

Palmiter, M. 1989. *Simulated evolution*. Bayport, NY: Life Science Associates.

Papert, S. 1980. *Mindstorms: Children, computers, and powerful ideas*. New York: Basic Books.

Papert, S. 1991. Situating constructionism. In Harel, I. & Papert, S. (eds.), *Constructionism*. Norwood, NJ: Ablex Publishing.

Resnick, M. 1990. MultiLogo: A study of children and concurrent programming. *Interactive Learning Environments*, *1*(*3*), 153-170.

Resnick, M. 1991. Animal simulations with StarLogo: Massive parallelism for the Masses. In Meyer, J.A. & Wilson, S. (eds.), *From animals to animats*. Cambridge, MA: M.I.T. Press.

Resnick, M. 1994. *Turtles, termites, and traffic jams: Explorations in massively parallel microworlds*. Cambridge, MA: M.I.T. Press.

Resnick, M. 1996. Beyond the centralized mindset. *Journal of the Learning Sciences*, *5*(*1*), 1-22.

Resnick, M. 1997. Learning through computational modeling. *Computers in the Schools*, *14*(*1*), 143-152.

Resnick, M., Bruckman, A., & Martin, F. 1996. Pianos not stereos: Creating computational constructions kits. *Interactions*, *3*(*5*), 41-50.

Reynolds, C. 1987. Flocks, herds, and schools: A distributed behavioral model. *Computer Graphics*, *21*(*4*), 25-36.

Richmond, B., & Peterson, S. 1990. Stella II. Hanover, NH: High Performance Systems, Inc.

Roberts, N., Anderson, D., Deal, R., Garet, M., & Shaffer, W. 1983. *Introduction to computer simulation: A system dynamics modeling approach*. Reading, MA: Addison-Wesley.

Simon, H. 1969. *The sciences of the artificial*. Cambridge, MA: M.I.T. Press.

Volterra, V. 1926. Fluctuations in the abundance of a species considered mathematically. *Nature*, *188*, 558-560.

Wilensky, U. 1993. *Connected mathematics: Building concrete relationships with mathematical knowledge*. PhD. dissertation, M.I.T.

6

An Object-Based Modeling Tool for Science Inquiry

Eric K. Neumann[1]

Wallace Feurzeig

Peter Garik[2]

OOTLs: The Object–Object Transformation Language

A major challenge of future secondary and undergraduate science instruction will be to help students learn to formulate, at an appropriate level of representation, mathematical models of physical phenomena for use with a computer simulation engine. Students will learn to investigate the behavior of these models and test their validity and scope of application. In order to make this leap in instruction to teaching model formulation, we have to confront the fact that students typically find it very difficult to express problems in the standard formal mathematical representations. The symbolic language of differential equations, for example, is very far removed from students' mental models of the objects and object interactions involved in problem situations. Another kind of representation language—mathematically equivalent and mechanically translatable to differential equations, but more natural and accessible to students—is needed to provide them with initial experiences in problem formulation. The transition to the standard formal language can be made later, after they have acquired the relevant insights. This chapter describes a modeling tool for expressing phenomena directly in terms of the characteristic interactions among the objects involved. This object-based representation facilitates the introduction of modeling ideas and activities in science education. At the same time, it offers science researchers a productive new approach for investigating complex phenomena.

We designed OOTLs (Object–Object Transformation Language), a computational modeling environment, to help students acquire experience and skill in formulating models of dynamic processes, expressed as objects and the in-

---

[1] Presently at NetGenics Inc., Cleveland, Ohio.
[2] BBN consultant, presently at Boston University.

teractions among them.[3] Events in OOTLs are conceptualized as interactions among the objects that are identified as the key players in the model processes. The OOTLs modeling language supports the description and simulation of phenomena for which the law of mass action holds: It applies to "well-stirred" systems composed of large numbers of dynamically interacting objects. OOTLs has application to a wide variety of phenomena in many areas of science, including epidemiology (the spread of contagious disease), population ecology (competition, predation, and adaptation), economics (market dynamics), physics (gas kinetics), chemical dynamics (reaction–diffusion equations) and traffic flow. OOTLs provides students with a parser to construct equations describing interactions between objects. The objects, which are represented as graphical icons, may represent chemical species, gas molecules, or humans. Objects interact with each other at specified rates. The equations describe the transformations that result from the object interactions. Objects may be created or consumed (for chemical reactions, there are sources and sinks for reactants; for a biological problem, birth and death of species; for a model of an economy, imports and exports or innovation and obsolescence).

In designing OOTLs, we have taken into account the visual representations that research scientists have found useful in formulating their problems. The mathematical science literature is filled with diagrammatic shorthands for equations where graphs embody the basic interactions and, at the same time, specify the equations to be solved to provide a numerical solution. Selection of an appropriate representation can greatly aid problem formulation and insight. For example, Feynman diagrams are a way to think of quantum electrodynamic processes that is conceptually simpler and clearer than an equivalent formulation in terms of the expansion of an integral equation. Similar diagrammatic techniques have been extended to many-body theory and to statistical mechanics.

When physicists, chemists, biologists, and engineers think about the time evolution of interacting systems, they often invoke similar mental models in their formulations. For a physicist the code words are *mean field*; for a chemical engineer or chemist, *stirred reactor*; for a population biologist, *high population density* or *rapid activation*. The common mental model is one of collisions between the interactants. The collisions give rise to specific products. In a chemical reaction, the result can be new species and the destruction of some of the reactants; in a predator–prey interaction, the death of the prey and the eventual birth of a new predator. The mental model is very con-

---

[3] This contrasts with modeling systems in which users describe phenomena in terms of differential equations. We believe that an effective environment for introducing the skills of modeling to a large student population must be founded on representations that depict directly—in a concrete, visual, dynamic fashion—the objects being modeled and their interactions rather than the mathematical machinery required to perform the simulation.

crete in both these instances, and the correspondence to the OOTLs formulation is direct.

OOTLs can be used for modeling both discrete and continuous phenomena in applications spanning a rich variety of science domains and levels of complexity. The system is capable of representing a wide variety of different types of model structures, including the following ones.

Aggregate behavior models describe the large-scale behavior of particles such as molecules and electrons. Models of this type typically assume random parallel motion of large numbers of particles. When the particles encounter each other, a number of interactions can occur under different conditions, such as agglomeration, recombination, rebounding, and splitting. When particles encounter a barrier, the set of interactions that are possible include penetrating the barrier, rebounding from it, and sticking to it. Some aggregate behavior models express the particles as objects with embedded behaviors and employ rate equations to describe the transitions of objects from one state to another. Aggregate behavior models are appropriate for describing phenomena in statistical mechanics, diffusion processes, and DNA replication, for example.

System dynamics models are designed to study systems of interrelated processes described in terms of causal relationships among the variables that characterize the component processes. The relationships among processes are often depicted graphically as a network of causal links relating the process variables by positive or negative connections to specify direct or inverse proportionality. Variable values can increase or decrease as a result of the effects of other variables linked to them, and there can be feedback loops among the process variables. Some implementations employ "semi-quantitative" values instead of algebraic expressions to represent the magnitude of the effect of one variable on another.

Cellular automata models are designed to study the behavior of discrete processes defined on a cellular grid. In the usual cellular automata operation, local rules for system behavior are associated with the cells of the grid and describe the change in the state of each cell at any time as a function of the states of the cells in its local neighborhood. Local behaviors within cells can give rise to emergent global effects. In some cellular automata models, the rules are associated with active objects rather than with the entire ensemble of cells. As these objects move through the grid, their behavior (for example, heading) is determined by the state (for example, color) of the cell they encounter at each step, and they can, in turn, modify the states of these cells. When active objects encounter each other, the rules determine the resulting action (such as annihilation, giving birth to a new object, or no effect). WorldMaker is an example of a cellular automata modeling system.

Events in OOTLs are conceptualized as interactions among the objects identified as the key players in the model processes. OOTLs is particularly well suited for modeling dynamic processes in terms of state transitions among the objects that are involved. The objects may represent chemical species, gas molecules, or humans. They may be represented visually. Changes in the objects' characteristics, such as the concentrations of the reactants in a chemical reaction–diffusion process, are also represented visually and can be shown dynamically by color change as the reaction proceeds. Visualizations can be richly informative in processes involving spatial diffusion, for example.

Equations are specified simply by dragging graphical icons into windows. This enables students to study the time behavior of the reactions before they have the mathematics necessary to understand the underlying differential equations. The number of coupled reactions and the number of participating objects are not limited.

Objects are assigned arbitrary colors—red, blue and green—which mix to form other colors on the screen. Thus, as the reactions progress, the color of the reaction products changes. Concentrations of all constituents, and any mathematical combinations of them, can be graphed in real time. OOTLs also models diffusion processes. Multiple reactors can be created and linked in linear or two-dimensional arrays. Diffusion constants can be specified, and the resulting dynamics can be displayed by means of animated colors. Because the diffusion constants of the different constituents need not be the same, the effects of variation in this important parameter are directly observable. OOTLs can function as a gateway to many different topics in various areas of science and mathematics. It provides a natural platform for building dynamic process models in a wide range of phenomena, including

- Chemical reaction dynamics: attractors; diffusion-induced Turing structures
- Population ecology: predator–prey relationships, mutual symbiosis
- Epidemiology: spread of disease, parasite distribution
- Immune response: lymphocyte activation and antibody selection
- Traffic flow and management: traffic waves, bus scheduling
- Developmental biology: growth and emergence of structures, tissue induction
- Economics: market competition, overutilization of limited resources
- Neurobiology: epileptic seizures, synaptic transmission
- Percolation: forest fires, aquifer formation, mineralization
- Animal behavior: swarming, termite tunneling
- Thermodynamics: crystallization, phase transitions

We have used OOTLs to model many such systems.

Model-Based Inquiry with OOTLs

The following example illustrates the use of OOTLs. The application describes a classic situation in epidemiology: the spread of disease in a large population concentrated in a local geographic area. A familiar example is mononucleosis (the "kissing disease") spread among students who live close to each other in university dormitories. The basic model assumes that most students will eventually contract the disease through contact with a student who is infected, and that each student who becomes infected will eventually recover and acquire immunity. Thus there are three subpopulations of students at any time: the *Susceptible* students, those who have not yet caught mononucleosis but who will catch it if they come in contact with an infected student; the *Infected* students, those who are currently ill; and the *Recovered* students, those who have been ill and are now immune.

The system of ordinary differential equations describing this dynamic model involves three populations of individuals and is defined as follows (where *a* is the transmission rate, the fraction of the individuals in the susceptible population that becomes infected per encounter per day; and *b* is the recovery rate, the fraction of the individuals in the infected population that recover per day).

$$dS/dt = -a^*S^*I = \text{change in Susceptible} \tag{1}$$
$$dI/dt = a^*S^*I - b^*I = \text{change in Infected} \tag{2}$$
$$dR/dt = b^*I = \text{change in Recovered} \tag{3}$$

For each susceptible individual who gets ill, S is decreased by the same amount as I is increased; thus the term a^*S^*I appears twice, once negative, once positive. The same applies to the recovery rate term, b^*I, though it is offset by only one equation. Our experience, and that of other investigators, is that most high school students are unable to formulate these rate equations.

This is how students might build the same spread-of-disease model using OOTLs. They begin by identifying the types of objects that are relevant. In this instance they identify two kinds of objects—individuals who are currently infected (denoted *I*), and those who are healthy but susceptible (denoted *S*.) They then describe the possible interactions between such individuals that can give rise to the observed behaviors—transmitting or "catching" the disease. In this case, the students identify a single interaction: "When a susceptible individual meets an infected one, the healthy individual becomes infected also." They specify an interaction rate, *a*. They next define and select the icons to specify susceptible and infected individuals and then arrange them to form the causal OOTLs interaction equation shown in Figure 6.1, which describes what occurs before and after the two types of individuals come into contact.

Once this transformation equation has been input via the OOTLs graphical interface, students can simulate the system on the basis of the initial conditions they choose. If they start with a small number of sick people and a large number of healthy ones, over time all the healthy individuals will "turn into" sick ones, reaching a stable final state, though the dynamics involved in attaining this are not trivial. Students are then asked whether this is the actual

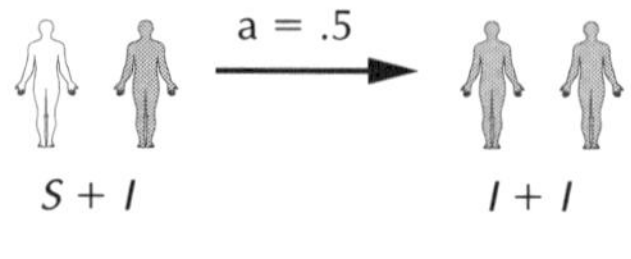

FIGURE 6.1 The First Interaction Equation.

FIGURE 6.2 The Second
Interaction Equation.

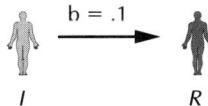

outcome that describes what happens in the real world. Their considered answer is "No. People do not stay sick forever. They get better."

The issue they now must address is how people stop being sick and how this is to be represented. One way to extend their model is simply to allow for sick individuals to become healthy again after a period of time. This requires creating a new type of object (denoted R), for individuals who have recovered and are immune to further infection. Accordingly, a second transformation equation, expressing recovery, is added to the model: Sick individuals eventually recover at some rate b (see Figure 6.2).

This is known as a first-order decay, and it produces exponential diminution over time. The result of simulations with this new two-equation model now yields a peak level of infection, with the number of infected dropping thereafter, followed by a new stable state in which not all the originally healthy (susceptible) individuals necessarily become sick. Students can extend the model by adding more transformations of increasing complexity, such as a rule to allow recovered healthy players to become susceptible to infection again over time. Alternatively, recovered individuals could still be carriers without exhibiting any outer symptoms, thereby infecting healthy individuals. And finally, students might incorporate population dynamics, allowing individuals to reproduce, die, and form subpopulations with different rates of growth and death.

In realizing these OOTLs models, the appropriate mathematics is handled by the OOTLs graphics language preprocessor. Note that whereas the differential equations representation employs three equations, one for each possible health state of the individual, in OOTLs only two process equations are required. The differential equations form is redundant. Beginning students are often confused by the significance of its terms. The dynamics of the differential equations are fully captured by OOTLs, as illustrated by the simulation output in Figure 6.3.

The dynamics resulting from this formulation display the classic onset and course of an epidemic, with the number of infected peaking at a certain time and then diminishing as the number of recovered increases asymptotically. Note, however, that not all susceptible individuals will necessarily get ill. If the rate of spread is less than the rate of recovery, then some individuals escape infection. However, decreasing the rate of recovery (lengthening the incubation–illness period) has the effect of ensuring that more individuals will get the disease. This important concept is very easily explored in the process-specific form embodied in OOTLs.

The OOTLs system provides its own differential equations simulation engine. However, OOTLs can also be used as a language front end to drive other simulation engines, including those that employ discrete and stochastic

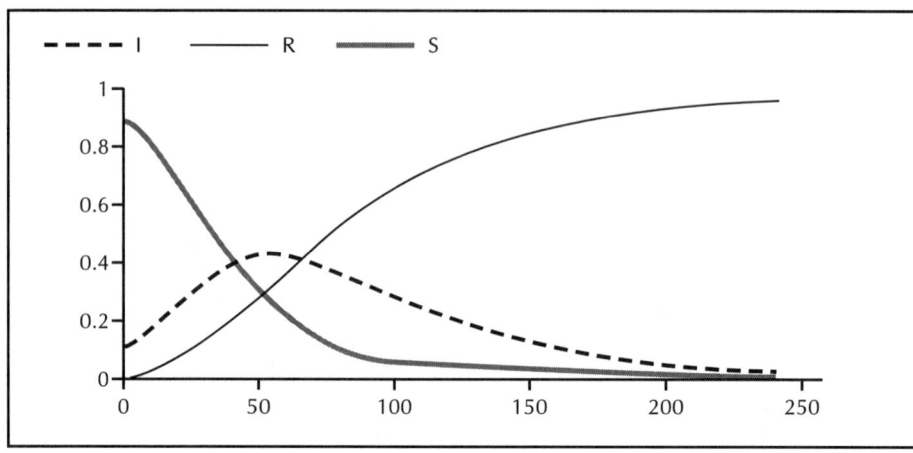

FIGURE 6.3 The Output of the Epidemiology Model.

mechanisms as well as those that employ continuous dynamics. This can have the advantage of expressing the model interactions that drive the simulation in a clearer and more straightforward fashion.

Using OOTLs to Drive the StarLogo Simulation Engine

The contagious disease phenomena expressed in OOTLs can also be implemented by using a discrete process simulation instead of the differential equations continuous dynamics simulation just described. OOTLs has been used in this way with the StarLogo simulation system.[4] Figure 6.4 shows the output of the OOTLs disease model coupled to a StarLogo simulation. Though the model is realized now through a stochastic and discrete dynamic mechanism, it clearly exhibits the same dynamics that we observed in Figure 6.3 for the OOTLs-driven differential equations simulation. In Figure 6.4, the colored StarLogo turtles shown in the left-hand window represent individuals from the three subpopulations. The outputs of the OOTLs-driven StarLogo simulation are shown visually both as a spatial embedding (the left-hand window) and as a graph of the level of the subpopulations across time (the right-hand window.)[5]

When OOTLs drives the StarLogo engine, the StarLogo code for the disease process (in Figure 6.4) is automatically generated by the OOTLs front end. For example, the StarLogo code generated for the state change from Infected

---

[4] See Chapters 5 and 7 for detailed descriptions of StarLogo and its use.
[5] It is important to make clear to students the nontrivial relation that holds between concentration variables in the differential equations and population numbers within finite spaces, because these factors affect the scaling of rates between continuous and discrete models.

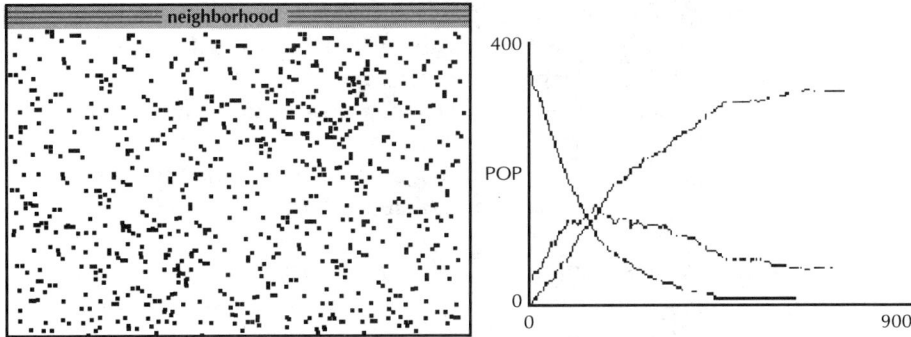

FIGURE 6.4 Epidemiology Model Outputs: OOTLs-Driven StarLogo Simulation.

to Recovered is as follows, where model specifics are shown in **bold** and health denotes the state of an individual:

if **(health = infected)** and **(b % > random 100)** **[set health recovered]**

This StarLogo procedure states that an infected individual has a probability b of recovery in the simulation's cycle.

Phenomena such as those expressed in the contagious disease model, where the population may be relatively small (thousands of objects), where the distributions of objects or the geometry of the environment may be non-homogeneous, and where global averaging masks a great deal of information, may best be simulated as discrete systems. Many problems in biology and engineering fall into this class. StarLogo's simulation engine, which permits the parallel interaction of many independent reactants, is an excellent vehicle for treating such problems.

OOTLs can be used with other modeling tools to generate both continuous and discrete implementations and both deterministic and stochastic implementations. Although OOTLs is a state transition modeling tool, it can be used to generate model inputs for driving cellular automaton models, system dynamics models, and aggregate behavior models. As the foregoing examples with continuous and discrete simulation engines illustrate, OOTLs presents students with a unified conceptual framework for thinking about processes of many different kinds, and it provides a generic language for expressing these processes in a precise way.

The Interplay Between Computer Modeling and Laboratory Experiment

There are three levels of modeling activities—formulating a model, reasoning about model behavior within the model world itself without reference to external reality, and extending the model to capture key aspects of the real-world phenomena addressed by the model that were missing or inappropriately expressed. The true power and relevance of science modeling comes from relating models to reality as a means of helping to make sense of the world. But students seldom experience this level of modeling. Often, they do not advance beyond learning to handle the formalisms and manipulations associated with a model (such as how one needs to think in order to build systems in STELLA® or to program in StarLogo.) As a result, they seldom experience the use of modeling as a rich vehicle for true scientific inquiry. The following example shows how a modeling investigation that does not go beyond the formal modeling level can miss a critical aspect of science inquiry, that of developing an understanding of the underlying mechanism that gives rise to the phenomenon.

Many dynamic systems problems involve processes that require a catalyst—a compound that is required for a reaction but not used up by it. Explorations of catalysis to determine a causal mechanism can engage students in experiments that involve the close coupling of modeling and laboratory work, with extensive interplay between the real-world and computer-world investigations. A representative example is an enzyme lab, where a substance (S) is broken down by an enzyme (E), to yield a product (P) that can be stained for quantification. The lab work begins with the student adding the substance to the reaction chamber. She can measure the amount of S used up and the amount of product P produced, and she can observe that the catalyst E remains constant over time. She is then asked to describe the process as simply as possible, using the OOTLs object model interface. She proposes the simple OOTLs model shown in Figure 6.5.

The model states that S goes to P at rate k. The reaction requires the catalyst E, though E is not consumed. The OOTLs model is then used to drive a simulation engine, such as StarLogo, which generates populations of S and E objects that come into contact to transform S into P. Here the first interesting phenomenon becomes apparent, one that is usually overlooked when only a numerical model is used: The number of collisions between S and E goes down over time, because the level of S decreases as it is transformed into P, and this results in slower P production. E remains constant as required—this fact is made more compelling by viewing the objects directly. Indeed, if the student considers the time required to convert half of S to P, she will also see that E stays constant for the subsequent series of half-way conversions (50% -> 25%, 25% -> 12.5%). This observation also provides the critical point for understanding exponential decay.

After simulations are performed with different starting amounts of S, the graphical outputs from the student's model will also show that the rate of P

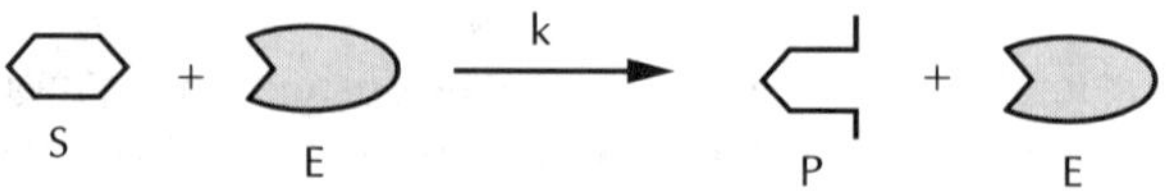

FIGURE 6.5 Initial Enzyme Model.

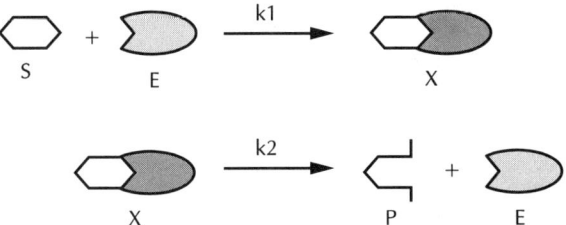

FIGURE 6.6 Extended Enzyme Model.

production increases linearly with increases in initial levels of either S or E, separately. There is no intrinsic limit to the production rate, a result that does not reflect reality. If the student performs the corresponding experiments in the laboratory world, she will discover important new phenomenon characterizing catalytic reactions: saturation. This is clearly shown experimentally as diminishing returns in P production even if S is increased indefinitely, to the point where no matter how much S is added, the rate of P synthesis becomes constant.

The student's original model is thus seen to be flawed. To duplicate the saturation effect seen in the experiment, she needs to modify or extend her model. She intuits that this can be done by postulating a new mechanism, the formation of an intermediate product. Thus she adds a step, the insertion of a new compound between the two sides. In her extended model, E and S coalesce at some rate k_1 to form the compound E*S, which she calls X. This compound then decomposes into P and E at a different rate, k_2. She inputs to OOTLs the pair of equations shown in Figure 6.6

This OOTLs process is then simulated—by driving StarLogo, for example. Initially, while E is bound in the compound E*S form, it is not available to catalyze the S -> P reaction. The E that is bound can be seen to be eventually released when P is produced by the k_2 reaction. Even so, a substantial fraction of E remains bound as E*S as it is recycled until all S is used up, at which time all E is finally released with the completion of P production.

By using differing initial amounts of S in the reaction, the student observes that the fraction of E bound as E*S depends on how much S is used. At some specific concentration of S, most of the enzyme becomes bound to E*S as it "waits" for the rate-limiting k_2 step to finish converting E*S to P + E. No matter how much more S is added, the total rate of P production cannot surpass a hard limit. To a first-order approximation, the student has captured the necessary relations in the process to explain a saturation effect in enzyme catalysis. Her model can now be used as a "mind's eye" into deeper comprehension and exploration of related mechanisms.

In addition, several factors that may influence the amount of saturation for S are now open for exploration: Does the amount of enzyme have an effect on the saturation limit? Does k_1 have an effect as well? How critical is the k_2 reaction constant? Does k_1 have to be greater than k_2? The student's OOTLs model can be used to generate quantitative simulations and analyses to help answer these and associated questions. Activities of these kinds can greatly enhance science education. The synergetic coupling of modeling and laboratory experimentation can help students get at what is really going on and make sense of observed phenomena by revealing and explaining the underlying processes.

Acknowledgments

This research was supported in part by the National Science Foundation under NSF Grant MDR-8954751, "Visual Modeling: A New Experimental Science."

Part 2

Model-Based Inquiry

The chapters in this section describe highly focused projects for integrating modeling into the secondary science curriculum. Although the modeling tools are different in each case, these four projects all focus on the use of modeling to help students develop scientific reasoning skills and think like scientists.

GasLab—an Extensible Modeling Toolkit for Connecting Micro- and Macro-properties of Gases, by Uri Wilensky, describes the work of learners engaged in substantial model-based investigations of stochastic phenomena. High school students and teachers used a version of the StarLogo language to study the behavior of gas molecules in a box. Particles in GasLab are modeled as Newtonian billiard balls colliding elastically with the walls of the box and with each other. Particles are color-coded by speed or energy. The model can support hundreds of thousands of particles. GasLab is embedded in a general-purpose modeling language; thus the user can extend the model and run new experiments by programming the molecules and the surrounding box with the desired properties. The chapter describes an intensive modeling investigation by a high school physics teacher, leading to the creation of a model for investigating the physics of gases under widely varying conditions. The chapter describes several examples of high school students' investigations with, and extensions to, the teacher's model. Wilensky advocates giving the students "expert-built" demonstration models as a starting point for encouraging them to extend their knowledge and test their ideas by creating their own models.

The thesis of *Designing Computer Models That Teach*, by Paul Horwitz, is that computer models can be extraordinarily effective in teaching science, but only if their design and use clearly address certain psychological and pedagogical requirements. He expresses the concern that learning accomplished entirely through interactions with a computer model may be effective in developing competence only within the computer context. He therefore calls for broadening the learning process so that students are made aware of what they are learning in the computer-modeling world and of how it applies in the world outside. He identifies some of the principles underlying the design of successful teaching models and discusses the problems that must be overcome before a computer model can become a practical classroom tool. He describes the development and classroom use of GenScope, a current science-modeling microworld that concretely exemplifies these design principles. GenScope enables students to undertake investigations of a wide variety of genetic phenomena. It represents and interconnects phenomena at several distinct levels of description, including the molecular, cellular, individual organism, family tree, and population levels. In a concerted effort to make GenScope an effective teaching tool, automated scripts are being developed to provide careful problem sequencing and selective scaffolding and to link activities on the computer to knowledge of the real world.

Modeling as Inquiry Activity in School Science: What's the Point? describes experiences of middle-school students in the use of Explorer Science computer models developed by science educators for teaching topics in wave phenomena, population ecology, and the human cardiovascular system. The research focus was on determining what conditions best support students in learning to conduct investigations using models built by experts. The modeling situations included open-ended exploration, structured classroom teaching, and guided inquiry approaches. The computer modeling investigations were combined with real-world observation and laboratory experimentation. Authors William Barowy and Nancy Roberts found that the key to effective students involvement and learning was in allowing students some measure of control over the modeling activities and agenda, rather than requiring them to work through a fixed preassigned lesson sequence.

Alternative Approaches to Using Modeling and Simulation Tools for Teaching Science chronicles several years of development of ThinkerTools, a microworld to teach Newtonian physics more effectively to precollege students. Authors Barbara Y. White and Christina V. Schwarz show how even relatively simple modeling and simulation tools, such as those embodied in the ThinkerTools software, can make possible a variety of instructional activities and approaches. The first iteration of the software presented a series of games that gradually increased in complexity and abstraction to help students develop an understanding of force and motion. Current iterations of ThinkerTools combine computer-based modeling activities with real-world experiments. The authors state that modeling tools are transforming the practices of science and engineering. Their study illustrates an approach to the use of modeling tools that, they believe, can facilitate even greater transformations in science education.

7

GasLab—an Extensible Modeling Toolkit for Connecting Micro- and Macro-properties of Gases

Uri Wilensky

Introduction: Dynamic Systems Modeling

Computer-based modeling tools have largely grown out of the need to describe, analyze, and display the behavior of dynamic systems. Recent decades have seen increasing recognition of the importance of understanding the behavior of dynamic systems—how systems of many interacting elements change and evolve over time and how global phenomena can arise from local interactions of these elements. New research projects on chaos, self-organization, adaptive systems, nonlinear dynamics, and artificial life are all part of this growing interest in system dynamics. The interest has spread from the scientific community to popular culture, with the publication of general-interest books about research into dynamic systems (Gleick 1987; Waldrop, 1992; Gell-Mann, 1994; Kelly, 1994; Roetzheim, 1994; Holland, 1995; Kauffman, 1995).

Research into dynamic systems touches on some of the deepest issues in science and philosophy: order vs. chaos, randomness vs. determinacy, analysis vs. synthesis. The study of dynamic systems is not just a new research tool or new area of study for scientists. The study of dynamic systems stands as a *new form of literacy for all*—a new way of describing, viewing, and symbolizing phenomena in the world. The language of the present mathematics and science curriculum employs *static* representations. Yet our world is, of course, constantly changing. This disparity between the world of dynamic experience and the world of static school representations is one source of student alienation from the current curriculum. The theoretical and computer-based tools arising out of the study of dynamic systems can describe and display the *changing* phenomena of science and the everyday world.

Dynamic Systems Modeling in the Connected Probability Project

The goal of the Connected Probability project (Wilensky, 1995a, 1995b, 1997) is to study learners (primarily high school students) engaged in substantial investigations of stochastic phenomena. As part of the project, learners are provided with access to a wide variety of modeling tools that they can use in pursuit of their investigations. They are particularly encouraged to use the StarLogo (Resnick, 1994; Wilensky, 1995a) modeling language to conduct their investigations.

StarLogo is one of a new class of object-based parallel modeling languages (OBPML). Chapter 5 in this book includes a detailed description of StarLogo. In brief, it is an extension of the Logo language in which a user controls a graphical turtle by issuing commands such as "forward," "back," "left," and "right." In StarLogo, the user can control thousands of graphical turtles. Each turtle is a self-contained "object" with an internal local state. Besides the turtles, StarLogo automatically includes a second set of objects, "patches." A grid of patches undergirds the StarLogo graphics window. Each patch is a square or cell that is computationally active. Patches have local state and can act on the "world" much as turtles do. Essentially, a patch is just a stationary turtle. For any particular StarLogo model, there can be arbitrarily many turtles (from 0 to 32,000 is typical in the StarLogo versions we have used), but there are a fixed number of patches (typically, 10,000 laid out in a 100×100 grid).

The modeling projects described in this chapter have run in several different versions of the StarLogo language on several different platforms. For simplicity of the exposition, all models are described in their reimplemented form in the version of StarLogo called StarLogoT1.0,[1] which is a Macintosh computer implementation—an extension and superset of StarLogo2.0.[2]

This chapter describes in detail the evolution of a set of StarLogo models for exploring the behavior of gases. We now call this collection of models GasLab. The original GasLab model was built, in the Connection Machine version of StarLogo, by a high school physics teacher involved in the Connected Probability project. He called the model GPCEE (**Gas Particle Collision Exploration Environment**). In the reimplementation of GPCEE for newer versions of StarLogo, the GPCEE model was renamed Gas-in-a-Box, and it is one of an evolving collection of models that constitute GasLab.

---

[1] Developed at Tufts University's Center for Connected Learning and Computer-based Modeling.
[2] Developed at the M.I.T. Media Laboratory.

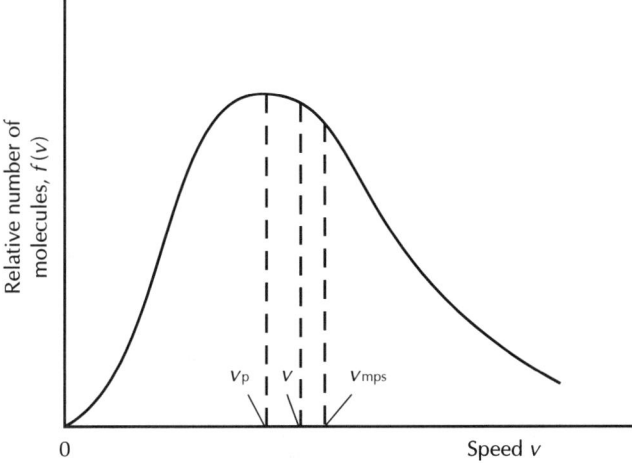

FIGURE 7.1 Maxwell–Boltzmann Distribution of Molecule Speeds (illustration from Giancoli, 1984).

The Creation of the Gas-in-a-Box Model—Harry's Story

In the context of the Connected Probability project, students were offered the opportunity to construct StarLogo models of phenomena of interest to them that involved probability and statistics. Harry, a high school physics teacher enrolled in an education class that I was teaching, had long been intrigued by the behavior of a gas in a sealed container. He had learned in college that the speeds of the gas molecules were distributed according to a famous law, the Maxwell–Boltzmann distribution law. This distribution had a characteristic right-skewed shape (see Figure 7.1). He had taught this law and its associated formula to his own students, but there remained a gap in his understanding. How/why did this particular distribution come about? What kept it stable? To answer these questions, he decided to build (with my help[3]) a StarLogo model of gas molecules in a box.

Harry built his model on the basis of certain classic physics assumptions:

• Gas molecules are modeled as spherical "billiard balls"—in particular, as symmetrical and uniform with no vibrational axes.
• Collisions are "elastic"—that is, when particles collide with the sides of the box or with other gas molecules, no energy is lost in the collision; all the energy is preserved as kinetic energy of the moving molecules.

---

[3] At the time Harry was building his model, StarLogo was not nearly so "user-friendly" as it is in current versions. This necessitated my working with Harry in constructing his model. Harry specified the behavior he wanted, and I did most of the coding. As StarLogo got more robust and easier to use, subsequent students were able to program the GasLab extensions themselves.

• Points of collision between molecules are determined stochastically. It is reasonable to model the points of collision contact between particles as randomly selected from the surface of the balls.

Harry's model displays a box with a specified number of gas particles randomly distributed inside it. The user can set various parameters for the particles: mass, speed, direction (see Figure 7.2). The user can then perform "experiments" with the particles. Harry's program was a relatively straightforward (though longish) StarLogo program. At its core were three procedures that were executed (in parallel) by each of the particles in the box:

go: The particle checks for obstacles, and if none are present, it moves forward (an amount based on its speed variable) for one clock tick.
bounce: If the particle detects a wall of the box, it bounces off the wall.
collide: If the particle detects another particle in its vicinity, the particles bounce off each other like billiard balls.

Harry was excited by the expectation that the macroscopic laws of the gas should emerge, spontaneously, from the simple rules, at the microscopic level, that he had written for the particles. He realized that he wouldn't need to program the macro-level gas rules explicitly; they would come "for free" if he wrote the underlying (micro-level) particle rules correctly. He hoped to

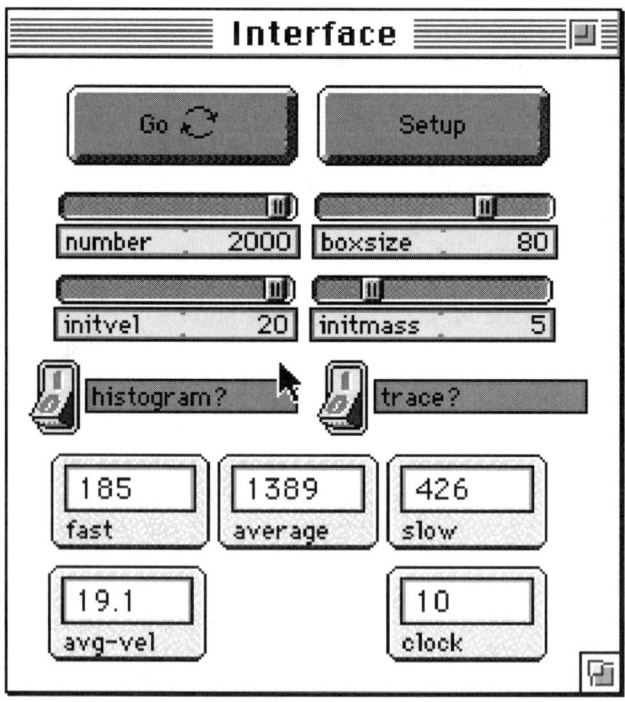

FIGURE 7.2 Gas-in-a-Box Interface Window.

gain deeper explanatory understanding of, and greater confidence in, the gas laws through this approach — seeing them emerge as a consequence of the laws of individual particles and not as some mysterious orchestrated properties of the gas.

In one of his first experiments, Harry created a collection of particles of equal mass randomly distributed in the box. He initialized them to start at the same speed but move in random directions. He kept track of several statistics for the particles on another screen. When looking at this screen, he noticed that one of his statistics, the average speed, was going down. This surprised him. He knew that the overall energy of the system should be constant: Energy was conserved in each of the collisions. After all, he reasoned, the collisions are all elastic, so no energy is lost from the system. The number of molecules isn't changing, so the average energy or

$$\frac{\text{total energy}}{\text{number of molecules}}$$

should also be a constant. But energy is just proportional to the mass and the square of the speed. Because the mass is constant for all molecules, the average speed should also be constant. Why, then, did the model output show the average speed to be decreasing? In Harry's words,

The IMPLICATION of what we discovered is that the average length of each of the individual vectors does indeed go down. PICTURE IT! I visualize little arrows that are getting smaller. These mental vectors are just that. Little 2 (or 3)-dimensional arrows. The move to the scalar is in the calculation of energy (with its v**2 terms). Doesn't it seem difficult to reconcile the arrows (vectors) collectively getting smaller with a scalar (which is a quantity that for a long time was visualized as a fluid) "made up" from these little vectors NOT getting less!

Harry was dismayed by this new "bug" and set out to find what "had to" be an error in the code. He worked hard to analyze the decline in average speed to see whether he could get insight into the nature of the calculation error he was sure was in the program.

But there was no error in the code. After spending some time unsuccessfully hunting for the bug, Harry decided to print out average energy as well. To his surprise, the average energy stayed constant.

At this point, Harry realized that the bug was in his thinking rather than in the code. To get a more visual understanding of the gas dynamics, he decided to color-code the particles according to their speed: Particles are initially colored green; as they speed up, they get colored red; as they slow down, they get colored blue. Soon after running the model, Harry observed that there were many more blue particles than red. This was yet another way of thinking about the average-speed problem. If the average speed were indeed to drop, one would then observe more slow (blue) particles than fast (red) ones, so this result was consistent with the hypothesis that the bug was in his thinking, not in the code.

Harry now began to see the connection between the shape of the Maxwell–Boltzmann distribution and the visual representation he had created. The color-coding gave him a concrete way of thinking about the asymmetrical Maxwell–Boltzmann distribution. He could "see" the distribution: Initially, all the particles were green, a uniform symmetrical distribution, but as the model developed, there were increasingly more blue particles than red ones, resulting in a skewed, asymmetrical spread of the distribution (see Figures 7.3, 7.4, and 7.5).

Even though Harry knew about the asymmetrical Maxwell–Boltzmann distribution, he was surprised to see the distribution emerge from the simple rules he had programmed. Because he himself had programmed the rules, he was convinced that this stable distribution does indeed emerge from these rules. Harry tried several different initial conditions, and all of them resulted in this distribution. He now believed that this distribution was not the result of a specific set of initial conditions but that any gas, no matter how the particle speeds were initialized, would attain this stable distribution.[4] In this way, the StarLogo model served as an experimental laboratory where the distribution could be "discovered." This type of experimental laboratory is not easily (if at all) reproducible outside of the computer-modeling environment.

But there remained several puzzles for Harry. Though he believed *that* the Maxwell–Boltzmann distribution emerged from his rules, he still did not see *why* they emerged. And he still did not understand how these observations squared with his mathematical knowledge. How could the average speed change when the average energy was constant?

Reflecting on this confusion gave Harry the insight he had originally sought from the GasLab environment. Originally, he had thought that because gas particles collided with each other randomly, they would be just as likely to speed up as to slow down, so the average speed should stay roughly constant. But now, Harry saw things from the perspective of the whole ensemble. The law of conservation of energy guaranteed, Harry knew, that the overall pool of energy was constant. Although there were many fewer red particles than blue ones, Harry realized that each red particle "stole" a significant amount of energy from this overall pool of energy. The reason: Energy is proportional to the square of speed, and the red particles were high-speed. Blue particles, in contrast, took much less energy out of the pool. Therefore, each red particle needs to be "balanced" by more than one blue particle to keep the overall energy constant. In Harry's words,

There have to be more blue particles. If there were the same number of blues as reds, then the overall energy would go up. Let's say 1000 green particles have mass 1 and speed 2, then the overall energy is equal to 2000 [ED—1/2 * m * V**2]. If half the

---

[4] In fact, it is a deep insight of mathematical physics that any set of objects that has randomized collisions and conserves energy will relax into a Maxwell–Boltzmann distribution.

FIGURE 7.3 8000 Gas Particles After 30 Ticks. Faster Molecules Are Red, Slower Molecules Are Blue, and Molecules Moving at Medium Speed Are Green. (See color slide in CD-ROM.)

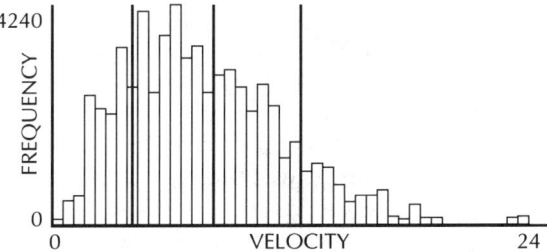

FIGURE 7.4 Dynamic Histogram of Molecule Speeds After 30 Clock Ticks.

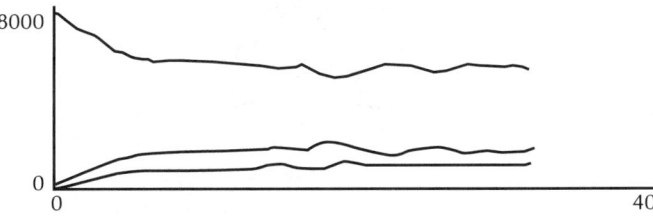

FIGURE 7.5 Dynamic Plot of Fast, Slow, and Medium-Speed Particles.

greens become red at speed 3 and half become blue at speed 1, then the energy of the reds is 500 * 1/2 * 9, which equals 2250. (Wow, that's already more than the total energy.) And the energy of the blues is 500 * 1/2 * 1, which equals 250. Oh, yeah, I guess I don't need the 500 there, a red is 9 times as energetic as a blue so to keep the energy constant we need 9 blues for every red.

Harry was now confident that he had discovered the essence of why the Maxwell–Boltzmann distribution arose. As particles collided they changed speeds, and the energy constraint ensured that there would be more slow particles than fast ones. Even so, he was still puzzled on the "mathematical side." He saw that the greater number of blue particles than red particles ensured that the average speed of the molecules would indeed decrease from the initial average speed of a uniform gas. But how did this square with the mathematical formulas?

Harry had worked on the classic physics equations when he felt sure there was a bug in the StarLogo code. He had worked on them in two different ways, and both methods led to the conclusion that the average speed should be constant. What was wrong with his previous reasoning?

In his first method, he had started with the assumption that momentum[5] is conserved inside the box. Because mass is constant, the average velocity as a vector is constant. And because the average velocity is constant, he had reasoned that its magnitude, the average speed, had to be constant as well. But now he saw that this reasoning was faulty:

[I] screwed up the mathematics—the magnitude of the average vector is **not** the average speed. The average speed is the average of the magnitudes of the vectors. And the average of the magnitudes is not equal to the magnitude of the average.

In his second method, he began with the assumption that the energy of the ensemble would be constant. This could be written $\Sigma_i 1/2 m v_i^2$ is constant. Factoring out the constant terms reveals that $\Sigma_i v_i^2$ is a constant. From this he had reasoned that the average speed, $\Sigma_i \mathrm{abs}(v_i)$, would also have to be constant. He now saw the error in that mathematics as well. It is not hard to show that if the former sum (corresponding to energy) is constant, then the latter sum (corresponding to speed) is at its maximum under the uniform initial conditions. As the speeds diverge, the average speed decreases, just as he "observed." For a fixed energy, the maximum average speed would be attained when all the speeds were the same as they were in the initial state. From then on, more particles would slow down than would speed up.

Although both of these bugs were now obvious to Harry and he felt that they were "embarrassing errors for a physics teacher to make," confusion between the vector and scalar averages still lurked in the background of his thinking. Once brought to light, it could readily be dispensed with through standard high school algebra. However, the standard mathematical formalism did not help Harry see his errors. His confusion was brought to light (and led to increased understanding) through his constructing and immersing himself in the Gas-in-a-Box model. In working with the model, it was natural for him

---

[5] A source of confusion in many a physics classroom: Why do we need these two separate quantities, energy $= mv^2$ and momentum $= mv$. The algebraic formalism masks the big difference between the scalar energy and the vector momentum.

to ask questions about the large ensemble and to get experimental and visual feedback. This also enabled Harry to move back and forth between different conceptual levels: the level of the whole ensemble, the gas, and the level of individual molecules.

Harry was now satisfied that the average speed of the ensemble would indeed decrease from its initial uniform average. The reasoning reflected in his comments above relieved his concerns about further investigations into the connection between the micro- and macro-views of the particle ensemble.

Harry was led inexorably to the question of why the particle speeds would spread out from their initial uniform speed. Indeed, why do the particles change speed at all? When teased out, this question could be framed as follows: "The collisions between particles are completely symmetrical. Why, then, does one particle change speed more than the other? To answer this question, Harry conducted further modeling experiments, repeating collisions of two particles in fixed trajectories. After seeing two particles collide at the same angle again and again but emerge at different angles each time, he remembered that "randomness was going on here." The particles were choosing random points on their surface to collide, so they did not behave the same way each time. By experimentally varying the collision points, he observed that the average speed of the two particles did not usually stay constant. Indeed, it remained constant only when the particles collided head-on.

It was not long from this realization to the discovery of the broken symmetry: "When particles collide, their trajectories may not be symmetrical with respect to their collision axis. The apparent symmetry of the situation is broken when the particles do not collide head-on—that is, when their directions of motion do not have the same relative angle to the line that connects their centers." See Figure 7.6.

Harry went on to do the standard physics calculations that confirmed this experimental result. In a one-dimensional world, he concluded, all collisions

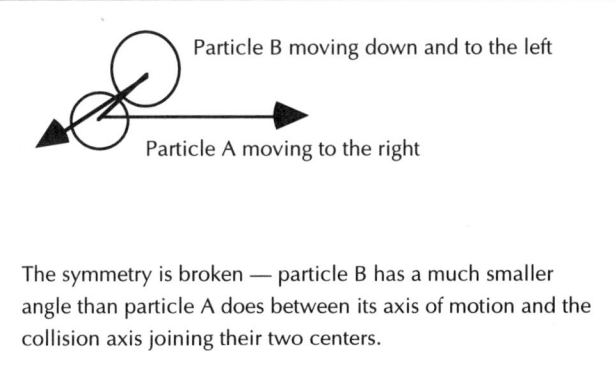

Particle B moving down and to the left

Particle A moving to the right

The symmetry is broken — particle B has a much smaller angle than particle A does between its axis of motion and the collision axis joining their two centers.

FIGURE 7.6 Broken Symmetry Leads to Changing Speeds.

would be head-on, so the average speed would stay constant[6]; in a multidimensional world, particle speed distributions become nonuniform, and this leads inevitably to the preponderance of slower particles and to the characteristic asymmetrical distribution.

Harry had now adopted many different views of the gas and used many different methods to explain the asymmetry of the particle speed distribution. Through connecting the macro-view of the particle ensemble with the micro-view of the individual particle collisions, he had come to understand both levels of description in a deeper way. Through connecting the mathematical formalism to his observations of colored particle distributions, he had caught errors he had made on the "mathematical side" and, more important, anchored the formalism in visual perception and intuition. Harry felt he had gained great explanatory power through this connection of the micro-view and macro-view. This connection was made feasible through the support offered by the StarLogo modeling language.

When asked what he had learned from the experience of building the Gas-in-a-Box model, Harry made one more trenchant observation. He had found that the average speed of the gas molecules was not constant. Upon reflection, he realized,

Of course the average speed is not constant. If it were constant, I'd have known about it. It isn't easy to be a constant and that's why we have named laws when we find constants or invariants. The law of conservation of energy guarantees that the energy of the gas is a constant. We do not have a law of conservation of speed.

Harry now understood the concept of energy in a new way. He saw that energy could be seen as a statistical measure of the ensemble that was invariant. He saw that there could be many statistical measures that characterize an ensemble. Each of them could lay claim to being a kind of "average"— that is, a characteristic measure of the ensemble. The idea of "average" is then seen to be another method for summarizing the behavior of an ensemble. Different averages are convenient for different purposes. Each has certain advantages and disadvantages, certain features that it summarizes well and others that it does not. Which average we choose or construct depends on which aspect of the data we see as important. Energy, Harry now saw, was a special such average—not (as he had sometimes suspected) a mysteriously chosen formula, but rather a measure that characterized invariantly the collection of particles in a box.

---

[6]Although this insight of Harry's is strictly true in the formal environment of Harry's thought experiment, in a real-world one-dimensional environment, there would be randomizing factors (such as nonsphericality of the particles) that would cause the distribution to relax into the highly asymmetrical one-dimensional Maxwell–Boltzmann distribution.

Creation of the GasLab Toolkit—Extensible Models

After Harry finished working with the Gas-in-a-Box model, I decided to test the model with students who had not been involved in its development. I contacted a local high school and arranged to meet 3 hours a week for several weeks with a few juniors and seniors taking introductory physics. The group was somewhat fluid; it consisted of three regular members and three or four others who sometimes dropped in. The students who chose to be involved did so out of interest. Their teacher described the three regular members as "average to slightly above average" physics students. I introduced the students to the Gas-in-a-Box model, showing them how to run the model and how to change elementary parameters of the model. I asked them to begin by just "playing" with the model and talking to me about what they observed. I describe below these students' experience with GasLab. I have introduced GasLab to dozens of groups of students (high school and collegiate) since that time. The details of their explorations differ in each case, but the overall character of the model-based inquiry is typified by the story related below.

The students worked as a group, one of them "driving" the model from the keyboard and the others suggesting experiments to try. One of the first suggested experiments was to put all of the particles in the center of the box.[7] This led to a pleasing result as the gas "exploded" in rings of color, a red ring on the outside, with a nested green ring and a blue ring innermost. The students soon hit on the same initial experiment that stimulated Harry. They started with a uniform distribution of 8000 green particles and immediately wondered at the preponderance of blue particles over red particles as the simulation unfolded. Over the next week, they went through much of the same reasoning that Harry had gone through, connecting the energy economy of the gas particle ensemble with the speed distribution of the particles.

But these students were not as motivated by this question as Harry was. One student, Albert, became very excited by the idea that the micro-model should reproduce the macroscopic gas laws:

What's really cool is that this is it. If you just let this thing run, then it'll act just like a real gas. You just have to start it out right and it'll do the right thing forever. We could run experiments on the computer and the formulas we learned would come out.

Albert went on to suggest that because this was a real gas, they could verify the ideal gas laws for the model. The group decided to verify Boyle's law—that is, to confirm that changing the volume of the box would lead to a reciprocal change in the pressure of the gas.

---

[7] To do this, they issued the simple StarLogo command 'setxy 0 0'. Though the StarLogoT code for doing this is quite simple, this is not an experiment that can be replicated in the laboratory—a case of the model as an instantiation of ideal gas theory rather than its real-world application.

Now the group was faced with creating an experiment that would test whether Boyle's law obtained in the Gas-in-a-Box model. Tania made a suggestion:

We could make the top of the box move down like a piston. We'll measure the pressure when the piston is all the way up. Then we'll let it fall to half way down and measure the pressure again. The pressure should double when the piston is half way down.

The group agreed that this was a reasonable methodology, but then they were stopped short by Isaac, who asked, "How do we measure the pressure"? This question was followed by a substantial pause. They were used to being given an instrument to measure pressure, a black box from which they could just read out a number. As Albert said for the group, "We have to invent a pressure-measure, a way of saying what the pressure is in terms of the particles." The group pondered this question. At their next meeting, Tania suggested the first operational measure:

We could have the sides of the box[8] store how many particles hit them at each tick. The total number of particles hitting the sides of the box at each tick is our measure of pressure.

They programmed this measure of pressure into the model. There ensued lots of discussion about what units this measure of pressure represented. At long last, they agreed that they did not really care what the units were. All they needed to know, in order to verify Boyle's law, was that the measure would double, so a unit scale factor would not affect the result of the experiment.

They created a "monitor" that would display the pressure in the box and ran the model. To their dismay, the pressure in the box fluctuated wildly. Tania was quick to point out the problem:

We only have 8000 particles in the box. Real boxes full of gas have many more particles in them. So the box is getting hit a lot less times at each tick than it should be. I think what's happening is that the number of particles isn't big enough to make it come out even.

Persuaded by this seat-of-the-pants "law of large numbers" argument, they made an adjustment to the pressure-measuring code. They calculated the number of collisions at each tick over a number of ticks and then averaged them. Trial-and-error simulations varying the averaging time interval convinced them that averaging over ten ticks led to a sufficiently stable measure of pressure.

Now that they had a stable pressure gauge, they were ready to construct the piston and run the experiment. But here again, they ran into conceptual difficulties. How was the piston to interact with the particles? Were they to

---

[8]They implemented this strategy by storing the numbers in the patches.

model it as a large, massive particle that collided with the small particles? In that case, how massive should it be? And if they did it that way, wouldn't it affect the pressure in the box in a nonuniform way? As Albert said,

If we do the piston, then the North–South pressure in the box will be greater than the East–West pressure. That doesn't seem right. Shouldn't the pressure in the box stay even?

This issue was discussed, argued, and experimented on for several hours. It was at this point that Tania suggested another approach.

I'm confused by the effect the piston is supposed to have on the particles. I have an idea. Why don't we start the particles out in half the box, then release the "lid" and let them spread out into the whole box. If we do that, we won't have to think about pistons and we can just see if the pressure decreases in half.

The group agreed that this was a promising approach and quickly implemented this code. They were now able to run the experiment that they hoped would confirm Boyle's law. Their experiment worked as they hoped. When they lifted the lid so that the box had double the volume, the pressure in the box did, indeed, drop in half. See Figures 7.7 and 7.8.

This confirming result could have led to an unfortunate acceptance of Tania's measure of pressure as accurate. (Indeed, experimental results with this isothermal version of Boyle's law could not have disconfirmed Tania's measure.) However, in time, the students did come to reject this measure on conceptual grounds. They reasoned that heavier particles ought to make more of a difference in the pressure than lighter ones. Similarly, they reasoned that faster particles should have more effect than slower ones. This led them to revise their pressure measure to the conventional physics definition: momentum transfer to the sides of the box per unit time.

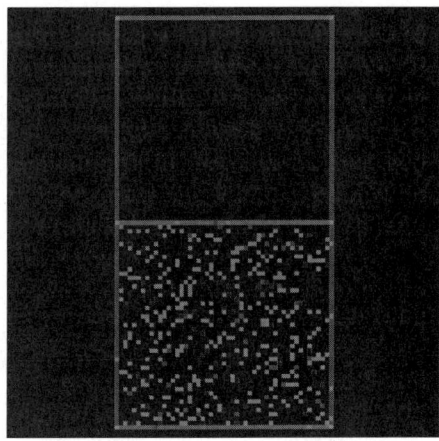

FIGURE 7.7 Box with Lid Down, Volume = 1200; Box with Lifted Lid, Volume = 2400. Lifting the Box Lid Proportionally Reduces Pressure.

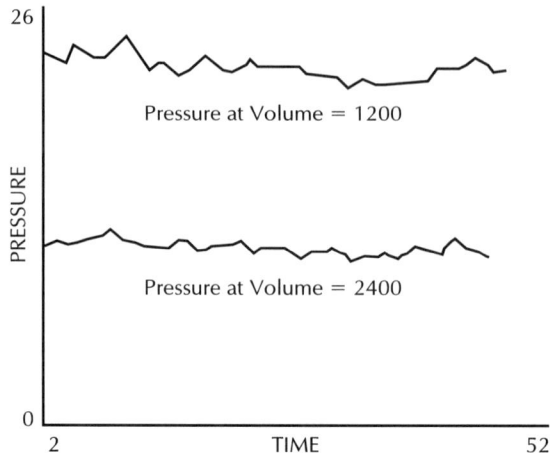

FIGURE 7.8 Plot of Pressure as Measured in the Box at the Two Different Volumes.

Their escapade with Tania's definition, however, did yield insights. As Tania later said,

I guess for Boyle's law to work, all that matters is how dense the molecules are in the box. With more space they're less likely to collide, so the pressure drops.

There is another incident of note surrounding the Boyle's law experiment. A week or so after completing the experiment, Isaac ran the model again with all particles initialized to be at the center of the box. While watching his favorite "explosion," Isaac noted that the gas pressure registered 0! Quickly, he realized that that was a consequence of their definition—no particles were colliding with the sides of the box. Somehow, this didn't seem right to Isaac, and he asked the group if they should revise their concept of pressure yet again. Argumentation ensued as to "whether a gas had internal pressure without any box to measure it." They realized that the experiment in question was not feasible in a real experimental setting, but nonetheless, it did seem that there should be a theoretical answer to the question. Isaac suggested various ingenious solutions to the problem, but in the end, the group did not alter their pressure gauge. The ingenious solutions were difficult to implement, and their current gauge seemed to be adequate to the experiments they were conducting.

One other noteworthy development was related to the emergence of the Maxwell–Boltzmann distribution discussed in the previous section. Albert came in one day excited about an insight he had had. The gas molecules, he said, can be thought of as probabilistic elements, like dice. They can randomly go faster or slower. But although there is no real limit to how fast they can go,[9] their speed is bounded below by zero. It's as if particles were con-

---

[9]He was ignoring the quite high speed limitation imposed by energy considerations.

ducting a random walk on the plane but there was a wall on the y-axis. Albert saw that this constrained random walk would have to produce a right-skewed distribution. I challenged him to go further: (a) Could he construct a StarLogo model to prove his theory? (b) Could he determine what particular probability constraints would produce a strict Maxwell–Boltzmann distribution? (c) Could he find other, seemingly unrelated phenomena that satisfied the same formal constraints and thus would also produce a Maxwell–Boltzmann distribution? Albert and his fellow students were up to these challenges.

These students (and subsequent groups of students) have conducted many more experiments with the Gas-in-a-Box model. Through revising and extending the model, they created a set of models that has since been expanded into the toolkit we now call GasLab. The set of extensions of the original Gas-in-a-Box model is truly impressive in its scope and depth of conceptual analysis. Among the many extensions they tried were heating and cooling the gas, introducing gravity into the model (and a very tall box) and observing atmospheric pressure and density, modeling the diffusion of two gases, allowing the top to be porous and seeing evaporation, relaxing elasticity constraints while introducing weak attraction and looking for phase transitions, introducing vibrations into the container and measuring sound density waves, and allowing heat to escape from the box into the surrounding container. Over the course of several weeks, these high school students "covered" much of the territory of collegiate statistical mechanics and thermal physics. Their understanding of it was deeply grounded both in the intuitive understandings gained from their concrete experience with the models and in the relations among the fundamental concepts.

GasLab provides learners with a set of tools for exploring the behavior of an ensemble of micro-level elements. Through running, extending, and creating GasLab models, learners were able to develop strong intuitions about the behavior of the gas at the macro-level (as an ensemble gas entity) and its connections to the micro-level (the individual gas molecule). In a typical physics classroom, learners usually address these levels at different times. When attending to the micro-level, they typically focus on the exact calculation of the trajectories of two colliding particles. When attending to the macro-level, they focus on "summary statistics" such as pressure, temperature, and energy. Yet it is in the *connection* between these two primary levels of description that the explanatory power resides.

Two major factors enable students using GasLab to make the connection between these levels: the replacement of symbolic calculation with simulated experimentation and the replacement of "black-box" summary statistics with learner-constructed summary statistics. The traditional curriculum segregates the micro-level and macro-level of description because the mathematics required to connect them meaningfully is thought to be out of reach of high school students. In the GasLab modeling toolkit, the formal mathematical techniques can be replaced with concrete experimentation with simulated objects. This experimentation allows learners to get immediate feedback about their theories and conjectures. The traditional curriculum hands learners summary statistics such as pressure as "received" physics knowledge. It is a "device" built by an expert, which the learner cannot inspect or question. Most fundamentally, the learner has no access to the design space of possibilities from which this particular design was selected. In the GasLab context, learners must construct their own summary statistics. As a result, the traditional pressure measure is seen to be one way of summarizing the effect of the gas molecules on the box—one way to build a pressure gauge. The activity of designing a pressure measure is an activity of doing physics, not absorbing an expert's "dead" physics.

The two factors described above (the ability to act on the model and to "see" its reactions, and the ability to create interpretations of the model in the form of new computational objects that, in turn, can be acted upon) make a significant difference in the kinds of understandings students can construct of the behavior of gas molecule ensembles. Through engaging with GasLab, high school students have access to the powerful ideas and explanations of statistical thermal physics. Yet by engaging in such activities, the students came to understand the gas as a concrete entity, much in the same way they experience physical entities outside the computer. These constructive modeling and model-based reasoning activities can give students a concrete understanding of, and a powerful way of apprehending, the physics and chemistry of gases—one that eludes even some professional scientists who learned this content in a traditional manner.

Implications for the Pedagogy of Modeling

Despite the rapid rate of infiltration of computer-based modeling and dynamic systems theory into scientific research and into popular culture, computer-based modeling has only slowly begun to affect education communities. Computer-based models are increasingly used in the service of pedagogic ends (Buldyrev *et al.*, 1994; Chen and Stroup, 1993; Doerr, 1996; Feurzeig, 1989; Horwitz, 1989; Horwitz *et al.*, 1994; Jackson *et al.*, 1996; Mandinach and Cline, 1994; Mellar *et al.*, 1994; Roberts *et al.*, 1983; Repenning, 1994; Shore *et al.*, 1992; Smith *et al.*, 1994; White and Frederiksen, 1998; Wilensky, 1997; Wilensky and Resnick, 1999), but there remains significant lack of consensus about the proper role of modeling within the curriculum.

Model Construction vs. Model Use

One tension that is felt is between students using already-constructed models of phenomena and students constructing their own models to describe phenomena. At one extreme is the use of preconstructed models purely for demonstration of phenomena. This use of modeling employs the computer to animate and dynamically display the structures and processes that describe the phenomena. It may permit students to modify the model's inputs and parameters, but it does not enable students to modify the model's structures, processes, or operation. At the other extreme, learners are involved in constructing their own models of phenomena *de nova*. Between these extremes are other kinds of modeling activities. One of particular interest is student use of preconstructed models as investigative tools for model-based inquiry—activities that may involve learner modification and extension of the initial models provided to them. Here, students are given starting models but are also involved in model design and development.

For the use of models to provide demonstrations, I employ the term *demonstration modeling*. Although such demonstration models can be visually striking, the experience is not very different from viewing a movie of the phenomenon in question. The computational medium is being used merely for delivery. From a constructivist point of view, this delivery model is unlikely to lead to deep learning, because it does not engage with the learner's point of entry into the phenomenon to be understood. Nor does this approach take advantage of the computer's interactivity to give the learner a chance to probe the model and get the feedback necessary to construct mental models of the phenomenon observed.

Constructivists might be happier with the "from scratch" modeling activity that requires the learner to start where she is and interact with the modeling primitives to construct a model of the phenomenon. That special breed of constructivist called constructionists (Papert, 1991) would argue that this externalized construction process is the ideal way to engage learners in constructing robust mental models. The learner is actively engaged in formulating a question, formulating tentative answers to her question, and (through an iterative process of reformulation and debugging) arriving at a theory of how to answer the question instantiated in the model. This process is an act of doing and constructing mathematics and science instead of viewing the results of an expert's having done the mathematics and science and handing it off to the learner. On the epistemological side, this understanding that mathematics and science are ongoing activities in which ordinary learners can be creative participants is an important meta-lesson of the modeling activity. The two sides of this debate are summarized in Table 7.1.

TABLE 7.1 Model Use vs. Model Construction

Model Use (Demonstration Models)	Model Construction (Model Based-Inquiry)
Passive	Active
Viewing a "received" mathematics and science	Constructing mathematics and science
Transmission of ideas	Expression of ideas
Dynamic medium used for viewing output of mathematical thought	Dynamic medium used as executor of mathematical thought
An expert's question	The learner's own question
An expert's solution	The learner's own tentative solution
Learning in a single step	Learning through debugging
Experts must anticipate relevant parameters for learning	Learners can construct parameters relevant to their learning

An argument on the side of using demonstration models is that the content to be learned is brought immediately and directly to the attention of the learner. In contrast, in the process of constructing a model, the learner is diverted into the intricacies of the modeling language itself and away from the content to be learned. Because there can be quite a bit of overhead associated with learning the modeling language, the model construction approach could be seen as very inefficient. Moreover, there is skepticism about whether students who are not already mathematically and scientifically sophisticated can acquire the knowledge and skills needed to design and construct models.

Selecting the Appropriate "Size" of Modeling Primitives

Like most tensions, this tension is not really dichotomous. There are many intermediate states between the two extremes. Demonstration models can be given changeable parameters that users can vary and, thereby, explore the effect on the behavior of the model. If there are large numbers of such parameters, as in the popular Maxis simulation software packages (Wright, 1992a, 1992b), the parameter space can be quite vast in the possibilities for exploration. This takes demonstration models several steps in the direction of model construction. On the other hand, even the most "from scratch" modeling language must contain primitive elements. These primitive elements remain black boxes, used for their effect but not constructed by the modeler. Not too many constructionist modelers would advocate building the modeling elements from the binary digits, let alone building the hardware that supports the modeling language. The latter can serve as an absurd reduction of the "from scratch" label. Thus even the die-hard constructionist modelers concede that not all pieces of the model need be constructed; some can be simply handed off.

I place myself squarely in the constructionist camp: The challenge for us is to construct toolkits that contain just the right level of primitives. In constructing a modeling language, it is critical to design primitives not so large-scale and inflexible that they can be put together in only a few possible ways. If we fail at that task, we have essentially reverted to the demonstration modeling activity. To use a physical analogy, we have not done well in designing a dinosaur modeling kit if we provide the modeler with three pieces, a *T. rex* head, body, and tail. On the other hand, we must design our primitives so that they are not so "small" that they are perceived by learners as far removed from the objects the learners want to model. If we fail at that task, learners will be focused at an inappropriate level of detail and so will learn more about the modeling pieces than about the content domain to be modeled. To return to the physical analogy, designing the dinosaur modeling kit to have pieces that are small metal bearings may make constructing many different kinds of dinosaurs possible, but it will also be tedious and far removed from the functional issues of dinosaur physiology that form the relevant content domain.

Thus modeling language (and model) designers must decide what characteristics of the primitive modeling elements to give to learners. Modeling language designers who choose to make their primitive elements on the large side we call demonstration modeling designers, whereas those who tend to keep their primitives small we call constructionist modeling designers. Demonstration modeling designers have no choice but to make the pieces, from which the models are built, semantically interpretable from within the model content domain. Constructionist modeling designers, however, can make the underlying model elements content-neutral,[10] thus creating a modeling language that is general-purpose, or they can choose modeling elements that have semantic interpretation in a chosen content domain, thus creating a modeling toolkit for that content domain.

General-Purpose vs. Content Domain Modeling Languages

Both of these choices—content domain modeling languages and general-purpose modeling languages—can lead to powerful modeling activities for learners. The advantage of the content domain modeling language is that learners can enter more directly into substantive issues of the domain (issues that will seem more familiar to them and to their teachers). The disadvantage is that the primitive elements of the language, which describe important domain content, are opaque to the learner. Another disadvantage is that use of the language is restricted to its specific content domain. That disadvantage may be nullified by designing a sufficiently broad class of such content domain modeling languages, though maintaining such a broad class may be challenging. The advantage of the content-neutral primitives is that all content domain structures, because they are made up of the general-purpose elements, can be inspected, constructed, and modified by the learner. The disadvantage is that the learner must master a general-purpose syntax before being able to make headway on the domain content. What is needed is a way for learners to be able to begin at the level of domain content but not be limited to unmodifiable black-box primitives.

In the Connected Probability project, the solution we have found to this dilemma is to build "extensible models" (Wilensky, 1997). In the spirit of Eisenberg's programmable applications (Eisenberg, 1991), these models are content-specific models that are built using the general-purpose StarLogoT modeling language. This enables learners to begin their investigations at the level of the content. Like the group of high schoolers described in the earlier section of this chapter, they begin by inspecting a prebuilt model such as Gas-

---

[10] This is a simplification. Even so-called content-neutral sets of primitives have affordances that make it easier to model some content domains than others. StarLogoT, for example, makes it much easier to model phenomena that can be viewed as accumulations of large numbers of elements, such as statistical and stochastic phenomena. Processes that are composed of a small number of larger elements are less naturally modeled in StarLogo.

in-a-Box. They can adjust parameters of the model such as mass, speed, and location of the particles and can conduct experiments readily at the level of the content domain of ideal gases. But, because the Gas-in-a-Box model is built in StarLogoT, the students have access to the workings of the model. They can "look under the hood" and see how the particle collisions are modeled. Furthermore, they can modify the primitives, investigating what might happen if, for example, collisions were not elastic. Finally, students can introduce new concepts, such as pressure, as primitive elements of the model and conduct experiments on these new elements.

This extensible modeling approach allows learners to dive right into the model content, but it imposes neither a ceiling on where they can take the model nor a floor below which they cannot see the content. Mastering the general-purpose modeling language is not required at the beginning of the activity but, rather, happens gradually as learners seek to explain their experiments and extend the capabilities of the model.

When engaged in classroom modeling, the pedagogy used in the Connected Probability project has four basic stages. In the first stage, the teacher presents a "seed" model to the whole class. Typically, the seed model is a short piece of StarLogoT code that captures a few simple rules. The model is projected through an LCD panel so that the whole class can view it. The teacher engages the class in discussion of what is going on with the model. Why are they observing that particular behavior? How would it be different if model parameters were changed? Is this a good model of the phenomenon it is meant to simulate? In the second stage, students run the model (either singly or in small groups) on individual computers. They engage in systematic search of the parameter space of the model. In the third stage, each modeler (or group) proposes an extension to the model and implements that extension in the StarLogoT language. Modelers who start with Gas-in-a-Box, for example, might try to build a pressure gauge, a piston, a gravity mechanism, or heating/cooling plates. The results of this model extension stage are often quite dramatic. The extended models are added to the project's library of extensible models and made available for others to work with as seed models. In the final stage, students are asked to propose a phenomenon, and to build a model "from scratch", using the StarLogoT modeling primitives.

Phenomena-based vs. Exploratory Modeling

When learners are engaged in creating their own models, two primary avenues are available. A modeler can choose a phenomenon of interest in the world and attempt to duplicate that phenomenon on the screen. Or a modeler can start with the primitives of the language and explore the possible effects of different combinations of rule sets. The first kind of modeling, which I call phenomena-based modeling (Wilensky, 1997; Resnick and Wilensky, 1998) is also sometimes called backwards modeling (Wilensky 1997) because the modeler is engaged in going backwards from the known phenomenon to a set of underlying rules that might generate that phenomenon. In the GasLab example, Harry knew about the Maxwell–Boltzmann distribution and tried creating rules that he hoped would duplicate this distribution. In this specific case, Harry did not have to discover the rules himself, because he also knew the fundamental rules of Newtonian mechanics that would lead to the Maxwell–Boltzmann distribution. The group of students who worked on modeling Boyle's law came closer to pure phenomena-based modeling as they tried to figure out the "rules" for measuring pressure. Phenomena-based modeling can be quite challenging, because discovering the underlying rule sets that might generate a phenomenon is inherently difficult—a fundamental activity of science practice. In practice, most GasLab modelers mixed some knowledge of what the rules were supposed to be with adjustments to those rules when the desired phenomenon did not appear.

The second kind of modeling, which I call exploratory modeling (Wilensky, 1997; Resnick and Wilensky, 1998) is sometimes called "forwards" modeling (Wilensky, 1997) because modelers start with a set of rules and try to work forwards from these rules to some as-yet-unknown phenomenon.

New Forms of Symbolization

In a sense, modeling languages are always designed for phenomena-based modeling. However, once such a language exists, it also becomes a medium of expression in its own right. In just such a way, we might speculate, natural languages originally developed to communicate about real-world objects and relations but, once they were sufficiently mature, were also used for constructing new objects and relations. Similarly, learners can explore sets of rules and primitives of a modeling language to see what kinds of emergent effects may arise from their rules. In some cases, this exploratory modeling may lead to emergent behavior that resembles some real-world phenomenon, and then phenomena-based modeling resumes. In other cases, though the emergent behavior may not strongly connect with real-world phenomena, the resulting objects or behaviors can be conceptually interesting or beautiful in themselves. In these latter cases, the modelers have created new phenomena, objects of study that can be viewed as new kinds of mathematical objects—objects expressed in the new form of symbolization afforded by the modeling language.

Aggregate vs. Object-based Modeling

In the previous section, we discussed the selection of modeling language primitives in terms of size and content-neutrality. They can also be distinguished in terms of the conceptual description of the fundamental modeling unit. Presently, modeling languages can be divided into two kinds: "aggregate" modeling engines, such as STELLA® (Richmond and Peterson, 1990), Link-It (Ogborn, 1994), Vensim, Model-It (Jackson *et al.*, 1996), and "object-based" modeling languages, such as StarLogo (Resnick, 1994; Wilensky, 1995a), Agentsheets (Repenning, 1993), Cocoa (Smith *et al.*, 1994), Swarm

(Langton and Burkhardt, 1997), and OOTLs (Neumann *et al.*, 1997). Aggregate modeling languages use "accumulations" and "flows" as their fundamental modeling units. For example, a changing population of rabbits might be modeled as an "accumulation" (like water accumulated in a sink) with rabbit birth rates as a "flow" into the population and rabbit death rates as a flow out (like flows of water into and out of the sink). Other populations or dynamics (the presence of "accumulations" of predators, for instance) could affect these flows. This aggregate approach essentially borrows the conceptual units—its parsing of the world—from the mathematics of differential equations.

Object-based modeling languages, by contrast, enable the user to model systems directly at the level of the individual elements of the system. For example, our rabbit population could be rendered as a collection of individual rabbits each of which has associated probabilities of reproducing or dying. The object-based approach has the advantage of being a natural entry point for learners. It is usually easier to generate rules for individual rabbits than to describe the flows of rabbit populations. This is because the learners can literally *see* the rabbits and can control the individual rabbit's behavior. In StarLogoT, for example, students think about the actions and interactions of individual objects or creatures. StarLogoT models describe how individual creatures (not overall populations) behave. Thinking in terms of individual creatures seems far more intuitive, particularly for the mathematically uninitiated. Students can imagine themselves as individual rabbits and think about what they might do. In this way, StarLogoT enables learners to "dive into" the model (Ackermann, 1996) and make use of what Papert (1980) calls "syntonic" knowledge about their bodies. By observing the dynamics at the level of the individual creatures, rather than at the aggregate level of population densities, students can more easily think about and understand the population dynamics that arise. As one teacher comparing students' work with both STELLA® and StarLogoT models remarked, "When students model with STELLA®, a great deal of class time is spent on explaining the model, selling it to them as a valid description. When they do StarLogoT modeling, the model is obvious; they do not have to be sold on it."

There are now some very good aggregate modeling languages, such as STELLA® (Richmond and Peterson, 1990) and Model-It (Jackson *et al.*, 1996). These aggregate models are very useful—and they are superior to object-based models in some contexts, especially when the output of the model needs to be expressed algebraically and analyzed via standard mathematical methods. They eliminate one "burden" of differential equations—the need to manipulate symbols—and focus instead on more qualitative and graphical descriptions of changing dynamics. But conceptually, they still rely on the differential equation epistemology of aggregate quantities.

Some refer to object-based models as "true computational models" (Wilensky and Resnick, 1999) because they use new computational media in a more fundamental way than most computer-based modeling tools. Whereas most tools simply translate traditional mathematical models to the computer (numerically solving traditional differential equation representations, for example), object-based languages such as StarLogoT provide new representations that are tailored explicitly for the computer. Too often, scientists and educators see traditional differential equation models as the *only* approach to modeling. As a result, many students (particularly students alienated by traditional classroom mathematics) view modeling as a difficult or uninteresting activity. What is needed is a more pluralistic approach that recognizes there are many different approaches to modeling, each with its own strengths and weaknesses. A major challenge is to develop a better understanding of when to use which approach, and why.

Concreteness vs. Formalism

Paradoxically, computer-based modeling has been criticized both as being too formal and as not being formal enough. On the one hand, some mathematicians and scientists maintain that computer models are insufficiently rigorous. As we noted in the previous section, it is somewhat difficult, for example, to get hold of the outputs of a StarLogoT model in a form that is readily amenable to symbolic manipulation. Moreover, there is not yet any formal methodology for verifying the results of a model run. Even in highly constrained domains, there is no formal verification procedure for guaranteeing the results of a computer program—much less any guarantee that the underlying assumptions of the modeler are accurate. Computational models, in general, are subject to numerical inaccuracies dictated by finite precision. Object-based models, in particular, are also vulnerable to assumptions involved in transforming a continuous world into a discrete model. These difficulties lead many formalists to worry about the accuracy, the utility, and (especially) the generality of a model-based inquiry approach (Wilensky, 1996). These critiques raise valid concerns that must be reflected on as an integral part of the modeling activity. As we recall, Harry had to struggle with just such an issue when he was unsure whether the drop in the average speed of the gas particles was due to a bug in his model code or to a "bug" in his thinking. It is an inherent part of the computer-modeling activity to go back and forth between questioning the model's faithfulness to the modeler's intent (as in seeking code bugs) and questioning the modeler's expectations for the emergent behavior (as in seeking bugs in the model rules).

Though the formalist critic may not admit it, these limitations exist in all modeling—even in using formal methods such as differential equations. Only a small set of the space of differential equations is amenable to analytic solution. Most modifications of those equations lead to equations that can be solved only through numerical techniques. The game for formal modeling, then, becomes trying to find solvable differential equations that can be said to map onto real-world phenomena. Needless to say, this usually leads to significant simplifications and idealizations of the situation. The classic Lotka–Volterra equations (Lotka, 1925), for example, which purport to describe the oscillations in predator–prey populations, assume that birth rates and death rates are numerically constant over time. This assumption, though reasonable to a first approximation, does not hold in real-world populations, so the solution to the differential equations is unlikely to yield accurate predictions. A stochastic model of predator–prey dynamics built in an object-based language will not produce a formal equation as a result, but it may give better predictions of real-world phenomena. Moreover, because object-based models are capable of refinement at the level of rules, adjusting them is also more clearly an activity of trying to refine content-based rules successively until they yield satisfactory results.

On the other hand, educator critics of computer-based modeling have expressed concern that the activity of modeling on a computer is too much of a formal activity, removing children from the concrete world of real data. It is undoubtedly true that children need to have varied and rich experiences away from the computer, but the fear that computer modeling removes the child from concrete experience with phenomena is overstated. Indeed, the presence of computer-modeling environments invites us to reflect on the meaning of such terms as *concrete experience* (Wilensky, 1991). We have come to see that those experiences we label "concrete" acquire that label through mediation by the tools and norms of our culture. Therefore, which experiences are perceived as concrete is subject to revision by a focused cultural and/or pedagogic effort. This is particularly so with respect to scientific content domains in which categories of experience are in rapid flux and in which tools and instruments mediate all experience. In the GasLab case, it would be quite difficult to give learners "real-world" experience with the gas molecules. A real-world GasLab experience would involve apparatus for measuring energy and pressure that would be black boxes for the students using them. The range of possibilities for experiments that students could conduct would be much more severely restricted and would probably be limited to the "received" experiments dictated by the curriculum. Indeed, in a significant sense, the computer-based GasLab activity gives students a much more concrete understanding of the gas, because they see it as a macro-object that emerges from the interactions of large numbers of micro-elements.

Concluding Remarks

The use of model-based inquiry has the potential to exert a significant impact on learning in the next century. We live in an increasingly complex and interconnected society. Simple models will no longer suffice to describe that complexity. Our science, our social policy, and the importance of an engaged citizenry require an understanding of the dynamics of complex systems and the use of sophisticated modeling tools to display and analyze such systems. There is a need for the development of increasingly sophisticated tools designed for learning about the dynamics of such systems and a corresponding need for research on how learners, using these tools, begin to make sense of the behavior of dynamic systems. It is not enough simply to hand learners modeling tools. Careful thought must be given to the conceptual issues that make it challenging for learners to adopt a system dynamics perspective. The notion of levels of description, as in the micro- and macro-levels we have explored in this chapter, is central to a system dynamics perspective, yet it is quite foreign to the school curriculum. Behaviors such as negative and positive feedback, critical thresholds, and dynamic equilibria pervade complex dynamic systems. It is important to help learners build intuitions and qualitative understandings of such behaviors. Side by side with modeling activity, there is a need for discussion, writing, and reflection activities that encourage students to reexamine some of the basic assumptions embedded in the science and mathematics curriculum—the assumption, for example, that systems can be decomposed into isolated subsystems and that causes add up linearly and have deterministic effects. In the Connected Probability project, we have seen the "deterministic mindset" (Wilensky, 1997; Resnick and Wilensky, 1998) prevent students from understanding how stable properties of the world, such as Harry's Maxwell–Boltzmann distribution, can result from probabilistic underlying rules.

A pedagogy that incorporates the use of object-based modeling tools for sustained inquiry has considerable promise as a means to address such conceptual issues. By providing a substrate in which learners can embed their rules for individual elements and visualize the global effect, it invites them to connect micro-level simulation with macro-level observation. By allowing them to control the behavior of thousands of objects in parallel, it invites them to see probabilism underlying stability and to see statistical properties as useful summaries of the underlying stochasm. By providing visual descriptions of phenomena that are too small or too large to visualize in the world, it invites a larger segment of society to make sense of such phenomena. By providing a medium in which dynamic simulations can live and a medium that responds to learner conjectures with meaningful feedback, it gives many more learners the experience of doing science and mathematics. A major challenge is to develop tools and pedagogy that will bring this new form of literacy to all.

Acknowledgments

The preparation of this paper was supported by the National Science Foundation (Grants RED-9552950 and REC-9632612). The ideas expressed here do not necessarily reflect the positions of the supporting agency. I would like to thank Seymour Papert for his overall support and inspiration and for his constructive criticism of this research in its early stages. Mitchel Resnick and David Chen gave extensive support in conducting the original GasLab research. I would also like to thank Ed Hazzard and Christopher Smick for extensive discussions of the GasLab models and of the ideas in this paper. Wally Feurzeig, Nora Sabelli, Ron Thornton and Paul Horwitz made valuable suggestions on the project design. Paul Deeds, Ed Hazzard, Rob Froemke, Ken

Reisman, and Daniel Cozza contributed to the design and implementation of the most recent GasLab models. Josh Mitteldorf has been an insightful critic of the subtle points of thermodynamics. Walter Stroup has been a frequent and invaluable collaborator throughout the GasLab project. Donna Woods gave unflagging support and valuable feedback on drafts of this chapter.

References

Buldyrev, S.V., Erickson, M.J., Garik, P., Shore, L. S., Stanley, H. E., Taylor, E. F., Trunfio, P.A. & Hickman, P. 1994. Science research in the classroom: *The Physics Teacher*, *32*, 411-415.

Chen, D., & Stroup, W. 1993. General systems theory: Toward a conceptual framework for science and technology education for all. *Journal for Science Education and Technology*, *2(3)*, 447-459.

Cutnell, J., & Johnson, K. 1995. *Physics*. New York: Wiley.

Daston, L. 1987. Rational individuals versus laws of society: From probability to statistics. In Kruger, Daston, L., & Heidelberger, M. (eds.), *The probabilistic revolution*, vol. 1. Cambridge, MA: M.I.T. Press.

Dawkins, R. 1976. *The selfish gene*. Oxford, England: Oxford University Press.

Dennett, D. 1995. *Darwin's dangerous idea: Evolution and the meanings of life*. New York: Simon & Schuster.

diSessa, A. 1986. Artificial worlds and real experience. *Instructional Science*, 207-227.

Doerr, H. 1996. STELLA®: Ten years later: A review of the literature. *International Journal of Computers for Mathematical Learning*, *1(2)*, 201-224.

Eisenberg, M. 1991. Programmable applications: Interpreter meets interface. *MIT AI Memo 1325*. Cambridge, MA: AI Lab, M.I.T.

Feurzeig, W. 1989. A visual programming environment for mathematics education. Paper presented 4th International Conference for Logo and Mathematics Education. Jerusalem, Israel, August 15.

Forrester, J.W. 1968. *Principles of systems*. Norwalk, CT: Productivity Press.

Gell-Mann, M. 1994. *The quark and the jaguar*. New York: W.H. Freeman.

Giancoli, D. 1984. *General physics*. Englewood Cliffs, NJ: Prentice-Hall.

Gigerenzer, G. 1987. Probabilistic thinking and the fight against subjectivity. In Kruger, L., Daston, L., & Heidelberger, M. (eds.), *The probabilistic revolution*, vol. 2. Cambridge, MA: M.I.T. Press.

Ginsburg, H., & Opper, S. 1969. *Piaget's theory of intellectual development*. Englewood Cliffs, NJ: Prentice-Hall.

Giodan, A. 1991. The importance of modeling in the teaching and popularization of science. *Trends in Science Education*, *41(4)*.

Gleick, J. 1987. *Chaos*. New York: Viking Penguin.

Hofstadter, D. (1979). *Godel, Escher, Bach: An eternal golden braid*. New York: Basic Books.

Holland, J. 1995. *Hidden order: How adaptation builds complexity*. Reading, MA: Helix Books/Addison-Wesley.

Horwitz, P. 1989. ThinkerTools: Implications for science teaching. In Ellis, J.D. (ed.), *1988 AETS yearbook: Information technology and science education*, pp. 59-71.

Horwitz, P., Neumann, E., & Schwartz, J. 1994. The Genscope Project. *Connections*, Spring, 10-11.

Jackson, S., Stratford, S., Krajcik, J., & Soloway, E. 1996. A learner-centered tool for students building models. *Communications of the ACM, 39(4)*, 48-49.

Kauffman, S. 1995. At home in the universe: The search for the laws of self-organization and complexity. Oxford, England: Oxford University Press.

Kay, A. C. 1991. Computers, networks and education. *Scientific American*, September, 138-148.

Kelly, K. 1994. *Out of control*. Reading, MA: Addison-Wesley.

Kruger, L., Daston, L., & Heidelberger, M. (eds.) 1987. *The probabilistic revolution*, vol. 1. Cambridge, MA: M.I.T. Press.

Langton, C., & Burkhardt, G. 1997. Swarm. Santa Fe, NM: Santa Fe Institute.

Lotka, A.J. 1925. *Elements of physical biology*. New York: Dover Publications.

Mandinach, E.B., & Cline, H.F. 1994. *Classroom dynamics: Implementing a technology-based learning environment*. Hillsdale, NJ: Lawrence Erlbaum.

Mellar *et al.* (1994). *Learning with artificial worlds: Computer based modelling in the curriculum*. London: Falmer Press.

Minar, N., Burkhardt, G., Langton, C., & Askenazi, M. 1997. The Swarm simulation system: A toolkit for building multi-agent simulations. http://www.santafe.edu/projects/swarm/.

Minsky, M. 1987. *The society of mind*. Simon & Schuster Inc., New York

Nemirovsky, R. 1994. On ways of symbolizing: Tthe case of Laura and the velocity sign. *Journal of Mathematical Behavior, 14(4)*, 389-422.

Neumann, E., Feurzeig, W., Garik, P., & Horwitz, P. 1997. OOTL. Paper presented at the European Logo Conference. Budapest: Hungary, 20-23 August.

Noss, R., & Hoyles, C. 1996. The visibility of meanings: Modelling the mathematics of banking. *International Journal of Computers for Mathematical Learning, 1(1)*, 3-31.

Ogborn, J. 1984. A microcomputer dynamic modelling system. *Physics Education, 19(3)*, 138-142.

Papert, S. 1980. *Mindstorms: Children, computers, and powerful ideas*. New York: Basic Books.

Papert, S. 1991. Situating constructionism. In Harel, I., & Papert, S. (eds.), *Constructionism*, pp. 1-12. Norwood, NJ: Ablex Publishing.

Papert, S. 1996. An exploration in the space of mathematics education. *International Journal of Computers for Mathematical Learning, 1(1)*, 95-123.

Pea, R. 1985. Beyond amplification: Using the computer to reorganize mental functioning. *Educational Psychologist, 20(4)*, 167-182.

Prigogine, I., & Stengers, I. 1984. *Order out of chaos: Man's new dialogue with nature*. New York: Bantam Books.

Repenning, A. 1993. AgentSheets: A tool for building domain-oriented dynamic, visual environments. Ph.D. dissertation, University of Colorado.

Repenning, A. 1994. Programming substrates to create interactive learning environments. *Interactive Learning Environments, 4(1)*, 45-74.

Resnick, M. 1994. *Turtles, termites and traffic jams. Explorations in massively parallel microworlds*. Cambridge, MA: M.I.T. Press.

Resnick, M., & Wilensky, U. 1995. New thinking for new Sciences: Constructionist approaches for exploring complexity. Presented at the annual conference of the American Educational Research Association, San Francisco, CA.

Resnick, M., & Wilensky, U. 1998. Diving into Complexity: Developing probabilistic decentralized thinking through role-playing activities. *Journal of the Learning Sciences, 7(2)*, 153-171.

Richmond, B., & Peterson, S. 1990. *Stella II*. Hanover, NH: High Performance Systems.

Roberts, N. 1978. Teaching dynamic feedback systems thinking: An elementary view. *Management Science*, *24(8)*, 836–843.

Roberts, N. 1981. Introducing computer simulation into the high schools: An applied mathematics curriculum. *Mathematics Teacher*, *74(8)*, 647–652.

Roberts, N., Anderson, D., Deal, R., Garet, M., Shaffer, W. 1983. *Introduction to computer simulations: A systems dynamics modeling approach*. Reading, MA: Addison-Wesley.

Roberts, N., & Barclay, T. 1988. Teaching model building to high school students: Theory and reality. *Journal of Computers in Mathematics and Science Teaching*, Fall, 13–24.

Roetzheim, W. 1994. *Entering the complexity lab*. Indianapolis, IN: SAMS Publishing.

Shore, L. S., Erickson, M. J., Garik, P., Hickman, P., Stanley, H. E., Taylor, E. F., and Trunfio, P. 1992. Learning fractals by "doing science": Applying cognitive apprenticeship strategies to curriculum design and instruction. *Interactive Learning Environments*, *2*, 205—226.

Smith, D. C., Cypher, A., & Spohrer, J. 1994. Kidsim: Programming agents without a programming language. *Communications of the ACM*, *37(7)*, 55–67.

Starr, P. 1994. Seductions of Sim. *The American Prospect*, 17, 19–29.

Thornton, R., & Sokoloff, D. 1990. Learning motion concepts using real-time microcomputer-based laboratory tools. *American Journal of Physics*, *58*, 9.

Tipler, P. 1992. *Elementary modern physics*. New York: Worth.

Tversky, A., & Kahneman, D. 1974. Judgment under uncertainty: Heuristics and biases. *Science*, *185*, 1124–1131.

Waldrop, M. 1992. *Complexity: The emerging order at the edge of order and chaos*. New York: Simon & Schuster.

White, B., & Frederiksen, J. 1998. Inquiry, modeling, and metacognition: Making science accessible to all students. *Cognition and Instruction*, *16(1)*, 3–118.

Wilensky, U. 1991. Abstract meditations on the concrete and concrete implications for mathematics education. In Harel, I., & Papert, P. (eds.), *Constructionism*. Norwood, NJ: Ablex Publishing, 193–204.

Wilensky, U. 1993. Connected mathematics: Building concrete relationships with mathematical knowledge. Ph.D. dissertation, M.I.T.

Wilensky, U. 1995a. Paradox, programming and learning probability: A case study in a connected mathematics framework. *Journal of Mathematical Behavior*, *14(2)*, 253–280.

Wilensky, U. 1995b. Learning probability through building computational models. *Proceedings of the Nineteenth International Conference on the Psychology of Mathematics Education*. Recife, Brazil, July.

Wilensky, U. 1996. Modeling rugby: Kick first, generalize later? *International Journal of Computers for Mathematical Learning*, *1(1)*, 125–131.

Wilensky, U. 1997. What is normal anyway? Therapy for epistemological anxiety. *Educational Studies in Mathematics*. Special Edition on Computational Environments in Mathematics Education, ed. R. Noss, (Ed.) *33(2)*, 171–202.

Wilensky, U. & Resnick, M. 1999. Thinking in levels: A dynamic systems approach to making sense of the world. *Journal of Science Education and Technology*, *8(1)*.

Wright, W. 1992a. *SimCity*. Orinda, CA: Maxis.

Wright, W. 1992b. *SimEarth*. Orinda, CA: Maxis

8

Designing Computer Models That Teach

Paul Horwitz

Introduction

Of all the species on earth, *Homo sapiens* is the only one, so far as we know, that uses models (Deacon, 1997). We invent models for many, often conflicting purposes: to provide parsimonious descriptions of observed phenomena, to predict what will happen under prescribed circumstances, and sometimes to explain why things happen the way they do. Models are the indispensable tools of modern science, and increasingly they run on computers, which enables us to predict, and to varying degrees control, the exact landing spot of a Mars probe, the three-dimensional configuration of a molecule, and the chance of rain tomorrow. Such uses of models, in fact, have given rise to a new kind of research, aptly described by the phrase *computational science*.

But whereas the research laboratory has embraced computer-based models as an aid to understanding, the same cannot be said for schools, where precollege science classes all too frequently concentrate on teaching facts, rather than scientific reasoning (Carey, 1986). The question naturally arises, then, whether the use of computational models in a school environment might not help students to think like scientists. Indeed, several efforts have been made to introduce, into the classroom, models similar or identical to those used in research.

This chapter will argue that computer models can indeed be extraordinarily effective in teaching science, but that for them to achieve this potential, they must be carefully designed and used. After a brief examination of various kinds of scientific models, I outline some of the pedagogic principles that underlie the design of successful teaching models. Using existing examples, I describe how a model may be linked in students' minds to the real-world phenomena it represents. Finally, I discuss some of the problems that must be overcome before the computer-based model can become a practical classroom tool.

Varieties of Models

Scientific models may vary quite dramatically across disciplines. As Ernst Mayr has pointed out (1989), a physics model, such as Einstein's theory of relativity (1905), is very different from a model in biology, such as Mendel's model for genetics (1866). As a result, scientists in different disciplines often differ considerably in what they consider a model and how they judge its utility. Physicists tend to set great store in the simplicity of a model, its fundamental nature, and its explanatory power. Particularly prized are those models that start from an axiomatic base (such as Newton's laws of motion, or the constancy of the speed of light)—particularly so if they can be shown to apply to a wide range of phenomena. Biological models, in contrast, are judged primarily on the basis of their explanatory power. They are expected to be approximate, to be somewhat *ad hoc*, and to admit of exceptions. An example may help to make the distinction clear.

In the early 1970s, high-energy physics was in a state of chaos and confusion. New particles were being discovered every other month, sometimes seen directly as tracks in a spark chamber photograph, more often inferred from "resonances" observed in various scattering reactions. But however they were observed, there was general consensus that there were entirely too many of them, and the central task of particle physics at the time was to classify them and impose some order on the experimental data. Many attempts were made to do this, but most were considered unsatisfactory, even by their inventors, because they were too complicated. (Of one new model it was rumored, only half in jest, that it had more free-floating input parameters than data points against which it could be tested!) The general feeling among physicists at the time was that "curve fitting" and "phenomenology" were never going to usher in any deep understanding. The true theory, if and when it ever arrived, would be, everyone was convinced, instantly recognizable by its simplicity and elegance and would not be judged solely on its ability to fit a large quantity of data.

At about this time, biology was going through a similar period, but in reverse. Even as the particle physicists thrashed about trying to find a model simple enough to be credible (a goal eventually achieved by Weinberg and Salam in the mid-seventies), the biologists were busy undermining one that was too simple! During the years immediately following the discovery of the double-helix shape of the DNA molecule, the "holy grail" of molecular biologists had been to work out the code that translates from the sequence of nucleotides strung out along the DNA to the sequence of amino acids on the resulting protein molecule. A few rash souls (mostly physicists!) ventured the opinion that once this code was broken and we could "read the book" of DNA, all the secrets of life would be laid bare. Nothing, of course, could have been further from the truth. In fact, the complications were just beginning. For example, although it is true that "codons" (triplets of nucleotides) correspond to amino acids in more or less the same way across all organisms, it does not follow that every protein encoded by the DNA will actually be produced within a particular cell. Furthermore, in many organisms, genes are often interrupted by long stretches of "noncoding" DNA that does not, according to our current understanding of the matter, appear to have any function at all. Moreover, genes are turned on and off in complex and poorly understood ways, either by other "control" genes or in response to conditions that affect the cell in which they reside. Even the assumption that every gene codes for a trait does not hold in general. So much for a simple, universal explanation of inheritance!

The moral of the story is simply stated: In physics, if a theory is too complicated and *ad hoc*, everyone distrusts it and looks for something simpler; in biology, if a theory is too simple and pat, everyone distrusts it and starts looking for exceptions. And both groups are right in reacting in these contrasting ways.

Teaching Models

Computational models used for research in whatever discipline are not necessarily much good for teaching. Consider, for example, a meteorological model that takes fixed input, such as topographic data, as well as readings of such time-varying quantities as temperature, barometric pressure, and wind velocity, at each of several million points within a volumetric grid. Using the applicable partial differential equations (such as various simplifications of the basic Navier–Stokes equations), the model churns out predicted values for each of the variables at each point over the next 24 hours (it typically requires several hours to do so), thereby making a prediction of the weather one day ahead, at each location within the grid.[1] A model of this kind, though it may be very useful for deciding whether to take an umbrella to work, is not appropriate for teaching students about the weather. It is far too complex, too opaque, and too far removed from the way humans think about the weather to be a useful guide. Trying to learn meteorology from such a model would be like taking chess lessons from Deep Blue.

So what do we look for in a good teaching model? It should be simple but not too simple, capturing the essence of the professionals' mental models of the domain but omitting unnecessary complications. It is also useful for the model to be modifiable—either by the teacher or by the students themselves—which may enable it, among other things, to change to meet the needs of students as they become better versed in the subject matter. At first, we may want certain aspects of the model to be inspectable by the students; later on, we may wish to turn this feature off, in order to force the students to make inferences indirectly by experimenting with the model. We shall see examples of this sort of thing later in this chapter.

---

[1] In general, additional information is obtained by interpolating between the grid points.

Note that a model that is intended to teach may be quite different from a simulation. Simulations are often prized for their realism—their ability to capture a sense of "being there." Thus simulations generally do not allow students to do anything they would not be able to do in real life, and conversely, anything that can be done in the real world can usually be done in the simulation as well. Teaching models frequently violate both these rules. In fact, as we shall see, much of their didactic power stems from their designers' freedom to decide what to allow the student to do.

The interface to a teaching model is often not user-friendly in the normal sense of the word. Usually, the designers of computer tools are keenly aware that they are designing something to be used by an expert. They study their intended users intensely to find out how the users think about the subject matter, and then they try to design software that matches the users' assumptions and reasoning patterns. But a model intended to teach is by definition written with naive users in mind. Clearly, students don't have the reasoning patterns of experts—the whole point is to help them acquire such reasoning patterns. Thus, as we shall see repeatedly in this chapter, educational models often have interfaces that are counterintuitive to students, precisely because they are designed to force the students to reason about the subject matter in unfamiliar ways.

Designing an Educational Model

The designing of a good teaching model starts with several simple but important questions. What exactly do we expect the students to do with it, and when they do it, what do we think they will learn? What do the students think they are doing when they use the model—what semantics or purpose do they associate with their manipulations of it? And, of course, the question one should ask of any piece of software: Why are we doing this on a computer? In this case, what educational value does the computer bring to the enterprise? What special role does it play that couldn't have been filled as well or better in some other way?

Fundamentally, all a computer program can ever do is show users things and let them do things. Thus, in designing any educational software, the two most important decisions one makes are what to show the students (the "representations" in the jargon of the trade) and what to enable the students to do (the "affordances"). To generalize slightly, there is usually a continuous spectrum in both dimensions. The software designer can choose how salient to make certain things and how carefully to hide others; certain features of the interface affect how easy or difficult it is for the user to perform various actions.

So far, so good. But surely such choices apply to any computer application. How do they play out specifically in the case of models intended to teach? The things a model shows the student are mostly in the form of objects, be they planets, fruit flies, or carbon atoms, that can be manipulated in various ways. But objects generally carry with them a certain semantics, and to the extent that the objects in the model are familiar to them, students assess the manipulations in terms of what these objects represent and associate causality with the consequences of those manipulations. In other words, the students form a mental model of what is going on—whether we want them to or not! Hence the power of the computer-based model to teach resides mainly in the *behaviors* of the various objects that we choose to include and in the reactions of these behaviors to the various manipulations that we allow the students.

It is worth noting, in this regard, that most students do not come to science class with a well-formed understanding that the physical world operates according to rules. Many, in fact, do not recognize that a "rule" can be descriptive in this way, rather than prescriptive. After all, the everyday rules with which most students are familiar are normative, generally stated in the imperative, and capable of being broken, though admittedly at some cost. "Wash your hands before coming to dinner," is a rule, as is "Speed limit 55 mph." Such rules describe what ought to be, not what is, and certainly not what cannot be otherwise. By contrast, the rule that "The energy of a closed system is constant" is an observed fact about the workings of the world, a fact that constrains physical systems—often in mysterious ways, as countless would-be inventors of perpetual-motion machines have discovered to their dismay. Computers, which are understood to operate in accordance with fixed internal programs, offer students an exemplar of a rule-based system that can guide them to a recognition of the operation of rules in nature.

A useful starting point for designing a computer-based model for teaching something is to choose the set of objects and manipulations that the model will incorporate. If we choose them carefully, these will be familiar and interesting enough to "jump start" the students' learning, but a formless and unstructured environment will not be enough to sustain the process. Often, we must impose a higher-level semantics and purpose on the model. It is not enough, in other words, that the students be able to manipulate the objects; they must have a *reason* for manipulating them—a reason that motivates their investigation and connects it to the science concepts we hope they learn. This semantic overlay can also serve to link the features of the computer model to their analogs in the real world—a crucial aspect of the learning process and, as we shall see, by no means an automatic consequence of students' interactions with the model.

Thinking along these lines, my colleagues and I at the Concord Consortium (and earlier at BBN Corporation) have created several game-like environments that pose problems to students and offer powerful computer-based tools with which to solve them (Horwitz and Feurzeig, 1994; Horwitz and Barowy, 1994; Horwitz, 1996). Each tool embodies an underlying model of a specific scientific domain, and each offers a set of representations and affordances appropriate to that domain. In each case, the student learns the domain by exploring the operation of the model. We call these open-ended exploratory environments "computer-based manipulatives" in order to emphasize their close pedagogic analogy with the mathematics manipulatives commonly used in the elementary grades.

In the discussion to follow, several existing and potential computer-based manipulatives are used as examples, but our main purpose is not to describe particular projects or products. Rather, we wish to propose a general methodology for designing computer models for teaching science, and the examples are intended for illustrative purposes only. We start with an examination of the factors that affect a designer's choice of representations and affordances.

Choice of Representations

In choosing what objects to represent, the educational software designer is not limited to those that are accessible in real life. On the computer we can show students many things that are ordinarily invisible, and we may choose, for pedagogical reasons, to hide other things that would normally be visible. Nor is it simply a matter of showing the user things that are too small to be seen with the naked eye or too difficult or hazardous to approach. Many scientific models include abstractions—for example, the center of mass of a collection of objects—that are invisible because they are not real but that are often more important for understanding the working of the model than the real objects themselves. In this situation it may be useful, from an educational standpoint, to show students the normally invisible object and perhaps to hide some or all of the visible ones.

But it is not enough to show them something: We must make the students care about it, too. One way to do this is by inventing a game. For example, we might present the students with a collection of point particles that they may affect by "landing on them" and firing little rocket engines to accelerate them in various directions. We then make the center of mass of the particles both visible and salient and create a game of "center-of-mass hockey," wherein the goal is to move the center of mass of the collection of particles into the opponent's territory by selectively moving the particles. In this way, though the students are able to affect only the real objects, their attention is focused on how such manipulations affect the position and velocity of the "unreal" center of mass.

Often we can get an educational advantage from hiding information that would normally be available to students. Referring to our collection of particles, suppose our goal were to help the students understand, at a qualitative level, the nature of collisions between the particles. We could simply confine the particles to a box (so as to keep them on the screen), have them bounce off the walls and each other according to an underlying model (conserving momentum and energy, for example), and tell the students to observe the motions of the particles very carefully and see whether they can figure out what is going on. This might work, but it would be a lot more motivating if we simply made one of the particles invisible and challenged the students to locate it by studying the motion of the visible particles. Every so often, one of these would bump into the invisible one and make a sharp turn. From a careful study of the motion—and a pretty detailed knowledge of the dynamics of the collision!—the students should be able to figure out where the invisible particle is and where it is going.[2]

Choice of Affordances

Often, for pedagogic purposes, we enable students to do things on a computer that they would not be able to do in real life. An obvious example is a simulation of a physical system in which the student can eliminate friction entirely in order to investigate the properties of a purely Newtonian world (White, 1993). Even further removed from real-world analogs are those environments that enable one to alter fundamental physical laws—letting the force of gravity vary as $1/r^3$ for instance—in order to see the effects of radically different models. (See Chapter 10.)

As with representations, the designer of a computer-based manipulative can sometimes teach a fundamental concept by not letting students do something they naturally want to do. For example, RelLab (Horwitz and Barowy, 1994), a computer-based manipulative for teaching special relativity, treats all reference frames as equivalent; every object constitutes a frame from which to observe the motion of all the others, and there is no such thing as a preferred rest frame. This "frame democracy" (which is the fundamental concept behind relativity) bothers many students because it violates their everyday perception of absolute motion (cars move, trees stand still, and so on). This came across in an especially illuminating moment at the end of one of our early trials of RelLab. After a 4-week relativity module, during which the students used RelLab nearly every day, we asked them to tell us what they liked and what they didn't like about the software. One of the more articulate students complained loudly and bitterly about how confusing it was, a condition she attributed to the absence of a privileged frame (she called it the "God's eye view") from which one could determine "What was really going on." Even as she said this, though, she caught herself and added, in a soft, wondering tone, almost as though she were speaking to herself, "But I guess that was the point, wasn't it?"

As a student grows in sophistication, the appropriate mix of affordances may vary. For example, in order to encourage students to reason indirectly from experimental evidence, we may choose to take away certain affordances in a progressive and systematic manner. Here is a detailed example of

---

[2]To dress the activity up and make it more fun, we could invent a tool that looks like a "butterfly net." Once a student has figured out where the invisible particle is, the object is to place the net over it and click the mouse button. This action renders the invisible particle visible and freezes all motion. If the invisible particle lies within the butterfly net, we award the student a point, create a new invisible dot at a random location with a random velocity, make the butterfly net just a wee bit smaller, and start the cycle over.

FIGURE 8.1 Examples of Dragons in GenScope. This species is used only in the early stages of a student's exploration of the model. Other available species include humans and labrador retrievers.

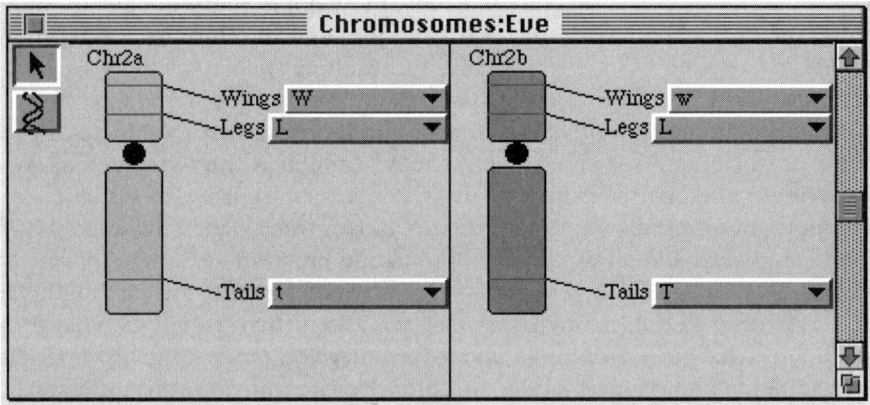

FIGURE 8.2 Chromosomes in GenScope represented as "popsicle sticks." The lines represent the locations of genes; the labels are the physical traits the genes control. The popup menus offer a simple method of altering the genes from one allele to another. When this is done, the organism changes its appearance in accordance with the classical Mendelian laws of inheritance. This is a purely informatic representation of the chromosomes (it focuses on their role as the carriers of genetic information) and offers no insight into their structure within the cell.

how this is done in GenScope, a computer-based manipulative we have designed to teach students about genetics (Horwitz, 1996).

GenScope offers students a multilevel view of genetics and enables them to move easily between the levels. Clicking on an organism (see Figure 8.1) with the "chromosome tool," for instance, brings up the textbook view of the organism's chromosomes, represented as short, fat rectangles like Popsicle sticks with lines across them representing the loci of various genes (these are shown in Figure 8.2). Another tool enables the student to see what those

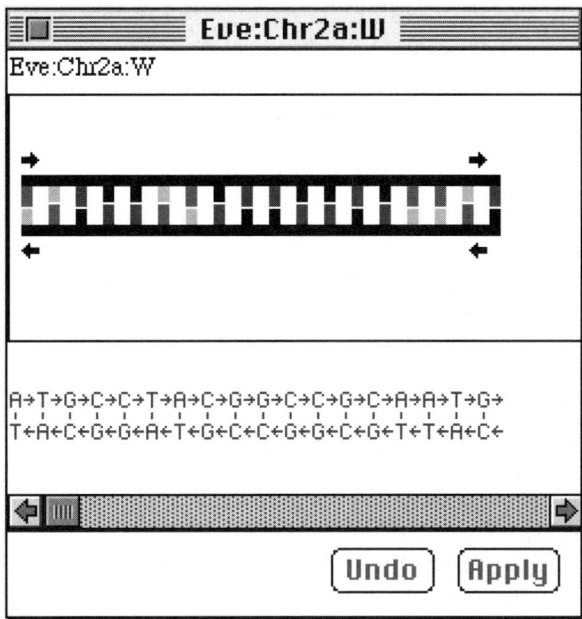

FIGURE 8.3 DNA in GenScope. This is the gene for wings—specifically, its dominant allele, *W*. It is much shorter than genes in real organisms (the main reason for starting with dragons is because we can make them very simple). The ladder-like structure at the top shows color-coded rectangles representing the four base pairs that are contained in the model: adenine, thymine, cytosine, and guanine. The sequence of base pairs is represented at the bottom of the window by their initial letters, which may be altered by placing the cursor between them and deleting or inserting letters, in a manner analogous to using a word processor.

genes are really "made of"—double strands of DNA in which the four different base pairs are represented both as colored beads and as sequences of letters (see Figure 8.3). At yet other levels, students may observe and manipulate cells, pedigrees, or whole populations of organisms that roam around the screen, mating with each other and maintaining a constantly changing gene pool.

Genes generally occur in multiple varieties, or *alleles*, and GenScope lets students alter an organism's genes in several different ways. In the early stages of a student's use of GenScope, the software is configured so that at the chromosome level, each gene is associated with a popup menu that enables the student to switch it from one allele to another. Just as the genes it carries affect the physical traits of an organism in the real world, so in Gen-Scope any alteration of a gene may affect[3] the appearance of the organism

_____

[3] Changes in genes do not always have an effect on the appearance of an organism. Mendel's well-known rules of inheritance are respected in GenScope, and organisms with two dominant alleles look the same as those with one dominant and one recessive allele.

that carries it. If the DNA window for the affected gene is open, it too reflects the change.

Obviously, in reality no one can alter a gene from one allele to another, nor would such a change, if it were possible, have any effect on an organism. Thus the operation of changing genes in no way simulates a laboratory or clinical procedure. The affordance is included in the software to allow students to discover Mendel's laws of inheritance for themselves by observing the consequences of these laws in a direct and motivating manner. This phase of exploration by direct manipulation of genes usually lasts 2 or 3 days and culminates, in our curriculum, with an activity designed to get students to articulate, discuss, and write down, using their own notation, the rules for inheritance of each trait.

In our introductory activities we generally use a simple—and fictitious—species: dragons. These are represented by deliberately cartoon-like graphics and have only three pairs of chromosomes and seven genes that code for six traits: presence or absence of horns, presence or absence of wings, shape of tail, number of legs, color, and whether or not they breathe fire.

Once students have explored the connection between the genotype (the set of alleles) of an organism and its phenotype (the set of traits), they are ready to investigate the reality of the genes themselves, in effect recapitulating much of the genetics research of the 1960s and 1970s. The representation of genes as loci on a symbolic chromosome conveys no information about what they really are—sequences of base pairs on an immensely long molecule—or about how they can have such a dramatic effect on all life forms. Although GenScope, in its present form, sheds no light on the latter mystery,[4] we can and do use it to teach students that genes are nothing more than segments of DNA.

The strategy for doing this is very simple. We challenge students to produce an organism with a particular trait by manipulating its genes. By this time the students have been altering genes and observing the consequences, and they have learned, for example, that giving a dragon two recessive *w* alleles will give the dragon wings. Starting with a wingless dragon, the students examine its chromosomes and determine that it is "heterozygous"; that is, it possesses two different alleles, one *w* and one *W*, for wings. They understand that in order to give the dragon wings, they must change the *W* into a *w*. But for this exercise we have taken away the students' ability to manipulate the wings gene directly. What are they to do? Using the DNA tool, they can examine *and alter* the two alleles. They will find that the *W* allele is represented in DNA by the sequence: ATGCCTACGGCCGCAATG.[5] The *w* allele,

---

[4] We are working on a more elaborate version of the software containing a full-fledged molecular level that will include representations of transcription (of DNA into RNA) and translation (of RNA into protein). This will offer the first link in the elaborate model of how genotype affects phenotype.
[5] Strictly speaking, this is just one strand of the DNA that makes up the *W* allele of the wings gene. The other, complementary strand is the same as the first, with A replaced by T, C by G, and vice versa.

in contrast, is represented by ATGCCTGCGGCCGCAATG. The difference is in the seventh base, which is an A (which stands for adenine) in the dominant, *W* allele, and a G (for guanine) in the recessive *w*. Because GenScope enables the students to delete and insert base pairs, it is a straightforward operation to change the *W* to a *w* and give the dragon wings, *but only if they understand the connection between the gene and the DNA*. When they do make the alteration, by the way, the gene's label changes at the chromosome level, and the dragon instantly sprouts wings.

Later still in the students' progression, we take away their ability to alter the DNA and challenge them to *breed* a dragon with wings, starting from two heterozygous parents (each of whom has a 1/2 probability of bequeathing that crucial *w* allele to its offspring, so that on average 1/4 of their offspring will have wings). This further reduction of the students' ability to manipulate the model forces them to reason in ways more closely analogous to those employed by professional geneticists. When, as a final challenge, we present them with an organism that has a previously unobserved trait (such as a dragon with horny plates on its neck), the students are now prepared to determine whether this trait is dominant or recessive and which chromosome it resides upon. At first they turn to the chromosome tool and seek to locate the gene that way. But we have hidden the gene, and it cannot be uncovered by such a simple and unrealistic technique. Instead, the students must investigate this new trait the way a real scientist would—by selectively breeding the unusual animal with other dragons of known genotype.

In summary, by a carefully sequenced set of moves that progressively *limit* students' interactions with the software until they are analogous to the reasoning available in the real world, we guide them bit by bit to reason in ways analogous to those of the professional scientist.

Evaluation for Redesign

It is not enough to design a model for teaching; One must also observe it in use, evaluate its effects, and modify it as required. Moreover, students are not simple, predictable robots. They do not bring identical attitudes and preconceptions to the learning process, and what they learn from working with an interactive model may differ, often dramatically, from what its designer intended. When this happens, it may suggest the need for substantial redesign of the model and/or its accompanying pedagogy. Again, a recent example from our research with GenScope may serve to make the point.

Students can get confused when the same thing is represented in more than one way. We ran into this problem early in the design of GenScope. GenScope represents chromosomes in two very different ways. At the chromosome level (Figure 8.2) they are represented as "popsicle sticks" with labels on them; at the cell level (Figure 8.4) they are seen as "spaghetti strands" that wiggle around and assort themselves randomly as the cell divides, via meiosis, into gametes. The first of these representations focuses on the informatic function of the chromosome—its role as the carrier of genetic information.

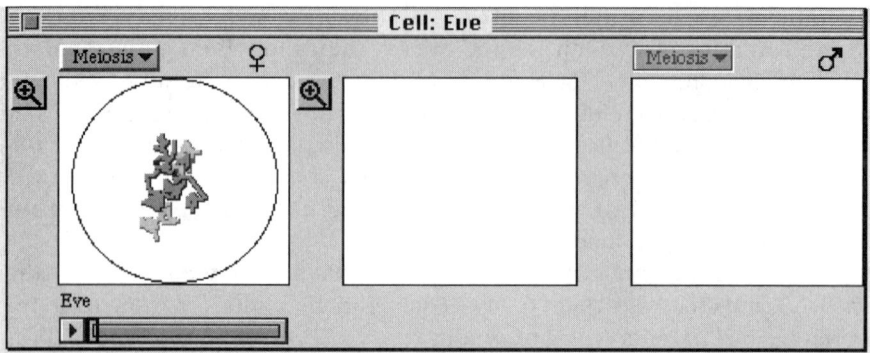

FIGURE 8.4 Cells in GenScope. This view of a cell (technically, only the nucleus of a cell) represents the chromosomes as they might appear in the laboratory. This representation makes no attempt to convey the information carried on the chromosomes, but when the student runs meiosis, the chromosomes move about on the screen and randomly assort themselves into gametes (sex cells). When one female and one male gamete are brought together in the middle pane of the window, fertilization takes place and a new organism is "born."

The second shows chromosomes more or less as they appear within cells. It emphasizes their structure and their behavior in meiosis, at the expense of not allowing the student to observe the genes directly.

When we first tried out GenScope in classrooms, we discovered to our dismay that students often did not understand that the spaghetti strands they observed and manipulated at the cell level were the same objects as the popsicle sticks whose genes they had observed and altered at the chromosome level. Our response was to add a bridging representation to GenScope—one in which a single cell is "zoomed out" so that it fills the entire screen (see Figure 8.5). In this representation the chromosomes become large enough so that the genes they carry can be labeled and observed, even while the cell undergoes meiosis (during which the chromosomes move about on the screen, carrying their allele tags with them and eventually randomly assorting themselves into gametes). This intermediate representation combines both the information and the structural aspects of chromosomes and helps students recognize that the two displays actually represent the same physical object.

Adaptive Models

Several of the examples we have used illustrate the need for a computer-based manipulative able to adapt itself to the changing needs of the student. In fact, as we have seen, quite often the curriculum we have created for GenScope, in particular, assigns students a series of similar tasks but progressively strips away their "privileges"—in terms of both representations and affordances—so that they are forced to rely more and more on indirect reasoning of the sort that scientists employ in the real world.

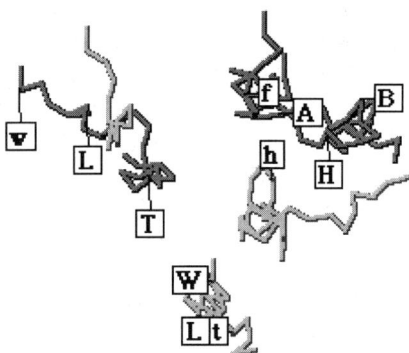

FIGURE 8.5 The "zoomed out" representation of chromosomes. The letters represent the alleles for each of the genes. This representation thus combines structure (physical layout of the chromosomes) and function (the information they carry) and forms a bridge between the representations depicted in Figures 8.2 and 8.4.

This is easy to implement if one is dealing with students in a controlled, laboratory environment where the researcher can intervene to alter the configuration of the software at strategic points in the student's learning. But how are we to arrange for such adaptive behavior in the classroom environment, in which at any given moment some 30 students may be working (usually in pairs) on a series of problems, each pair following a different route to the solution?

The management of inquiry-based classrooms poses problems unrelated to the use of computer-based manipulatives. Open-ended exploration that enables students to "construct their own knowledge" is a powerful teaching tool, but in practice it can be a very inefficient process, as students persevere though laboring under a misconception or "play around" for a significant fraction of the class time without making visible progress. It is all right— some might argue that it is essential—for students to struggle in this way, but if the struggle goes on too long, they become frustrated and "turn off." Ideally, a tool for open-ended inquiry should help the teacher to intervene at just the right moment.

Moreover, the designer of a computer-based manipulative must bear in mind that students have very different learning styles, just as teachers have different teaching styles. In some situations it may be appropriate to let the student loose to explore a model with little or no direction, but in different circumstances a more structured and linear approach may be called for. What is needed is a way to control, to some degree, how the software interacts with the student. We would like the computer-based manipulative itself to communicate with students, monitor their progress, and adapt itself to their needs. And would like it to do all this following the wishes of the teacher or curriculum developer. In software terms, the computer-based manipulative must become *scriptable*.

Scripts are not a new technology. Most business applications are scriptable, allowing the user to write simple programs that cause them to perform a specified sequence of often-used functions with a single mouse click. Scripts also provide a convenient medium of communication between applications, enabling a spreadsheet, for example, to obtain input from a database and hand its results to a word processor. Nor is the idea of altering the quality and quantity of help available to students particularly new. Such flexible support, often known as *scaffolding*, has been available for many educational applications for many years.

What *is* new is the notion of giving teachers and curriculum developers control over the form that such scaffolding takes at various stages of the student's learning. In addition to controlling an underlying computer-based manipulative, scripts must be capable of displaying information to the student in the form of text, animations, audio, or video material. They must also be able to gather information from the student, in the form of text entry or mouse clicks, and to receive updates from the computer-based manipulative itself. And most important, the scripts must be written in a high-level language that is inspectable and editable by teachers, researchers, and curriculum developers. We are in the process of creating this capability in GenScope.

What will happen when a student starts up the scripted version of GenScope? It depends, of course, on the script. To take a specific example, a prototype script that we have written starts by opening up a window that challenges the students to mate two parent organisms and, by selecting the appropriate egg and sperm to fertilize, create a baby with a specific set of traits. When the student clicks "OK" the script closes the text window, brings up GenScope, and configures it for the activity, reading in a file that creates the female and male organisms, locking their chromosomes so that students cannot alter the genes, and eliminating certain other unnecessary features. Then the script "goes away" and leaves the students to solve the puzzle. It is not entirely passive, however.

A major feature of scripts is their ability to monitor the students' actions. By constraining the problem very precisely, a curriculum developer can use this monitoring capability to identify "teachable moments" and can tell the script to intervene when such opportunities present themselves. For example, in the context of the fertilization activity just described, the prototype script we have designed remains in the background, keeping track of (and logging) certain critical student actions but not intervening until the students click the button to run fertilization. At that point, if the gametes have the right genes and the students have actually looked at them (which it can tell from the log it has created), the script waits until the baby appears and then congratulates the students and suggests that they call their teacher over and explain how they solved the problem. If, however, the students have not examined the gametes to determine which genes they carry, the script gently reminds them to do that and starts the cycle over.

But what if the task requires a special arrangement of the chromosomes, one that is unlikely to happen by chance? How can a script handle that situation? It can do just what a teacher might do. The first time the students try and fail, the script can simply say something like "Too bad—try again." The second time, it can give them a hint ("Does this baby look like the first one you made?") and then suggest that they "Run meiosis again and watch carefully how the chromosomes move." By the time the third offspring comes off the assembly line with the wrong chromosomes, the script can reconfigure GenScope to give the students control over how the chromosomes move and ask them to give it "one more try." The objective here is that the students, in their mounting frustration, realize why the babies are not coming out with the desired collection of traits and why the assortment of chromosomes during meiosis, ordinarily random, is so important.

Will this strategy work? We don't know. We are just now building into Gen-Scope the ability to write interactive scripts of this kind, and we have not yet had a chance to try them out with students.

Linking Models to the Real World

Models, by definition, are not real, and it is not always obvious how they connect to real things. The most carefully crafted computer model designed to teach some important scientific concepts may come across to students as just another video game. As a result, what they learn from the computer may be nothing more than the skill required to "win" the game. In particular, it may not extend to reasoning about real-world examples. Furthermore, many scientific discoveries carry with them important implications for society. Consider, for example, the legal, ethical, and moral dilemmas that seem to arise almost daily from scientific advances in genetics. In a world increasingly confronted with such issues, it is unacceptable to teach science without encouraging students to consider its social implications.

Although this linking of model to real-world analogs can be accomplished through classroom discussions, slide shows, field trips, or laboratory experiments, we have found that students do not easily relate their experience on the computer to the other things that happen to them in and out of school. Thus it is important to forge these links on the computer itself.

We are currently exploring the possibility of creating multimedia material, in the form of video, audio, animations, and text, and linking it to an underlying computer-based manipulative via the same scripting technology that we originally created to handle the adaptive scaffolding problem. One example of how this might be done is to create an "interactive documentary" that links video sequences with computer-based manipulative activities. Actors playing the roles of researchers (or patients with a genetic disease, or forensic scientists trying to determine the identity of a criminal) would be used to establish a context for student investigations, posing problems and appearing at important points to offer advice or encouragement. Alternatively, the results of computer-based manipulative investigations could be linked and compared to laboratory data (such as DNA sequences) available over the Internet or stored on a CD-ROM.

Assessment

In our research, particularly with GenScope, we have found that students often fail to transfer knowledge gained on the computer to performance on paper-and-pencil assessments (Horwitz and Christie, 1998). In their interactions with the computer, they may show evidence of multilevel reasoning and of using the model for abstract problem solving, but when confronted with a paper-and-pencil test, they do not score significantly higher than comparable students who have been taught genetics without the use of the computer. There are at least two possible explanations for this effect: (1) The students aren't really learning significantly more on the computer than they learn via lectures and textbook, or (2) they are learning more, but the knowledge doesn't transfer to the assessment.

If we accept the first hypothesis, we must explain how the students are managing to solve difficult problems on the computer without really understanding the underlying model of genetics. Are they just "playing around" until they get the answer they want, or do they really understand what they're doing? In other words, are they constructing a mental model of genetics, or are they merely mastering a complicated and counterintuitive computer interface? We may be able to shed light on this question by narrowing the gap between the assessment items and the computer-based activities. If we find, for example, that students who have used GenScope are able to answer questions about dragons but fail to perform at the same level when presented with homologous questions about, say, dogs, we will take it as evidence of shallow and superficial understanding. We may, however, discover that students are able to answer such "transfer" questions orally but fail to do so on paper—a finding that would have profound implications for the very nature of assessment itself.

In future research, we hope to use the scripting technology described here to probe more deeply into the nature of students' learning with computer-based manipulatives. We believe this can be done in various ways. For instance, we will be able to collect data unobtrusively during the students' interactions on the computer, logging their significant actions, and at critical points asking them probing questions aimed at characterizing their reasoning process. Sometimes these questions can be asked directly on the computer, but at other times we will instruct the students to explain, to us or to their teacher, what they are doing. Of course, in addition to shedding light on students' cognition, this kind of "embedded assessment" can also have a significant positive instructional effect and may thus serve to improve the students' scores on written tests.

If the second hypothesis appears to be correct, then we will need to find out why we are not achieving transfer between the two modalities. Is it, for instance, because the pencil-and-paper modality "turns the students off" and thus reduces their ability to demonstrate what they know? Is it a consequence of the specialized vocabulary, or other abstract representations, used on the written assessment? Or is a latency effect at work (a period of days to weeks may elapse between the computer activity and the written test)? Is performance undermined by the psychological stress inherent in taking a scary-looking written assessment? Our ability to embed the assessment function within the software will enable us to explore these hypotheses. By asking the students gentle questions every once in a while, we will be better able to judge what they are thinking as they work their way through the puzzles we give them. We may then be able to devise written tests that are more appropriate. Alternatively, by collecting a portfolio of students' work and using it to assess their progress, we will attempt to compensate for any "test phobia" that may be undercutting their performance.

Conclusion

Models, whether on or off the computer, aren't "almost as good as the real thing"—they are fundamentally different from the real thing. From an educational standpoint, they are neither better nor worse than "hands on" methods. Rather, the two approaches are complementary, and neither works very well in isolation. We have concentrated in this chapter on a particular kind of computerized model, the computer-based manipulative, as an example of one way to use computers to teach science. We have examined the design of such computer-based manipulatives, paying particular attention to such issues as selective scaffolding, careful sequencing of problems, and linking activities on the computer to knowledge of the real world. As we have emphasized, the computer-based manipulative paradigm, powerful though it may be, must be brought to bear in the context of conjunction with many other tools—"wet" labs, textbooks, and classroom activities—that can help students to link the various features of the computer-based manipulative to the real-world facts, phenomena, and procedures that they represent.

The most important question that still confronts us in the use of computer-based manipulatives is "What are the students learning?" The discrepancy we have observed between students' performance on the computer, captured in observation notes and on videotape, and their scores on written tests points out the importance of addressing this question.[6] We do not lay the "blame" for this discrepancy on the tests themselves, which have been designed to assess what we think the students are learning. Rather, it appears that learning accomplished entirely within the context of interactions with a computer-

---

[6] For simplicity, we have limited ourselves in this chapter to GenScope. We have, however, also observed this effect with RelLab, a computer-based manipulative for special relativity.

based manipulative may become learning about that computer-based manipulative, rather than generalizing to learning about the domain. It is very important, therefore, to broaden the learning process so that students are made aware of the model underlying the computer-based manipulative and of its application to real-world phenomena. This broadening process has implications for the teacher, the curriculum developer, and the software designer. We hope that the scripts we are currently designing will go a long way toward making students conscious of what they are learning when they explore and solve problems on the computer and of how what they are learning applies in the world outside the classroom.

References

Carey, S. 1986. Cognitive science and science education, *American Psychologist*, *41(10)*, 1123–1130.

Deacon, T. W. 1997. *The symbolic species: Co-evolution of language and the brain*. New York: Norton.

Einstein, A. 1905. Elektrodyamik bewegten Körper, *Annalen der Physik*, *4(17)*, 891–921.

Horwitz, P. 1996. Linking models to data: Hypermodels for science education. *The High School Journal*, *79(2)*, 148–156.

Horwitz, P., & Christie, M. T. 1998. Computer-based manipulatives for teaching scientific reasoning: an example. In Jacobson, M. & Kosma R. (eds.), *Learning the sciences of the 21st century: Research, design, and implementing advanced technology learning environments*. Hillsdale, NJ: Lawrence Erlbaum (in press).

Horwitz, P., & Feurzeig, W. 1994. Computer-aided inquiry in mathematics education, *Journal of Computers in Mathematics and Science Teaching*, *13(3)*, 265–301.

Horwitz, P., & Barowy, W. 1994. Designing and using open-ended software to promote conceptual change, *Journal of Science Education and Technology*, *3(3)*, 161–185.

Mayr, E. 1989. *Toward a new philosophy of biology: Observations of an evolutionist*. Cambridge, MA: Harvard University Press.

Mendel, G. 1866. Versuche über Pflanzen-Hybriden. *Verhandlungen des naturforschenden Vereines, Abhandlungen, Brünn*, 4, 3–47.

White, B. Y. 1993. "ThinkerTools: Causal models, conceptual change, and science education." *Cognition and Instruction*, *10*, 1–100.

9

Modeling as Inquiry Activity in School Science: What's the Point?

William Barowy

Nancy Roberts

Introduction

"What's the point?" Asked a middle-school student who was given the task of exploring modeling software with her partner in a clinical interview. Having no prior exposure to the computer model, and having been given no other directions than what they needed to run a simulation, she and her partner questioned the authenticity of the moment. They had just met the interviewers, and a camcorder was located behind them, pointed over their shoulders at the computer screen. In the process of designing the interview to explore student inquiry with computer models in the least invasive way, we as researchers created a context that made no sense to the students.

"What's the point? I don't get it," echoed a sophomore a year later when asked to sit in front of a computer running modeling software in her high school biology class. In contrast to the middle-schooler, this student was sitting with the rest of her class. She was working on a lesson specifically designed to fit into the fourth day of a constructivist conceptual change unit on plant nutrition. The unit had been modified so that the students would be working with the *Explorer: Photosynthesis*[1] computer model (Duffy and Barowy, 1995) as an alternative to lecture for presenting the "school science view" (Driver and Bell, 1986; Driver, 1987). As we talked with the student, it soon became clear to us that she too was confused about what we expected, although the lesson sheet included an introduction to the computer model and described a purpose—*the teachers' purpose*—for the lesson.

This class was not like her usual experiences with her teacher, who describes himself as "traditional." She and her classmates had no prior experience in the computer lab or with the new software. Yet, after careful and de-

---

[1] Explorer software is available from LOGAL®, 125 Cambridge Park Drive, Cambridge, MA 02140.

liberate discussion with one of us about computer models and how this task was related to her other learning experiences in the science lab, this student became thoughtfully and enthusiastically immersed. We later observed her working diligently with the computer model and conversing with another student.

What is the point? Although both students used similar science-modeling software, their participation in two disparate systems of activity with different goals, tasks, and social interactions contributed to substantially different consequences. The high school student became engaged when we intervened, whereas the middle school student became engaged only sporadically during the clinical interview. In this chapter we explore middle-school student interactions across several situations that shed light on how students can learn with expert-built science computer models—models designed by scientists and built by professional programmers for educational use. We observe different forms of student reasoning and talk emerging with distinct systems of (negotiated) goals, expectations for participation, computer-modeling tools, and co-constructions of mathematical graphs and system diagrams.

The Issue

We began our work with a simple question: What can students learn with computer models? As we sought to answer this question critically, we encountered theoretical and practical issues in learning that underlie the use of modeling for science and math education. Generative-approach advocates of *model-building* (Papert, 1990) insist that student externalization of knowledge is essential to their learning. Efforts therefore should be directed at building tools to help students construct their own models. The construction of models by learners has turned out to be a difficult goal to attain, however. Mandinach and Thorpe (1988) and Riley (1990) have reported difficulties not only with students being able to build models, but also with teachers being able to adopt and manage model-building activities in the classroom.

Historically, two approaches have been pursued. The first has been to develop more ability-appropriate model-building software such as Model-It (Jackson *et al.*, 1996 and Chapter 3) and StarLogo (Resnick, 1994 and Chapter 5), which, generally speaking, offer students more support in building models. The second approach has been to provide expert-built models that students may investigate in order to understand the behavior of the system being simulated (Richards *et al.*, 1992). The distinction between the two approaches is somewhat blurred by the continuum of expert-built microworlds along this dimension, which facilitate student construction of simulations that conform to the rules imposed by particular knowledge domains. For example, Interactive Physics and RelLab (Horwitz and Barowy, 1994) are two environments that support student constructions in Newtonian physics and special relativity, respectively.

We have chosen to explore how middle-school students learn from expert-built, student-investigable models. Through our work, we have refined our initial question as follows: What conditions best support student investigation with expert-built models? What processes can and will students engage in when these conditions are met? What knowledge or skills must students already have acquired to learn effectively with computer models? We explore these questions with and without software in four different situations: an open-ended exploration, the classroom, guided inquiry, and the Nature of Models interview. We discuss aspects of social interactions and re-presentational tools (which include computer model tools, graphs, and causal diagrams) that, in our present theoretical view, provide some clarity about what happens when students either build or manipulate models.

Approach

This study is an exploration of the thinking and learning of middle-school students as they worked with computer models and simulations. Because our primary goal was to seek optimum contexts for learning with scientific computer models, the study did not attempt to control variables—an approach that presumes a theoretical framework to define which variables to control. Instead, in our effort to understand student interactions emerging from a variety of learning and problem-solving situations, we created and varied the conditions in a manner *approaching* the experimental-developmental method of Vygotsky (1978), or design experiment methodology (Brown, 1992; Collins, 1992; Hawkins and Collins, in press), or the positive-critical method of cultural psychology (Cole, 1997, 1996) that emerged from the integration of experiment with naturalistic observations (Cole and Scribner, 1974). These methods for research into cognitive processes share the use of an intervention shaped by a theoretical framework.

Our study falls between these and grounded theory (Glaser and Strauss, 1967). Like many other experimenters before us, as we approached the issue of learning with models, we were guided by our own prior experience designing and using models, learning from them, and teaching with them. It was this craft knowledge that shaped the design of our interventions. We began building systems models to explain our observations (Roberts *et al.*, 1996). Activity theory retroactively emerged as a framework that concisely encompassed the results of our inquiry into the prerequisite skills and knowledge and the types of settings in which students begin to use scientific models much as scientists do: as explanatory constructs and vehicles of investigation (Leont'ev, 1975).

The subjects of our study were seventh-grade students from an inner-city, Massachusetts school. The students were recruited by a participating teacher to represent a wide spectrum of abilities. The same students participated in all aspects of the study. These students joined our study as an opportunity to learn science and have fun. Some students were clearly supported by their

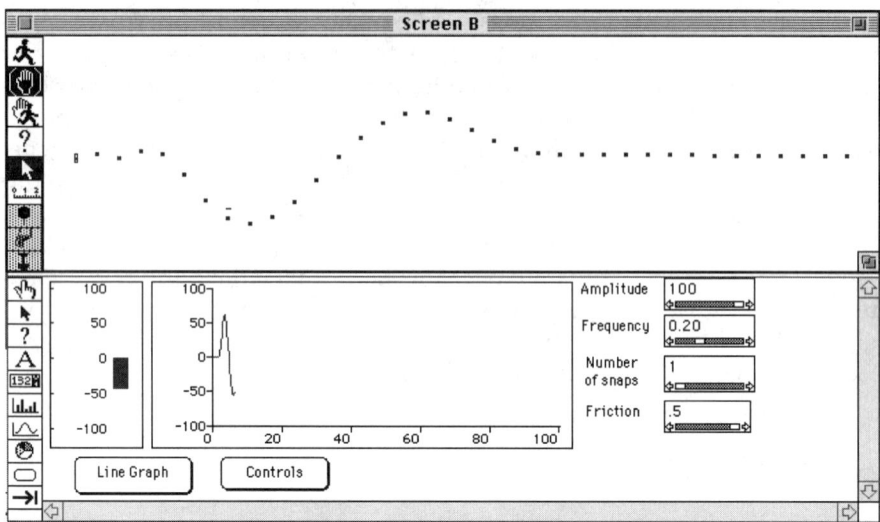

FIGURE 9.1 Screen from the Explorer: Waves Simulation. One complete wave cycle has been created [number of snaps] with initial amplitude of 100 and frequency of 0.2, in arbitrary units.

parents, who wanted to meet with us to learn more about the study. We have the distinct impression that some were strongly encouraged by the teacher despite their having some initial reservations, but we have no tangible evidence to substantiate this claim.

We studied four situations. First, we began and ended this study by conducting the Nature of Models clinical interview with individual students. The interview assesses student understanding of models and how they are used in science. The Nature of Models interview is described in detail later in this chapter (Grosslight *et al.*, 1991). Second, we conducted baseline clinical interviews using the Explorer: Waves model, in which pairs of students were asked to explore a highly abstract model of wave phenomena with minimal intervention. Third, we engaged the students in classroom teaching on equilibrium in aquatic populations, using the Explorer: Population Ecology model. Unlike the prior investigations, the ecology teaching component was conducted in a school with the students' teacher. Fourth, we investigated the combination of causal diagramming, graphing, experimentation, and the Explorer: Cardiovascular model. During this last component, half of the students participated in a section run by their teacher in school, and half participated in a section run solely by the investigators in their research offices.

Open-Ended Exploring

We selected the Waves model, a highly abstract representation of wave phenomena, as the basis for a clinical interview. The initial interface to the model is shown in Figure 9.1. We developed the interview as a precursor to understanding the use of a computer model as an explanatory construct. The students were asked to explore the Waves model with minimal intervention from the interviewer. We were interested in exploring the notion that science students bring knowledge acquired prior to instruction (Clement, 1982; Clement *et al.*, 1991; Saxe, 1991). The Waves model poses an extreme case: The middle-school students had not previously studied wave phenomena with this software.

We felt that our results would suggest what types of learning might occur in similar, "Exploratorium" settings. We have encountered the approach of letting students explore software on their own in some schools, where the social ecology of work sometimes precludes teachers from learning about the software first (Barowy and Laserna, 1996; Saferstein and Souviney, 1996). Yet success with this approach presumes that students have mastered experimentation skills or that they are able to view the computer model as an analog. Unable to identify what phenomena the software is modeling, what sense would students make of the model? What patterns of experimentation, if any, would emerge when they were given the task of exploring the model. Would they be able to generate analogies to explain what they observed?

During clinical interviews, the interviewer refrains from providing information relevant to the students' problem at hand in order to learn about the students' abilities without adult intervention. Within this structure, the goals of exploration are determined by the students. Interviewers ask questions only to clarify student comments or encourage them to vocalize their thinking. The format affords a useful glimpse into the types of reasoning students can apply without adult intervention. Because the interviewer gives students no guidance, they must make their own decisions about how to proceed.

At the beginning of the interview, the students were introduced to the software. We explained to them that we were interested in what they were thinking as they explored it. They were shown how to run, stop, and restart a simulation. Most students had previous experience with operating the Macintosh computers on which the software ran. They were familiar with the click and drag interface and with the use of buttons and pull-down menus. We encouraged them to ask questions concerning the operation of the software, such as how to operate a slide control, which we answered directly when asked. Otherwise, researchers simply asked clarifying questions, such as "Can you tells us what you are doing/thinking now?" in order to encourage students to explain their actions.

We found, in an observation consistent with Schauble *et al.* (1991), that when left to their own initiative, our seventh-grade students often applied an "engineering" model of experimentation in this exploratory setting. Students with an engineering orientation manipulate the model to produce a desired outcome, which may be an instance that contrasts highly with a previous outcome. Students simultaneously test many variables they believe to be causal, while trying to make causal inferences. In comparison, students oriented toward a "scientific" model of experimentation apply the strategy of testing one variable at a time (controlling variables) to achieve greater understanding. The latter strategy *per se* (not necessarily a full model of experimentation) is often addressed in school science. In the following dialogue, "S1" and "S2" are student comments, and "R" is a researcher. The interview took place in the students' school. One student explains:

S2: She [S1] changed all of them [variables], and if you wanted to see what the difference was, then you keep only one variable because then you don't know what's causing it to be different.

R: Is that something you've learned here?

S1, S2: Yeah, in science.

Later, the students' comments revealed that a tension had emerged between the two students: S1 wished to experiment in the pattern of the engineering model, whereas S2 wished to control variables. Without close scrutiny of both her actions and her explanations, S1 would appear to have been engaged in random probing, which a RAND study of students interacting with microworlds labeled *thrashing*, wherein "one issue would take center stage, another would arise to replace it, and later the first issue would resurface" (McArthur and Lewis, 1991). However, key utterances (such as "make it look like . . .") indicate her goal orientation and distinguish her engineering approach from random exploration. Her persistence also supports a goal-oriented explanation: S1 makes *several* attempts to vary two parameters in the model despite her earlier recognition of the strategy of controlling variables.

S2: So that's forty . . . [amplitude]

S1: Yeah, try to make it look like . . .

S2: So let's not fool around with the forty since we have a good guess. Keep it at forty . . .

S1: Well, let's try and make it go like . . .

S2: No. Let's fool around with something we are not really sure of, like, try friction. [sets friction to the midpoint in the range]

S1: But let's see if it [amplitude] goes to 30 this time [is changing friction value at same time]

S2: But you don't want to test two things at the same time because then you don't know what's causing it [sets amplitude back to forty and runs simulation].

All students treated the simulation primarily as a novel phenomenon. Most of the subjects' explorations were consistent with Schauble's engineering model, although recovering students' purposes for making experimentation decisions was difficult and uncertain. In one other exceptional interview, two students consistently displayed an ability to generate analogies to explain the animation on the computer screen. Like the other students, they changed variables in a manner consistent with the engineering model, yet they often expressed the purpose of trying to generate wave patterns that were similar to those they had seen in other contexts, such as in the wave tank at the local science museum. These students appeared to apply minimally abstracted experiences, or *p-prims* (diSessa, 1988), to explain the animation as they changed the model variables. Sutton (1993) calls this activity *figuring*, "applying something we already have in order to make sense of the relatively unfamiliar."

When left on their own, as in this exploratory interview, our students struggled to experiment in a meaningful way. That is to say, they had difficulty accomplishing tests that provided them with results with which they could make sense of their situation. The transcript provided earlier is our best evidence of what knowledge the students were able to bring to the interview. The two students brought recognition of the strategy of controlling variables but were unable to apply it consistently. Apparently their goal orientation—their tendency to think in terms of producing a desired outcome—militated against changing one variable at a time. If otherwise the best that students can do is generate a disparate collection of analogies to experiences they have had earlier in life, then they will not reliably see patterns emerging during their exploration. The variables and interrelationships programmed into a model may be obscured by the students' haphazard testing of variables.

It is quite possible, within the design of the Explorer software, to provide a scaffolding so that students have access to one or a few variables at a time. The obvious approach is to design a lesson in which the students are carefully led through manipulation of the parameters of the model. Students tend to find this dull and unmotivating, and furthermore, they are not given a chance to exercise investigation skills. Instead, we prefer to consider *systemically* the open-ended design of the software as coupled with social interactions that are specific to the content, skills, and motivations to be learned. One example is the use of Explorer software for *coordinating theoretical explanations* with complex physical phenomena that the students have experienced (Richards, *et al.* 1992). Another is the support for students *creating thought experiment* resolution of cognitive dissonance (Horwitz and Barowy, 1994) in relative motion.

Classroom Experiments

The classroom experiments explore activity in which students were provided with a structured environment for learning with a computer model. The small-group teaching sessions focused on the development of the students' understanding of the ecological behavior of populations. Guided model investigations were followed by a hands-on experiment in which the students built an aquarium containing organisms that constituted a three-level food chain: guppies, *Daphnia*, and algae. The hands-on experiment provided a measure for whether students' understanding of the computer model would spontaneously transfer to the design of environmental conditions for stable equilibrium of the aquarium.

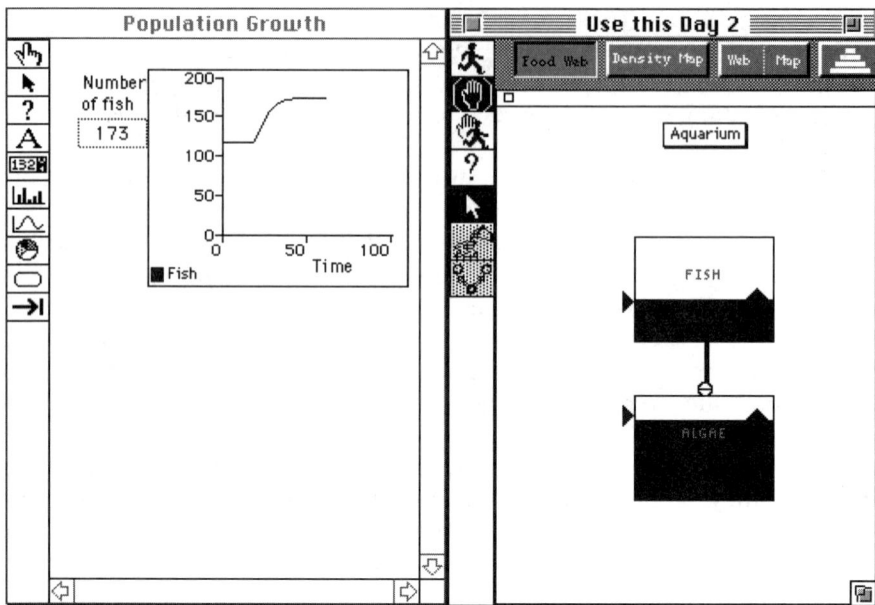

FIGURE 9.2 The Population Ecology Model.

The Population Ecology model simulates the growth and interaction of populations of organisms in a controlled ecosystem. The ecosystem can be divided into regions and habitats to examine any of several imaginary combinations of species, observing noncompetitive, competitive, and predatory relationships. The model generates "real-time" graphical presentations of data for each population. Students are able to adjust initial population levels of species and then run the simulation to observe the model behavior over time.

The students with whom we worked had been studying ecology in science with the aid of a classroom aquarium. Our intervention drew on the work by White and Horwitz (1988) with the Thinkertools software (Chapter 10), in which students built a causal model of force and motion through a series of increasingly complex microworlds. We first presented students with instructions to manipulate the model to recreate particular graphs. Students were given a set of challenges paraphrased as "Manipulate the conditions to create this kind of graph." In contrast to open-ended exploring, this approach provides carefully structured but non-negotiable experimentation goals for the students.

This intervention seemed plausible for learning about population ecology, where both direct observation and experimentation are difficult. (Observation of ecological systems requires the investment of significant resources in quantifying participating species, and the characteristic lifetimes and rates of interactions among species rule out experimentation within the time constraints in schools.) Understanding whether a computer tool could circumvent these difficulties and build student understanding seemed a worthwhile pursuit.

Students were exposed both to transient and long-term equilibrium behavior of two-level and three-level food chains. They were given target graphs of population levels to recreate by modifying variables and running simulations. Written lessons focused on how changes in population and supporting habitat at one level would affect that same level and other levels in the system. We asked students to make semi-quantitative predictions of model behavior before they made any changes in the model prior to running simulations, and we asked them to justify their predictions verbally and in writing.

Formative evaluation of each lesson influenced the design of the next lesson. This process enabled lesson designs to be sensitive to students' difficulties or insights with the population ecology content. The lessons generally included work sheets on which students were asked to make and explain predictions, to describe their results, and to indicate how these results compared to their predictions.

The teacher worked with us in developing the lessons and was responsible for interacting with the students during the experimental teaching sessions. We had selected this teacher in part because she described herself as an inquiry teacher. Together with the teacher, we explicitly negotiated the goal of developing students' abilities to inquire with the model. We developed the plan for students to contribute more fully to the direction of the work as they grew more proficient with the software, applying a cognitive apprenticeship metaphor (Collins *et al.*, 1989). We *coached* the teacher in how to use the software and made suggestions for interacting with the students. We *modeled* for the teacher a style of interacting with the students that we believed would support inquiry.

We noted sustained teacher-centered discourse in the sessions, described by Stubbs (1983) as IRF (for Initiate, Respond, Follow-up or Feedback). The teacher *Initiates* discursive interaction with the students by posing a question, to which the students *Respond*. The teacher *Follows-up* or provides *Feedback* to the students, which may be a form of *Evaluation* of the students' responses. The latter, labeled IRE, is a common discourse genre in school settings (Cazden, 1988; Lemke, 1990; Wells, 1996). The following dialogue is an example of the discourse we observed.

T: Now what do you think time means? What do you think the time part means?

S1: How long it takes.

T: Usually how long it takes for fish to reproduce. . . . It could be years, it could be weeks, it could be months, for guppies it could be days. All right so you have a lot of information there . . . it depends on what species of fish you are talking about.

T: Now, you have how many variables that you can play with right now?

S3: Three.

S1: Three.

T: Three? Think about it. How many little arrows or diamonds do you have?

S1-3: Four.

T: Four. So, how many variables do you have?

S1: Four.

T: Four variables, OK.

Implicit in the above discussion is the teaching of control of variables—the objective the teacher later articulated to us. We also observed the teacher to use the word "So" to draw conclusions, and the word "Well" to begin expostulation, which offers clues to the students about whether they are responding with answers she expects. When the students became unresponsive, the teacher's approach sometimes became more tutorial-like, in which she hinted about her expectations to the students. In the following discussion, her voice softens. She helps the students focus their attention on features of the display that are central to the task: the number of algae, the shape of the graph. In explaining the shape of the graph, the teacher indirectly suggests possible actions for the students. She then directs attention to essential features of the software.

T: OK, well you've got the first . . . you got almost opposite of what it should be doing. You want it to come out and up. You've got the curve going one way. Now how do you think you might get it down to here?

S1: Really lower the algae.

T: OK!

S1: And the fish.

T: Well, you can change both of them if you feel that's the way you would do it. OK, now compare . . .

S1: Looks about the same.

T: Looks about the same . . .

S1: One started a little lower than the other one.

T: OK now, where do we want the fish roughly to start out from?

S1: The bottom.

T: At the bottom.

S1: Decrease . . .

S2: You want us to make it look like that?

T: Ah hmm.

S2: ALL RIGHT!

When the teacher directed the lesson, the students did not contribute to the learning goals but instead followed the teacher's lead. ("You want us to make it look like that?") To our surprise, the goals that we made explicit in the series of challenges we provided for the students helped to sustain the teacher-centered social ecology. As the students struggled with manipulating variables to create the target graphs, the teacher took control. After 6 or 7 years of public schooling and perhaps 12 or 13 years of adult authority, the students expected this. They looked to the adults to set the tasks, and when we asked them simply to explore (in other words, to set their own goals) as in the open-ended explorations, they would sometimes ask if they were doing that right, if they were doing what we wanted them to do.

It is widely noted that students' and teachers' participation in teacher-directed investigations can hinder the opportunities for students to engage in the processes considered to be scientific inquiry (NRC, 1996; AAAS, 1993). So what did the students learn? Supported by the teacher's suggestions, and by questions directing the students' attention to features of the software related to the challenges, students became adept at manipulating the model's independent variables to recreate graphs of dependent variables over time. This in itself is a significant learning outcome. Can it transfer to other situations? Although we did not specifically design the lessons for transfer, we did test whether transfer would occur spontaneously. When it came time to design the actual aquarium, the computer model was made available to the students. No students spontaneously tried to use the model to determine the initial populations of algae, *Daphnia*, and fish.

During the entire school-based teaching experiment, we observed that the students played, experimented, and investigated their own questions in the intervals between teacher-led interactions. The teacher allowed these episodes as motivation for the students to complete the software challenges that they found difficult and discouraging. Clearly, the teacher is not solely responsible for the social structure of the classroom teaching experiments. In the classroom example, the curriculum also contributed by imposing a non-negotiable goal structure that the students and teacher were compelled to carry out. The teacher, intuitively recognizing the need to motivate her students, provided ways around the curriculum to keep the students interested.

Guided Inquiry

We conducted the teaching experiment on the cardiovascular system with half the middle-schools students in the BBN labs, while the classroom teacher worked with the other students in their school. We will describe only the BBN sessions, because only this component was qualitatively different from the classroom teaching experiments. Preliminary analysis of the school sessions conducted by the teacher indicates that these sessions were qualitatively no different from the classroom teaching experiments in ecology.

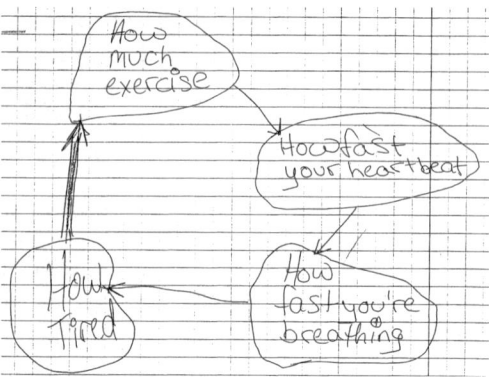

FIGURE 9.3 Anne's Causal Diagram.

Each session lasted one and one-half hours. During the course of these sessions, we introduced dynamic feedback systems thinking (Roberts, 1978), encouraged hands-on experimentation, and provided a simulation of the human cardiovascular system. The discussion began between two students, Anne and Joanna, and the two of us, as the students described a set of experiments they did in their regular science class with their teacher the week before.[2] Prior to their participation with us, the students had determined their pulse rates while resting, walking, running, and doing jumping jacks. The teacher had discussed with them the relationship between level of effort and heart rate.

As we began by discussing what topic to pursue with the students, they expressed an interest in continuing the learning about the cardiovascular system that they had started in school. Thus, in our first session, we introduced the students to causal diagramming of the effects of exercise on heart rate and breath rate. As we discussed the cardiovascular system with the students, we asked them to draw a diagram of how exercise affected heart rate. With help from Joanna and us, Anne drew the diagram shown in Figure 9.3.[3]

The diagram is interpreted in the following manner, working clockwise from the upper left: Increasing how much exercise you do increases how fast your heart beats. An increase in how fast your heart beats increases how fast you are breathing. An increase in how fast you are breathing increases how tired you are. An increase in how tired you are *decreases* how much exercise you do—hence a *different* arrow.

The second session included students Peter, John, and Anne. Schedule changes allowed the boys to participate, but Joanna's family moved away,

---

[2] Only two students attended the first session because of scheduling difficulties.
[3] Where possible, we have provided the students' drawings. Reproduction quality and the difficulty of producing drawings recorded on videotape or poster paper, preclude our presenting them all.

and she was no longer able to attend. We asked Anne to draw and explain her diagram to John and Peter. As she recreated her diagram, John interjected.

John: Wait a minute, you forgot to put the question mark in.

Anne: No, we didn't use question marks.

John: Well, it's not proper English. Mr. Driscoll would be disappointed.

Anne: Well, Mr. Driscoll's not here.

We learned that Mr. Driscoll was the students' English teacher. As a new-comer, John had not yet experienced the new ways of participation we had begun to develop with Anne and Joanna. Instead, he brought knowledge appropriate to another activity system he participated in, his school experience. Perhaps because he had no researcher expectations "laid out" for him, John displayed those expectations that he transferred from school. John read the phrase as a question, requiring a question mark, and he summoned the authority of his teacher, Mr. Driscoll, to bolster his interpretation.[4]

In contrast, Anne's use of the phrase was to express a quantity, as she had learned during the first lesson. With her rebuttal "Mr. Driscoll's not here," Anne expressed her understanding that the BBN lessons were different from school. In other words, the new lessons constituted a system of activity with different conventions for participation than those that prevailed in school. The episode with John is representative of the tension we initially encountered as we began to engage the students in the new teaching experiment with new expectations for participation.

Anne, John, Peter, and the two of us began a discussion of Anne's original diagram. During the discussion, both John and Anne modified the diagram, which was drawn on a white-board. Their co-construction of the systems model led to a disagreement between John and Anne, both of whom were trying to express their own views. Their final version appears in Figure 9.4. The diagram was not consensual: The students did not agree about several aspects of its form, and the diagrams the students drew individually were all different from the one on the white-board. Peter's and John's diagrams showed some causal links reversed, but the correctness of their diagrams is secondary. More important, the disagreements became focal points for learning during subsequent discussion. The students' performance in resolving these disagreements is in striking contrast to the performance of the very same students in interviews and in the classroom.

John, and soon Peter, disagreed with Anne about two aspects of the diagram. In the first, Peter maintained that "How much exercise" could be reduced to the constituents "How hard exercise" and "How long exercise." Anne expressed that "how much exercise" was correct—that Peter's interpretation was not right. "How much exercise" appears undifferentiated in

---

[4]Christine Théberge Rafal identified and interpreted this interaction.

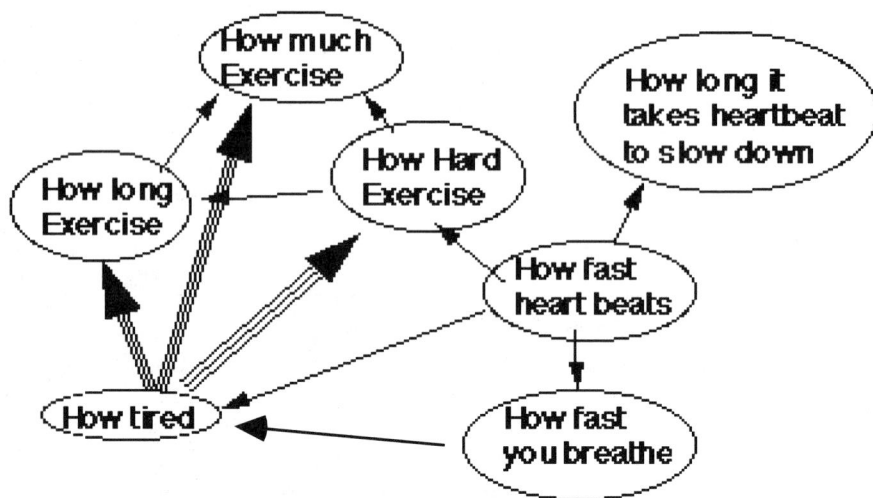

FIGURE 9.4 Anne, John, and Peter's Co-Constructed Diagram.

Anne's original model. "How hard exercise" and "How long exercise" were distinctions contributed by John and Peter.

We were unable to facilitate resolution, and the discussion reached an impasse. Anne then refocused work on another relation in the causal model, the relationship between how long it takes the heartbeat to slow down and how fast the heart beats. She stated that the amount of time required for the heart to make the transition from its running equilibrium rate to its resting equilibrium rate would be the same as going from running rate to walking rate. John and Peter again disagreed with her. We then asked students to make predictions about how heart rate would change over time and to draw what they were saying in terms of graphs of heart rate vs. time. Anne and Peter expressed competing predictions.

Anne: If you walk, it doesn't matter if you walk or you rest, it's gonna be the same, so they should be about, in the same line.

Peter: But I'm saying though, say your heart rate like slows down this fast, it'll go there sooner than it'll go to there.

Peter drew his graph (Figure 9.5) and pointed out that in order for the heart rate to drop from the running rate to the resting rate, it must first pass through the walking rate. Peter's logical argument became an opportunity to resolve the different predictions that Anne and Peter were making at a theoretical level. The basis of Peter's argument is similar to what Clement has called *rational evaluation*. In a study of professional scientists' construction of models, Clement (1989) describes rational evaluation as a process through which hypotheses are supported or disconfirmed depending on their consis-

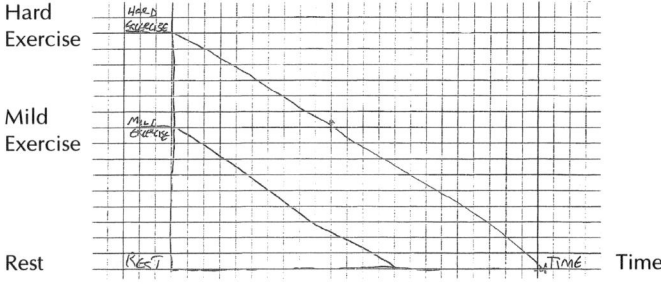

FIGURE 9.5 Peter's Graph.

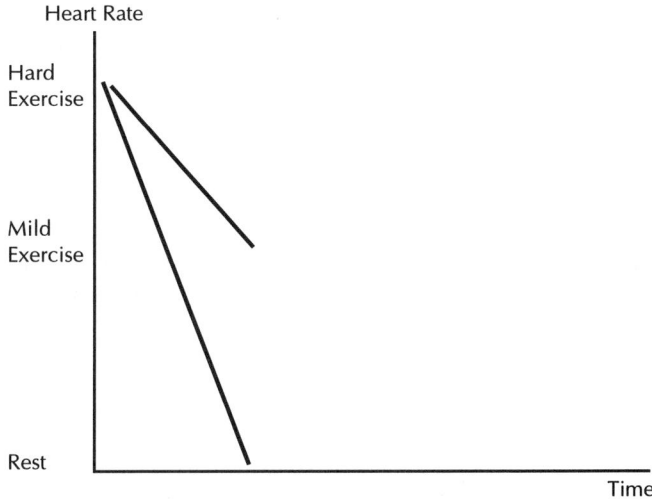

FIGURE 9.6 Peter's Graphical Interpretation of Anne's Argument.

tency with other established theories. In contrast, *empirical testing* is the process by which models are compared to experimental evidence.

Peter's argument rested upon his mathematical knowledge of graphing. He drew lines showing heart rate decreasing linearly with time for both hard-exercise and mild-exercise initial conditions. He placed, on the line for hard exercise, a mark that coincides with the rate for mild exercise, supporting his claim that as the heart relaxes to its resting rate, the heart rate will pass through the value for mild exercise. Insofar as Peter's argument draws not on experimental evidence but on an understanding of mathematics, its form is consistent with Clement's notion of rational evaluation.

Anne disagreed and claimed that the time for the heart to relax would be independent of the starting heart rate. After some verbal discussion that led to no resolution, Peter drew a graph to interpret what Anne was saying (Figure 9.6).

R2: Anne is saying it'll get to the same, the walking and resting heart rate, at the same time.

Peter: Anne is saying it'll go to there, say, and like that.

Anne: Say . . . no, that's not . . . I am saying that it will take about the s . . . I can't really express it in a graph.

Peter: You're saying it will take the same amount of time, right . . .

R1: OK, well . . .

R2: Maybe your graph isn't set up right.

Anne: Not set up . . .

Peter: So running . . . rest . . .

R1: Go ahead . . .

Anne: I don't think the graph is set up right. It doesn't make sense. It doesn't show what I'm saying.

Anne objected to the graph Peter drew when she echoed R2's comment "I don't think the graph is set up right." To us, however, Peter's graph seemed to show precisely what Anne was saying. Anne continued to protest the graph, looking to the researchers to support her, as though to appeal to their authority. Anne's earlier performance with graphing tasks indicated to us that her objection could have arisen because she had not mastered graph interpretation. Peter made several more attempts to reconcile the situation.

Peter drew a graph and said, "You sound like you're saying that. . . ." Peter was trying to make sense of Anne's argument in a graph and was proposing another re-presentation of her view. Anne continued to protest the graph: "I don't like that graph. It doesn't express what I am saying." We attempted to reconcile by suggesting the students do an experiment, but the conversation had a momentum of its own. Unlike the earlier teacher-centered discourse that appeared in the classroom experiments, we adults seemed to have far less influence on the direction of this conversation. As we tried to focus on empirical testing to resolve the conflict, Peter and Anne continued to ignore us and stayed focused on the graphing issue. As the conversation continued, we became convinced that Anne's disagreement was stemming from her difficulty interpreting Peter's graphs.

Anne: But I don't know what kind of graph to do. I don't think a graph really expresses it . . . that curve doesn't work.

Failing to redirect the conversation, one of us (R1) attempted to describe how the graph that Peter drew was consistent with Anne's verbal argument. R1's description of the slope of the graph appears understandable to Peter, but not to Anne.

R1: . . . hard exercise to walking and this is going from hard exercise to rest, so your pulse rate, here, as time's going on, your pulse rate's dropping.

Peter: Well, yeah.

R1: And so if it takes the same amount of time . . .

Peter: So you think it's dropping slower when it's going to . . .

R1: . . . then your heart rate's dropping faster when you're going from hard exercise to rest. Does that make sense?

R1: So it's slowing down, when you, you're doing this hard exercise, and when you stop and rest, your heart beat slows down faster than if you go from hard exercise to mild exercise. That's what that graph is saying. It's different from that graph.

Peter: That's what it sounds like to me she's saying.

Anne: That's not what I'm saying.

R1: OK what do you say?

Anne: I'm saying that it doesn't matter, it's gonna take, I say it doesn't matter if you go to mild exercise or to rest it's still going to be the same amount of time.

Peter: Yeah I know, but that's what this says. This is the amount of time and the amount of time is the same.

R1: Right.

Anne: I suppose.

Anne's comment "I suppose" carries a sense of resignation. One might ask whether it was Peter's final reinterpretation of the graph in a manner that Anne could understand, or perhaps the researcher's agreement, that persuaded her, but that would be missing the essential features of an emerging system of activity in which the motivation and ideation emerged jointly between the adults and the students.

The social and cognitive dissonance that started with the creation of the causal diagram brought both adults and students to participate together in changing Anne's point of view. No one conducted rational evaluation alone. The process was distributed through the group and mediated by the graph. It is in moving from this description of activity as a social system to Anne's acceptance of the graph that we begin to understand how social activity is related to individual learning. Perhaps Anne is finally convinced because she understands Peter's logical argument. Or perhaps the authority of the researcher as an adult led her to relinquish her stance.

The discussion focused on the details of the mathematical graph during the entire episode and did not appear confrontational to us, yet there was a resistive tone in Anne's voice. R1 finally tried to rearticulate Peter's point about how this graph represents Anne's argument.

R1: See the amount of time is the same. Because that's what, this graph, I should really label the graph. This is heart rate. This is heart rate, that's the number that's on the graph and over here is time. So if you, so you're starting over here where you're exercising hard and you say OK, let's make the change right away. And time's proceeding and you're checking your pulse rate and your pulse rate's dropping. Here's where your. . . . This is the curve where you go from hard exercise to mild exercise . . .

Anne: Right.

R1: . . . but it takes the same time as going from hard exercise to resting.

Anne: Yeah. That's what I'm saying.

Peter has already expressed how Anne's argument is conveyed by his redrawn graph, yet it is not until R1 rephrased what Peter said that Anne finally consented—we doubt that she was convinced. Anne may have been following a convention she brought from participating in other activity systems, home and school, that students learn from an authoritative figure rather than from another student. This view, and/or her inability to interpret graphs, may have prevented Anne from considering and accepting at first the graph that Peter drew for her.

Peter's initiative afforded the group an opportunity to engage collectively in rational evaluation, reflecting on the validity of both Anne's and John's opinions. This can be identified in the phrases where the researchers and the students are saying, "I think this is what you are saying." Would any of the students have engaged on their own in this reflective process, or in any of the other processes in Clement's model? It does not seem likely, although we hope to have provided a plausible *existence proof* that the processes can be distributed socially, with the participation of adults.

We finally intervened successfully, to ask the students if their ideas could be tested, and then engaged them in designing experiments to test the relation in question. The students drew on the cardiovascular experiments done earlier in school. These were exercises in which they first determined the equilibrium resting, walking, and running heart rates of the human subject. In experiment 1, the subject runs for a period of time and then walks. The students measure the time it takes for the pulse rate to drop to the walking rate. In experiment 2, the subject runs for a period of time and then rests. The students then measure how long it takes for the heart rate to drop to the resting rate. For each case, the students waited for the heart rate to drop to the resting pulse rate before determining the next rate.

Just prior to the experiments, we prompted the students to make predictions about how heart rate would change over time in terms of graphs of heart rate vs. time. Surprisingly, Anne drew a graph (Figure 9.7) nearly identical to the Peter's earlier interpretation of her argument! In the 1-week interval between the discussion of the graph and the development of the experiments, Anne appeared to have accepted Peter's graphical interpretation. Yet the two students still disagreed in their predictions.

A fourth student, Brian, attended for the first time. Not having been in previous lessons, Brian was unable to participate equally with the other students. Recognizing this, the students made him the subject of the experiments the others had designed the lesson before. The students used stethoscopes and stopwatches, as they did in their science class, to make their measurements. The students had some difficulty when they realized that

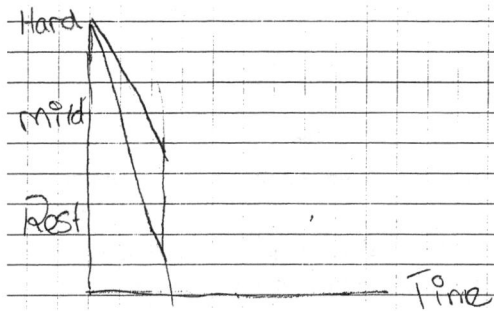

FIGURE 9.7 Anne's Final Graph.

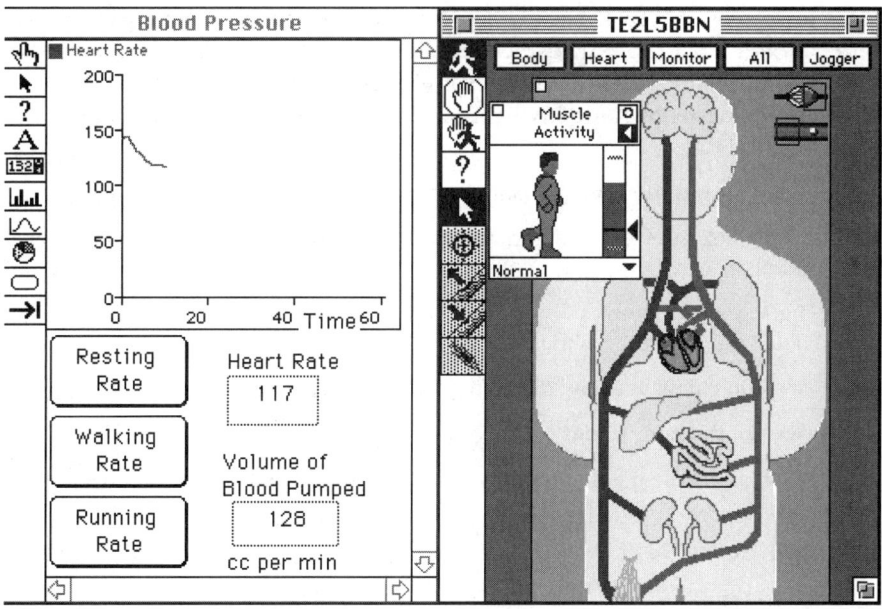

FIGURE 9.8 Explorer: Cardiovascular System Model.

Brian's heart rate was dropping significantly over the 15-second interval they were using to measure his heart rate. Peter and John became disappointed with the experiments.

The students reached another impasse. We adults knew that shortening the time interval would not solve the problem, because the measurement uncertainty would increase as fewer heartbeats were counted. Without access to more sophisticated tools such as an electrocardiogram, a computer model seemed appropriate for continuing the investigation. Thus we provided the Explorer: Cardiovascular model to investigate the phenomena further. The simulation includes a theoretically calculated heart rate that is displayed continuously and thus does not have the limitations that appeared in the students' measurements. The simulation interface was modified to look like that shown in Figure 9.8.

R1: How do you guys think . . . Does it seem possible to do the experiment that we were doing before this way?

All: Um huh.

R1: What do you think we have to do now to test?

Anne: Go from running to walking and then go from running to resting and see how long it takes.

In contrast to the students' earlier experiences with computer models in the clinical interviews and in the classroom teaching experiment, the students' use of the Cardiovascular model appeared purposeful and meaningful —the model was being used to test the hypotheses they had developed together earlier. The students controlled the variables, but not because they were told to do this by their teacher. Instead, the students came to the computer model with the purpose of resolving a disagreement in their predictions. When Anne found that the simulation did not support her point of view, the students become involved in a discussion about the validity of the experimental data:

Anne: I don't know, this one [the simulation] I think is better than Brian's.

R1: Do you think this one is better than Brian's, even though this one doesn't do what you predicted?

Anne: I guess so, it doesn't do what anybody predicted, really. Nobody predicted it would do all different times. Brian did what I predicted.

Peter: We don't actually know if he did what Anne predicted.

Anne: We got the same thing.

John: What about in an experiment you can't be sure about some data on some experiments.

R1: Why is that?

Anne: Why John?

John: Because in most experiments Brian's heart could have slowed down quickly in the 15 seconds. . . . You know it has to do with timing of the stuff that things could go wrong that you know and you record data that might be wrong. So, the good thing to do in an experiment is to do it more than once.

We observe that the students do not distinguish between testing their predictions through experiment and testing their predictions with the computer model. Is it reasonable to expect students to make these distinctions? Kuhn (1989) has shown that it is hard for students to delineate between theory and evidence at the middle-school level. Our results are puzzling and ambiguous, because the students *did* differentiate between the predictions of their causal models and the data, as indicated by the foregoing dialogue. Furthermore, John recognized the limitations imposed by the finite time interval over which they accumulated their heart rate data. Could the students have made the epistemological distinction between the computer model and evidence?

To do so would mean understanding the nature of the computer model—that it is constructed in congruence with scientific theory. We did not make explicit attempts to help students understand computer models in this way.

Technically, the students' comparison of their predictions, based on their causal model, to the Explorer: Cardiovascular Model is an example of rational evaluation. That is, the students compared their predictions to those made from culturally acquired theoretical knowledge as programmed in the model and played out through the simulation. We view the model as a valuable cultural artifact to help the students learn the theory through interacting with it, but our view is not generally shared within the science education community.

Although modeling is recognized across many disciplines, such as physics (Hestenes, 1992) and biomedicine (NRC, 1985), and has spawned modern computational science (Kaufman and Smarr, 1993), neither rational evaluation nor (more generally) modeling is considered "real science" in many classrooms (Barowy and Laserna, 1997). Some science teachers consider computer simulations vicarious experiences. We suggest that this is because the populace views science as proceeding primarily by testing theories against experimental evidence. The belief has roots in science methodology (Popper, 1959; Lakatos, 1978) and appears as the "scientific method" in textbooks, which are used much more commonly in schools than are computer models.

Nature of Models

We began and ended this study by conducting the Nature of Models interview with individual students in a clinical interview setting to assess their conceptions of models and of how models are used in science. The full Nature of Models interview is described in detail elsewhere (Grosslight *et al.*, 1991) and so is only briefly described here.

The interview begins by asking students general questions about the nature and purpose of science. What is the reason for doing science? How do you think scientists do that? Students are then given a series of questions about models, which they answer in their own words. Are there different kinds of models? What are models for? Can you ever change a model? Next, students are presented with four examples and asked if these are models. We presented students with a toy airplane, a graphic depicting a house, a map of the local public transportation rail, and a diagram of the earth's hydrogeologic cycle. Finally the students are asked: Can you use models in science? Do scientists make up models?

By conducting the interview before and after the classroom experiment and guided inquiry, we were able to gauge whether the students' knowledge of models improved as a result of our interactions with them. The students' views did not change; they maintained a simple *copy epistemology*. Students with a copy epistemology express the view that a model is a copy, albeit sometimes smaller, of the original. Models are merely physical copies of reality that are intended to "show off stuff." This is to be compared to the more expert view that models are constructed re-presentations that may embody different theoretical perspectives. Not surprisingly, the *copy* view is consistent with the way most students presently come into contact with models—what they see in science museums, in science classroom demonstrations, or in science education manipulables.

Reviewing the videotapes of the interview induces discomfort. The students sometimes sit with their hands or arms folded, or hunched over, and they often glance at the hands and face of the interviewer, as though looking for clues that will help them answer the questions. Granted, it is not the purpose of the interviews to make the participants comfortable. Yet, even though an interviewer may try to act relaxed and take actions such as sitting on the same side of the table as the student, the design of the interview for a one-way flow of information creates an environment that seems artificial to all participants through the entire process.

The following dialogue is indicative of the discourse style that is typical of the interview.

R: How about in science. Do you think scientists can have more than one model for the same thing?

S: I think so.

R: Could he have different models for the same thing?

S: I'm pretty sure he could.

R: How could they be different?

S: I don't know.

Participating in the interview also feels far from being natural. The discourse style is that of repeating blocks of [I-R], the researcher *Initiating* a question to which the student *Responds*. It is not clear to us whether the responses of the student represent minimal knowledge of models or the student's attempt to minimize the chances of giving a wrong answer to an authority figure. Similar strategies are employed in classroom contexts. In *Talking Science*, Lemke (1990) notes that in classroom IRE (triadic) dialogue, students often *call out* answers to questions posed by a teacher, before the teacher selects them to give an answer. Doing so "tends to reduce anxiety about speaking formally to the class and being on the spot if you're wrong."

Our interviews with these middle-school students are consistent with Grosslight's findings. Yet the students were able to apply system and computer models, much as scientists use models, during guided inquiry. Without being prompted, they engaged in discussion about the validity of the causal model with which they were working, and they engaged in testing it. These students were also able to use the expert-built Cardiovascular model to investigate situations where they had difficulty in obtaining direct data.

The activity system in which a student participates—with the associated resources, routines, expectations, and adult support—shapes his or her actions. An interview constitutes a relatively barren activity system. There are no tools to support student thinking, and the design prohibits the adult participation that might suggest possibilities for the students to pursue. Consequently, the interview should not be viewed as a measure of students' ability to engage in modeling in other contexts.

Discussion

We have conducted an exploratory investigation of science student learning mediated by expert-constructed computer models and student-constructed conceptual models across several situations. We have looked carefully at the interdependence among motives, technological artifacts, and social interactions. We see these dimensions coming to play through influencing what students are able to do and what they know. Student interest in learning, the actions afforded by the computer tools, and social conventions come together as a system of interacting, interdependent elements, shaped by a history of participating in other systems of activity, and with tools shaped by a history of scientific endeavor.

In the guided inquiry experiment, where the goal of investigation was negotiated with the students, we were able to achieve the most powerful results by creating a system of activity that nurtured student investigation. Although the students did not recognize explicitly, nor could they articulate, the epistemological status of models, as in the Nature of Models interview, they were able to build and use models in effective and sophisticated ways. It is an untested hypothesis that if students could better understand the nature of models, then they would be able to benefit even more from interacting with computer models.

In contrast, students had difficulty forming coherent goals for investigation alone, as in the open-ended investigations. Furthermore, when adults provided complete goals for students, as in the classroom experiments, the students participated marginally in an authority-based and teacher-centered structure. Consequently, the classroom provided students with experience in controlling variables, a skill understood to be important by many science educators, yet the students essentially did what they were told. When left to their own, as in the open-ended investigations, they were unable to demonstrate these skills completely, never having practiced them in an authentic context.

Our results indicate that the ability to develop a causal model and to use a scientific model did not arise from the talent of any individual student. Rather, the determination of causal model validity occurred in the interaction between the students and adults within a system of activity that the adults helped to establish. The system enabled students to make the moves they were unable to make in the classroom activity or to demonstrate as individuals in the interviews. The open-ended interviews and the Nature of Models interviews provide data about what the students knew and could do alone — but these were interview situations, not settings in which the students were truly supported in pursuing their own goals.

Although this phenomenon is captured in the core idea of zone of proximal development (or zoped), we have attempted to explicate how one defines and establishes a zoped as a system of activity. Scientific inquiry in the classroom is not the same as "discovery learning." In our example of guided inquiry, the goals, the questions, and the means to pursue them are the purview of neither the adult nor the student alone but, rather, are effectively established in the nexus of social interaction and the tools available. Adult intervention influenced the students to test their ideas with experiments and computer models — forms of empirical testing and rational evaluation in our eyes, though certainly not in the eyes of the students.

When disagreement arose from different expressed understandings, or socio-cognitive dissonance, the adults did not use their authority to suppress it but instead sought to explore and resolve those differences through scientific inquiry. Although cognitive conflict is a term associated with Piaget, we prefer socio-cognitive dissonance, because it emphasizes reconciliation in the social plane, rather than on an individual level. In addition, conflict carries an antagonistic meaning, when indeed we see the respectful consideration of differences in understanding as a useful precondition for engaging in inquiry.

The replacement of "authority based on power" with "authority based on competence" in guided inquiry affords essential elements of Piaget's theory of peer interaction in synergy with Vygotsky's apprenticeship learning model.

An important adult contribution in guided inquiry was the introduction to the students of causal diagramming. Causal diagramming provided artifacts that facilitated and mediated discussion. The artifacts included not only the drawings themselves but also the conventions for drawing them – a way of thinking. The generative capacity of causal diagramming, along with the ease with which it is learned, helped students formulate and articulate their beliefs, which then became the focal point for investigation.

Although the Explorer expert-built Cardiovascular model did not offer the same generative capacity as causal diagramming, it was sufficient for testing ideas as a form of rational evaluation. This tool does not support the ideation that we have found can engender socio-cognitive dissonance and that can subsequently lead to guided scientific inquiry. As an artifact that manifests cultural knowledge, however, it can be valued for the information that it contains and for the ways in which information may be sought interactively. Students may check their ideas against what is known, much as a scientist does when researching other scientists' work, or as modern computational scientists do in playing out the consequences of a model of a complex phenomenon.

Students did not attempt to evaluate the computer model per se. Why should they? We, as researchers, made the computer model available when the experiment provided unsatisfactory results. The authority of the model, like the authority of researchers as adults, was perhaps historically unquestionable. Even as Anne's prediction disagreed with the model, she did not wish to test the computer model itself. The transfer of status, expectations, routines, and scripts from the activity system of schooling, as seen in the Mr. Driscoll dialogue, can interfere with the emergence of new standards of activity. These rules for participation and their connections to authority and the use of language in schools are described in detail by Lemke (1990).

Furthermore, despite the students' experience in modeling, it must be recognized that they did not learn about modeling processes, as indicated in the concluding Nature of Models interview. This is not surprising. The adults unsuccessfully attempted to pay attention to this topic during the guided inquiry phase, but the students were focused on the differences between their predictions and were not responsive. We wish to pursue Grosslight's contention that students need more experience discussing the role of models in scientific inquiry.

Conclusions

This chapter has presented an exploratory study of science student learning with modeling tools. We provide evidence that meaningful building of models, and use of expert-built science computer models for student learning, can occur during guided inquiry, despite contrary evidence in other situations. We envision the design of activities and social interactions *using* computer models, together with the design of the computer models *per se*, for learning science as a *system*. The system forms a set of conditions and capabilities that we have observed to be useful in stimulating scientific inquiry with middle-school students. Inquiry learning with scientific models, along with assessment of that learning, includes the negotiation of learning goals and the distribution of cognitive effort. Like Cole and others (LCHC, 1989), we have pursued learning as "complex activity . . . in meaningful contexts."

The complex activity and meaningful contexts emerge in a system in which motive is negotiated. Students practice scientific inquiry when the questions and goals emanate from their efforts, as *guided and supported by adults*. We have found that facilitating student ideation, which we did in this study with causal diagrams, helps students express their personal knowledge of the system being studied. Because knowledge and prior relevant experiences differ from one student to the next, individual creations are distinct, and socio-cognitive dissonance emerges. These differences in understanding provide strong motives for students to engage in inquiry about the topic, if that activity is part of an established system of expectations and routines that provides the requisite cognitive and affective support.

The role of adults, or of more expert participants generally, is to recognize promising socio-cognitive dissonance; nurture it; suggest methods, approaches, and appropriate tools to resolve it; and facilitate an inquisitive, creative, and respectful atmosphere. Adults can guide discussion to ensure that it is meaningful, is purposeful, and leads to significant learning, even if not to resolution of the problem. Several approaches, such as "Convergent conceptual change" (Rochelle, 1992), "sense-making" (Newman *et al.*, 1994), and "paradox resolution" (Horwitz and Barowy, 1994), suggest this direction. Despite students' admirable performance during our guided inquiry sessions, the notable absence of expressed understanding about models and of any systematic treatment of models as provisional re-presentations, indicates that adults would also do well to address the nature of models in their discussions with students.

What is the point? The performance of the students during guided inquiry stands in stark contrast to their performance in the classroom experiment, the latter being closer to most practices of school science. It should be noted, however, that we have observed classroom interactions similar to any of those we have presented. In our experience, much work is to be done. Computer tools to support student inquiry should be created not alone, but together with contexts and activities that are meaningful to the students. Sarason (1972, 1996, 1997) frames the objective generally as the "creation of settings." That some essential precursors are the negotiation of goals and the distribution of cognition means that the latter cannot be done for classrooms but must *include students and teachers*.

Furthermore, engaging in the iterative constructing of models, as scientists do, is contrary to the traditional notion of curricula as material to be delivered. Indeed, we hope we have presented compelling evidence of how traditional curricula can militate against student involvement in inquiry.

Creating settings such as guided inquiry will require significant changes in science education practice and will create major challenges. We can begin by convincing all involved that these processes are legitimate. We can increase awareness and understanding of many matters, including how to foster scientific inquiry through modeling. Indeed, we can all better understand what scientific inquiry itself really is.

Acknowledgments

It is a pleasure to thank Christine Théberge Rafal, John Richards, George Blakeslee, and Lorraine Grosslight, who worked on the project. We are grateful to Cliff Konold and Rolfe Windward for extensive discussion of the text, both face-to-face and virtual. We appreciate the time Katherine Goff, Phillip Allen White, Jerry Balzano, Eva Ekeblad, David Dirlam, and Judy Diamondstone took to read the manuscript and offer many valuable comments, which we have tried to address in the available time.

This project was supported, in part, by the National Science Foundation, Grant No. 9153871. Opinions are those of the authors and are not necessarily those of the Foundation.

References

American Association for the Advancement of Science, Project 2061. 1993. *Benchmarks for science literacy*. New York: Oxford University Press.

Barowy, W., & Laserna, C. 1997. The role of the Internet in the adoption of computer modeling as legitimate high school science. *The Journal of Science Education and Technology, 6(1)*: 3-13.

Brown, A. 1992. Design experiments: Theoretical and methodological challenges in creating complex interventions in classroom settings. *The Journal of the Learning Sciences, 2(2)*, 141-178.

Carey, S., Evans, R., Honda, M., Jay, E., & Unger, C. 1989. An experiment is when you try it and see if it works: A study of grade 7 students' understanding of the construction of scientific knowledge. *International Journal of Science Education, 11*, 514-529.

Cazden, C. 1988. *Classroom discourse: The language of teaching and learning*. Portsmouth, NH: Heinemann.

Clement, J. 1982. Students' preconceptions in introductory mechanics. *American Journal of Physics, 50*, 66-71.

Clement, J. 1989. Learning via model construction and criticism. In Glover, G., Ronning, R., & Reynolds, C. (eds.), *Handbook of creativity: Assessment, theory and research*. New York: Plenum Press, pp. 341-381.

Clement, J., Brown, D.E., & Zietsman, A. 1989. Not all preconceptions are misconceptions: Finding "anchoring conceptions" for grounding instruction on students' intuitions. *International Journal of Science Education, 11* (Special Issue), 554-565.

Cole, M. 1997. Private communication.

Cole, M. 1996. *Cultural psychology*. Cambridge, MA: The Belknap Press of Harvard University Press.

Cole, M., & Scribner, S. 1974. *Culture and thought: A psychological introduction*. New York: Wiley.

Collins, A. 1992. Toward a design science of education. In Scanlon, E., & O'Shea, T. (eds.), *New directions in educational technology*. Berlin: Springer-Verlag, 1992.

Collins, A., Brown, J. S., & Newman, S. 1989. Cognitive apprenticeship: Teaching the crafts of reading writing and mathematics. In Resnick, L. B. (ed.), *Knowing, learning and instruction: Essays in honor of Robert Glaser*. Hillsdale, N.J.: Lawrence Erlbaum.

diSessa, A. 1983. Phenomenology and the evolution of intuition. In Gentner, D., and Stevens, A. (eds.), *Mental models*. Hillsdale, NJ: Lawrence Erlbaum, pp. 15-34.

Driver, R. 1987. *Approaches to teaching plant nutrition*. Children's Learning in Science Project, Centre for Studies in Science and Mathematics Education, University of Leeds.

Driver, R. 1989. Students' conceptions and the learning of science. *International Journal of Science Education, 11*, 481-490.

Driver, R., and Bell, B. 1986. Students' thinking and the learning of science: A constructivist view. *School Science Review, 67(240)*, 443-456.

Duffy, M., and Barowy, W. 1985. Effects of constructivist and computer-facilitated strategies on achievement in heterogeneous secondary biology. NARST 1995 Annual Meeting, San Francisco.

Glaser, B.G., and Strauss, A. L. 1967. *The discovery of grounded theory*. New York: Aldine De Gruyter.

Grosslight, L., Unger, C., Jay, E., & Smith, C. L. 1991. Understanding models and their use in science: Conceptions of middle and high school students and experts. *Journal of Research in Science Teaching, 28(9)*, 799–782.

Hawkins, J., & Collins, A. (1999). Design experiments: Evaluating the role of technology in supporting school change. In Hawkins, J., & Collins, A. (eds.), *Design experiments: Using technology to restructure schools*. New York: Cambridge University Press.

Hestenes, D. 1992. Modeling games in the Newtonian world. *American Journal of Physics, 60(8)*, 732–748.

Horwitz, P., & Barowy, W. 1994. Designing and using open-ended software to promote conceptual change. *The Journal of Science Education and Technology, 3(3)*, 161–185.

Jackson, S. L., Stratford, S. J., Krajcik, J. S., & Soloway, E. 1996. Making dynamic modeling accessible to pre-college science students. *Interactive Learning Environments, 4(3)*, 233–257.

Lakatos, I. 1978. *The methodology of scientific research programmes*. Cambridge, England: Cambridge University Press.

LCHC. 1989. Kids and computers: A positive vision of the future. *Harvard Educational Review, 59(1)*, 50–86.

Lemke, J. 1990. *Talking science: Language, learning, and values*. Norwood, NJ: Ablex Publishing.

Leont'ev, A. N. 1975. The problem of activity in psychology. *Soviet Psychology, XIII(2)*, 4–33. (Translation from *Voprosy filosofii*, 1972, no. 9, 95–108.)

Mandinach, E., & Thorpe, M. 1988. *The systems thinking and curriculum innovation project: Technical report*. Cambridge, MA: Educational Technology Center, Harvard Graduate School of Education.

McArthur, D., & Lewis, M. 1991. Overview of object-oriented microworlds for learning mathematics through inquiry. *A RAND Note*. Santa Monica, CA: RAND.

Mehan, H. 1979. *Learning lessons*. Cambridge, MA: Harvard University Press.

National Research Council. 1985. *Models for biomedical research: A new perspective*. Washington, DC: National Academy Press.

National Research Council. 1996. *National science education standards*. Washington, DC: National Academy Press.

Newman, D., Morrison, D., & Torzs, F. 1993. The conflict between teaching and scientific sense-making: The case of a curriculum on seasonal change. *Interactive Learning Environments, 3(1)*, 1–16.

Papert, S. 1990. *Introduction to constructionist learning: A fifth anniversary collection of papers*, ed. I. Harel. Cambridge, MA: M.I.T. Media Laboratory.

Popper, K. R. 1959. *The logic of scientific discovery*. London: Hutchinson.

Resnick, M. 1994. *Turtles, termites, and traffic jams: Explorations in massively parallel microworlds*. Cambridge, MA: M.I.T. Press.

Richards, J., Barowy, W., & Levin, D. 1992. Computer simulations in the science classroom. *Journal of Science Education and Technology, 1*, 67–79.

Riley, D. 1990. Learning about systems by making models. *Computers in Education 15(1–3)*, 255–263.

Roschelle, J. 1992. Learning by collaborating: Convergent conceptual change. *The Journal of the Learning Sciences, 2(3)*, 235–276.

Roberts, N. 1978. Teaching dynamic feedback thinking: An elementary view. *Management Science, 24(8)*, 836–843.

Roberts, N., Blakeslee, G., & Barowy, W. 1996. The dynamics of learning in a computer simulation environment. *Journal of Science Teacher Education*, *7(1)*, 41-58.

Saferstein, B., & Souviney, R. 1996. Secondary science teachers and the Internet: Community of Explorers project. Submitted to *The Journal of Computing and Teacher Education*.

Sarason, S. B. 1972. *The creation of settings and the future societies*. London: Jossey-Bass.

Sarason, S. B. 1996. *Revisiting "The Culture of the School and the Problem of Change."* New York: Teachers College Press.

Sarason, S. B. 1997. Revisiting the creation of settings. *Mind, culture and activity*, *4(3)*, 175-182.

Saxe, G. B. 1991. *Culture and cognitive development: Studies in mathematical understanding*. Hillsdale, NJ: Lawrence Erlbaum.

Schauble, L., Klopfer, L.E., & Raghavan, K. 1991. Students' transition from an engineering model to a science model of experimentation. *Journal of Research in Science Teaching*, *28*, 859-882.

Schwartz, D. 1993. The construction and anological transfer of symbolic visualizations. *Journal of Research in Science Teaching*, *30(10)*, 1309-1326.

Stubbs, M. 1983. *Discourse analysis: The sociolinguistic analysis of natural language*. Chicago: University of Chicago Press.

Sutton, C. 1993. Figuring out a scientific understanding. *Journal of Research in Science Teaching*, *30(10)*, 1215-1227.

Vygotsky, L. S. 1978. *Mind in society: The development of higher psychological functions*. Cambridge, MA: Harvard University Press.

Wells, G. 1996. Using the tool-kit of discourse in the activity of learning and teaching. *Mind, Culture and Activity*, *3(2)*, 74-101.

Wertsch, J. 1991. *Voices of the mind: A sociocultural approach to mediated action*. Cambridge, MA: Harvard University Press.

White, B. 1993. ThinkerTools: Conceptual change and science education. *Cognition and Instruction 10*, 1-100.

White, B., and Horwitz, P. 1988. Computer microworlds and conceptual change: A new approach to science education. In Ramsden, R. (ed.), *Improving learning: New perspectives*. London: Kogan Page.

10

Alternative Approaches to Using Modeling and Simulation Tools for Teaching Science

Barbara Y. White
Christina V. Schwarz

Introduction

Computer modeling and simulation software are transforming the way science and engineering are done. They make possible analytic and conceptual tools that allow scientists to employ new forms of analysis, engage in new kinds of thought experiments, and create new types of theories. In this chapter, we illustrate how such computer-based tools can also transform the practice of science education. We describe how modeling and simulation tools, such as those embodied in our ThinkerTools software, facilitate a variety of instructional approaches that attempt to realize the increasingly ambitious and varied goals being advocated for modern science education. These goals include engaging young students in authentic scientific inquiry in which they learn about the nature of scientific models and the processes of modeling. They also include enabling students to learn abstract and complex subject matter at increasingly younger ages.

Over the past 20 years, we have been developing and evaluating alternative, computer-enhanced approaches to science education. Our approaches have evolved to achieve differing pedagogical goals, which have been motivated by changing visions of what is important in science education and the role that technology can play in this reform process. In what follows, we present three successive versions of our ThinkerTools software and curricula in which the emphasis shifts from developing students' understanding of the subject matter to enhancing their inquiry expertise and finally to refining their expertise in modeling.

Early ThinkerTools Research: Intermediate Causal Models, Conceptual Change, and a Middle-Out Approach to Science Education

The focus in our early research was primarily on conceptual change and on enabling young students to acquire relatively abstract and complex theories of force and motion phenomena (Horwitz, 1989; White, 1981, 1984, 1993a; White and Horwitz, 1988). Our goal was to counteract the commonly held view that learning physics is beyond the reach of most students. We argued that students' difficulties in understanding physics arise from deficiencies in the traditional definition of knowing physics, with its focus on algebraic equations and solving quantitative problems, and from deficiencies in traditional approaches to teaching science in which students listen to lectures and solve quantitative problems for homework. We further argued that reconceptualizing what it means to understand physics, as well as how it is taught, can help make the subject accessible and interesting to a wide range of students, including younger and lower-achieving students.

Computer models and simulations were important in our reconceptualization process. What the first author calls "intermediate causal models" played a central role (White, 1989, 1993b). Intermediate causal models can be characterized as applying the laws of physics, in causal form, to predict what happens as events occur. They employ various visual representations to depict the resulting sequence of behaviors at an intermediate level of abstraction. This type of model can be embedded in a computer simulation that can explain its reasoning verbally (using synthesized speech) and can depict its behavior visually and dynamically. Intermediate causal models can also be internalized by students in the form of mental or conceptual models. Acquisition of such conceptual models enables students to step through time and events while using laws and representations to predict and explain what will occur.

Figure 10.1 shows an intermediate causal model from the domain of force and motion. This model is useful for predicting how forces affect the motion of an object in a one-dimensional world with no friction. It "reasons" by stepping through time and analyzing what forces are acting on the object. To do so, it uses the laws of physics that have been shown to determine the changes in velocity produced by those forces. The model also uses various diagrammatic representations (as illustrated in Figure 10.1) to encode these changes in velocity.

We find that intermediate causal models provide an excellent vehicle for reforming our conception of what it means to understand and teach science. They enable us to change our view of scientific understanding by providing formal models of reasoning processes, such as causal discrete-state reasoning, and representational forms, such as diagrams. These can play a crucial role in scientific theorizing, but they have not previously been accorded the

Basic Force-and-Motion Principle:

When no impulses are applied to an object, its speed stays the
same, because there is nothing to make it change. But whenever
an impulse is applied, it causes the object to change speed.

Prediction Law:

If the impulse is in the same direction in which the object is moving,
it adds 1 to its speed (+1);
An impulse applied in the opposite direction subtracts 1 from its speed (−1).

A Model-Based Prediction:

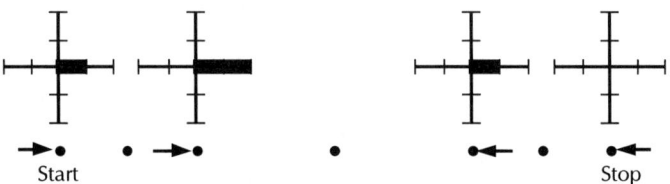

Start   Stop

FIGURE 10.1 A model-based prediction of the effects of a sequence of four impulses
on an object's motion. The "dotprints" show the position of the object as time passes.
The arrows indicate the direction and timing of each impulse. The "datacross" is used
to calculate and show the effect that each impulse has on the object's velocity.

respect they deserve. Intermediate causal models can thus serve as target
conceptual models embodying an important type of expertise that students
need to acquire. They can also change our view of science education, because enabling students to create and experiment with computer simulations
of this type allows for alternative approaches to science education that have
the potential to make difficult subjects, such as physics, interesting and accessible to a wide range of students.

We believed that focusing instruction on intermediate causal models instead of algebraic abstractions or real-world experiments has a number of advantages. First, even young children can make sense of such models, because
it is a small step from this form of representation and reasoning to how one
naturally reasons about real-world phenomena. Furthermore, intermediate
causal models provide an efficient starting point for instruction, because it is
also a relatively small step from this form of representation and reasoning to
algebraic abstractions and constraint-based reasoning, which are useful for
solving many types of problems. Working with intermediate causal models
thus provides a bridge between real-world phenomena and mathematical formalisms. In addition, they are useful in their own right and are particularly
good for predicting and explaining real-world phenomena.

Accordingly, we developed an instructional approach in which the emphasis is on experimenting with computer models with the goal of creating explicit, written-down conceptual models (like the one shown in Figure 10.1). As illustrated in Figure 10.2, this can be characterized as a "middle-

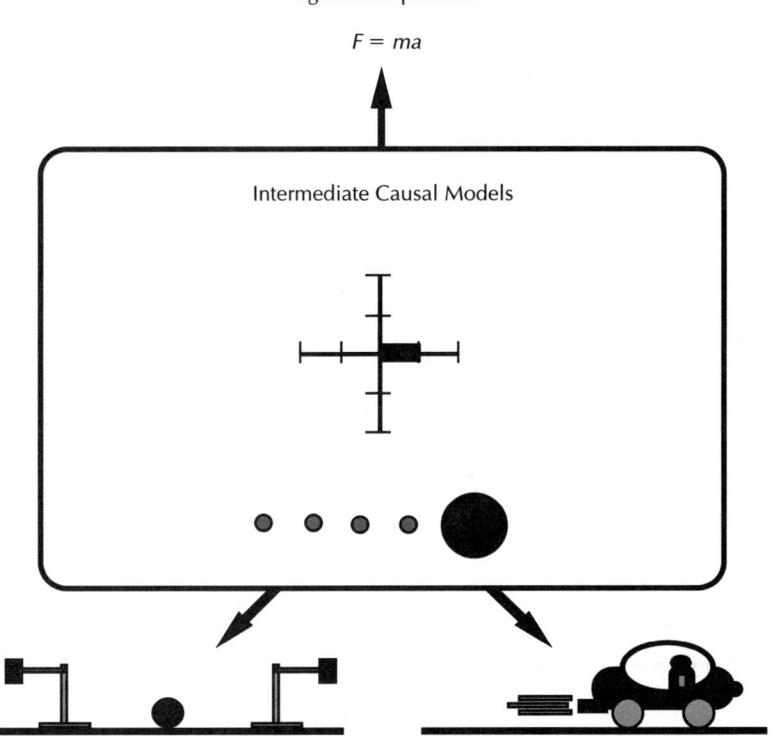

FIGURE 10.2 Casual models that portray sequences of events at an intermediate level of abstraction provide a bridge between real-world phenomena and abstract mathematical formalisms.

out" approach to science education, in which students start by working with and understanding intermediate causal models and then connect these models to real-world phenomena *and* to more abstract formalisms. This middle-out approach places far less emphasis than traditional instruction on doing real-world experiments or solving quantitative problems.

The ThinkerTools Computer Microworlds

In our early ThinkerTools curricula, the children's primary conceptual tools are a sequence of interactive microworlds that embody increasingly sophisticated models for how forces affect motion. The children interact with this sequence of microworlds, which incorporates intermediate causal models that gradually increase in complexity. They begin with a simple one-dimensional world in which there is no friction or gravity. Working with this microworld, the children discover that they can use scalar arithmetic to model force-and-motion phenomena. They then progress to a two-dimensional

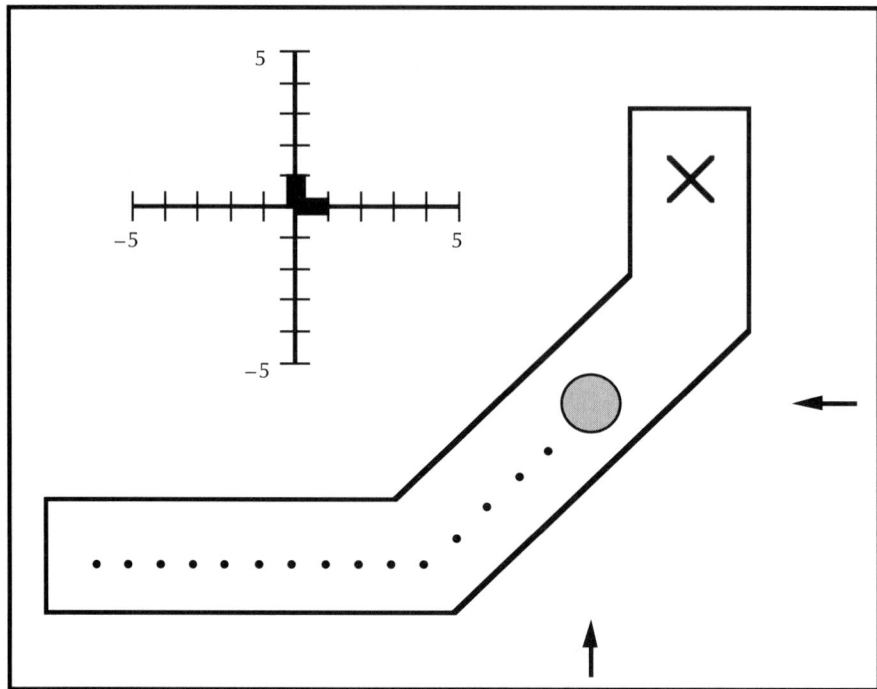

FIGURE 10.3 This game helps students develop a conceptual model of force-and-motion phenomena that incorporates a simple form of vector addition. The screen shot illustrates the representations of motion employed within the microworlds (that is, the moving dot and its dotprints, arrows, and datacross).

world in which they work toward creating a conceptual model that incorporates simple vector addition. Next, they go to a one-dimensional world with gravity. By doing a limit analysis, they learn to model continuous forces, such as gravity, as a series of small impulses closely spaced in time. Finally, they progress to a two-dimensional world in which gravity is acting. Working with this microworld, the children use their conceptual model to analyze and explain the trajectories of objects. This involves stepping through time and applying their laws and representations to predict and encode what will occur.

In working with the microworlds, children are frequently engaged in game-like activities and experiments in which they try to control the motion of an object by applying forces to it. (This object is introduced as a generic object called a dot, which is the pictorial equivalent of a variable; students can map it onto different objects, such as spaceships or billiard balls at different times.) The activity shown in Figure 10.3, for example, is set in the context of a two-dimensional microworld in which there is no friction or gravity. In this game, students apply impulses to the dot so that it navigates the track and stops on the target. (Impulses are forces that act for a limited time, like a hit or a kick.)

The ThinkerTools microworlds incorporate multiple levels of abstraction. For example, there is the dot itself, which moves dynamically around on the screen. A more abstract representation of the dot's motion is provided by "wakes" or "dotprints." These are little dots that the big dot leaves behind at fixed time intervals. They show, by their position and spacing, where the dot has been and how fast it was going. At a higher level of abstraction, the students are encouraged to think of the dot's motion in terms of its components—that is, its speed in the horizontal dimension and its speed in the vertical dimension. (Components are useful for reasoning about complex situations, such as analyzing trajectories.) The components of the dot's velocity are illustrated dynamically by arrows that are constantly pointed at the dot. They are represented more abstractly by a vector representation we call a datacross (shown at the top left of Figure 10.3), which indicates the magnitude of horizontal and vertical velocities.

In the early stages of working with these microworlds, students typically focus on the behavior of the dot and ignore the other representations. The teacher then gives them tasks to illustrate the utility of the more abstract representations. For example, in one such activity the dot is off the screen, and the only way the students can determine its velocity is to look at its datacross and see what effect the impulses they apply have on its velocity. By focusing on the datacross and applying impulses, the students can bring the dot back onto the screen and stop it on the target "X." Through activities like this, students become familiar with the power and utility of such abstract representations.

The Middle-Out Instructional Approach

In creating the original ThinkerTools curriculum, we developed a four-phase instructional cycle that is repeated with each increasingly complex microworld. The four phases in this cycle correspond to cognitive stages in the development of subject-matter expertise: motivation, conceptual evolution, formalization, and transfer. Students start by engaging in thought experiments to motivate their explorations. They then work with a given microworld and try to develop a conceptual understanding of it. Next, they formalize their understanding by creating an explicit rule or model. Then they transfer its use to several real-world situations by generating model-based predictions and explanations for various phenomena. By applying this cycle in conjunction with the microworlds, the curriculum embodies a middle-out approach to science education.

In the first phase of the instructional cycle, the *motivation phase*, the children are asked to do thought experiments in which they imagine what will happen as forces are applied to an object. For instance, in conjunction with the first microworld (which is a one-dimensional world with no friction or gravity), they are asked to predict what will happen in the following situation:

There is an object resting on a table. The table is very smooth, so there is no friction. A blast of air is applied to the object. Then, as it is moving along, a blast of air, the same size as the first, is applied in the opposite direction. What will the second blast of air do to the motion of the object?

The teacher simply listens to the children's answers and reasoning without commenting on their correctness or incorrectness. Some say that the second blast of air will cause the object to turn around and go in the opposite direction. Others say that the second blast of air cancels the first and the object will stop. Still others say that it simply slows the object down but that the object keeps moving in the original direction. The children thus discover that different people may have different hypotheses about what will happen in such situations. These disagreements motivate them to find out who has the correct beliefs about how forces affect motion.

The children then go on to the second phase of the instructional cycle, the *evolution phase*, in which they work in pairs with the computer games and inquiry activities. The games are designed to help them to determine the laws that are governing the behavior of the microworld (White, 1984). The children are asked to predict what they think is going to happen within this world. Then, by playing the games and doing the experiments, they see what actually happens. On the basis of these experiences, they write down laws that they think govern this microworld.

To facilitate this process, the children then proceed to the third phase of the instructional cycle, the *formalization phase*. In this phase, they work together in larger groups to evaluate candidate laws for describing the behavior of the microworld. These may include rules such as "Whenever you apply an impulse to the dot, it changes speed" or "Whenever you give the dot an impulse to the left, it slows down." Of course, the proposed laws vary in accuracy, precision, and range of applicability. The children's task is to decide which of the laws are right and which are wrong. For the laws they believe to be wrong, they have to prove to the rest of the class that those laws are wrong and explain why. For the laws they believe to be correct, they have to decide which are the most useful and defend their choice. Through such activities, the children come to realize that a useful scientific law is something that makes precise predictions that apply across a wide range of circumstances.

In the fourth phase of the instructional cycle, the *transfer phase*, the teacher introduces the real world. In this phase, the children do activities in which they try to determine whether the conceptual model they have developed is accurate and useful for explaining what happens in the everyday world. In these activities, students are asked to "step through time" and use their laws and representations to predict and explain what will happen to an object's velocity (as illustrated in the example presented in Figure 10.1). They then experiment with the real world to see what actually happens. Inevitably, they discover that their conceptual model has limitations. For example, their model may predict that an object maintains a constant velocity after it has been hit, but in the real-world experiment, they observe that the object slows down. To resolve such discrepancies, they carry out additional modeling activities, such as going back to the microworld and putting in friction. They can then see that a microworld with friction behaves according to some of their original predictions and is a better model for many real-world situations. In this way, the children's conceptions of force and motion phenomena evolve.

This four-phase instructional cycle is repeated with each new microworld. Each time it is repeated, more of the inquiry process is turned over to the students. For example, in the early microworlds they evaluate laws that we have created, but in the later microworlds they have to create and evaluate their own laws.

Instructional Trials of the Middle-Out Approach

In order to determine the effectiveness of this approach, we arranged for this instructional cycle, centered on the four microworlds, to be implemented in sixth-grade, public-school classes by a teacher who did not know any physics and had never before used computers in her classroom. What were the results?

At the end of the ThinkerTools curriculum, most of the children successfully answered questions that assessed their understanding of the alternative representations of motion and their knowledge of the laws of the microworlds. To assess their understanding of the representations, they were asked to choose which pattern of dotprints corresponds to statements such as "the dot is accelerating," and they were asked to translate between the datacross and dotprint representations ("Draw the datacross that goes with this dotprint pattern"). To assess their knowledge of the laws, they were asked to predict the path of the dot (by drawing its dotprint pattern) under various conditions (as it was given impulses in microworlds with and without friction or with and without gravity). On a test consisting of such questions, the children averaged 77% correct with a standard deviation of 19%. The evidence thus indicates that most understood the laws and representations used in the microworld.

Could the children take their conceptual model and apply it to real-world situations? To assess this, we constructed a test consisting of problems, such as the one shown in Figure 10.4, that involve predicting what will happen in simple, real-world situations. On these real-world transfer questions, children who had taken the ThinkerTools curriculum performed dramatically better than a control group of their peers (65% compared with 44%). They also performed significantly better than high school physics students in the same school system who were taught force and motion with a traditional physics text (and who averaged only 58% correct). This was a surprising finding, because the high school physics students not only were older but were also a much more select group (few students choose to take physics in high school).

How did the children reason when they answered such problems? To find out, we interviewed some of them at the conclusion of the program. As an example, consider what occurred when we asked one student about throwing a ball upward: "Imagine that we throw a ball straight up into the air.

Imagine that you kick a ball horizontally (→) off a cliff.
Drawn below are three paths that someone might think the
ball would take as it falls to the ground.

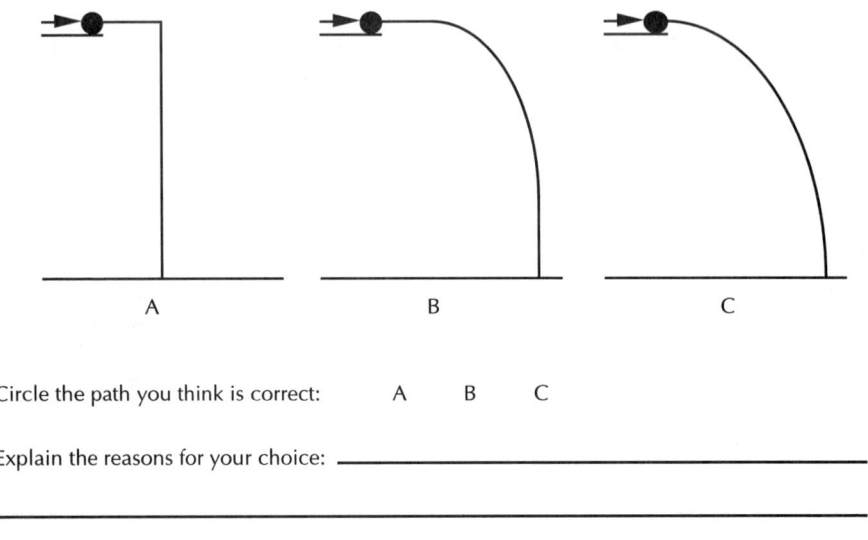

Circle the path you think is correct: A   B   C

Explain the reasons for your choice: _____

_____

_____

FIGURE 10.4 A sample problem from a physics test.

Describe what happens to the motion of the ball, and then explain why that
happens." First of all, the student described the motion of the ball in a step-
by-step manner: "It will start going up at the speed you threw it up, and then
it will gradually slow down, and there will be a second when it is stopped in
the air, and then it will start coming down, and it will gradually speed up."
Next, he gave a causal account of those events, using the conceptual model
that he had developed: "Because going up, the gravity keeps pulling, adding
another impulse down, and that will eventually stop the ball, and then going
down it keeps adding another impulse down which makes it go faster and faster." This quote demonstrates that he was able to do a causal, step-by-step
analysis of what would happen, and he showed no evidence of any of the
common misconceptions about force and motion.

We found that many students could also engage in a more precise and abstract form of reasoning. For example, we asked the students to draw and explain what would happen if the following sequence of impulses were applied
to an object in a frictionless environment: right, down, left, and up. One student drew the diagram shown in Figure 10.5. Each time an impulse was applied, she drew a datacross to indicate the object's new velocity. Then, on the
basis of the velocity shown in the datacross, she determined the position of
the object's next dotprint. In this way, she stepped through time and analyzed events to determine the object's velocity and position.

Thus classroom observations, interviews, and the results of our physics

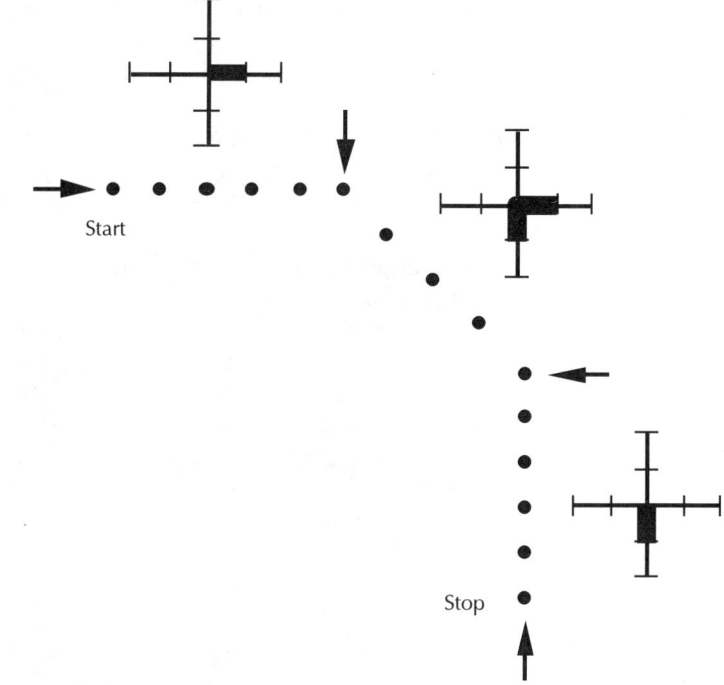

FIGURE 10.5 An example of a diagram drawn by a student to portray how a sequence of impulses would affect an object's motion.

transfer test reveal that many students were indeed able to generalize the use of conceptual models derived in the microworld contexts to a variety of simple real-world contexts. In the formalization phase of the instructional cycle, they had successfully abstracted what they learned from the computer microworld into a set of laws. In the transfer phase, these students had learned how to map the laws onto different real-world problem-solving situations. Students also saw how these laws had to be refined as they were generalized to increasingly complex circumstances, through the increasingly complex progression of microworlds.

Problems in the Development of Inquiry Expertise

When we examined what the students did when they formulated laws and designed experiments for themselves, we found limits to their scientific inquiry skills. For instance, students were asked to experiment by putting different amounts of friction into the first microworld to determine what would happen. We expected groups of students to come up with simple, qualitative laws such as "The more friction you put in, the faster the dot slows down" or "The faster the dot goes, the more friction slows it down." They did indeed derive such laws. One group of students surprised us, however, by discovering a law that we had not discovered ourselves (see Horwitz, 1989). We call it the "linear friction law:" In the microworld, the effect of friction is linearly proportional to the speed with which the object is moving. The consequence is that when students apply a sequence of impulses to the dot, it does not matter whether they apply them in quick succession or widely spaced in time; in either case, the dot will come to rest at the same point. The students discovered this fact, but they did not fully explore its implications (such as whether it means the dot can stop only at certain locations on the screen), nor did they go on to investigate whether it is an accurate model of real-world friction (such as that which affects rolling balls and sliding hockey pucks).

As a further example of limited inquiry skills, in the fourth unit on analyzing trajectories, the students were asked to design an educational experiment. They had already determined the principles that affect trajectories and were asked to develop a game-like experiment, using objects such as ping-pong balls, blowguns, and buckets, that would help someone else learn about trajectories. The activities they created were, for the most part, entertaining games, not good instructional experiments. Also, when evaluating each other's activities at the end of the period, their criteria for a good activity had more to do with its being enjoyable than with whether it was instructive.

If one reflects on the instructional approach, these limitations in the students' inquiry skills are understandable. The students were never given examples of good and bad experiments, as they were with laws. In fact, their primary perception of what they were doing in the microworlds was "playing fun games," not doing experiments. This orientation was revealed in their answers to the opinion survey taken at the end of the course. For example, in response to the question "What did you like about the course?" many students said, "I learned physics while playing fun games." In response to the question "How could the course be improved?" many said, "More games and less talk." Further, students were never explicitly told that the four phases of the instructional cycle correspond to steps in the scientific method or that they represent a particular kind of systematic inquiry. For instance, the students who discovered the linear friction law did not investigate the implications or generality of their law. Although they had practiced generalizing laws in the transfer phase of the instructional cycle, it was not made explicit to them that they were testing the explanatory power of a law and that the generalization process plays a crucial role in scientific discovery. Therefore, it is perhaps not surprising that the students did not spontaneously explore the generality of laws they discovered.

Recent ThinkerTools Research: Scientific Inquiry, Modeling, and Metacognition—Creating a Classroom Research Community

Such problems in the development of the students' inquiry expertise, as well as an increasing focus on inquiry skills within both the cognitive science and the science education communities, led us toward an increased emphasis on enabling students to learn about inquiry goals, processes, and strategies (Frederiksen and White, 1998; White and Frederiksen, 1998). We worked with teachers from urban classrooms to develop new educational tools and instructional approaches and have been investigating whether they do indeed make physics and scientific inquiry accessible to all students. This effect has included research on the use of various metacognitive tools and activities designed to enable students to create explicit models of both subject-matter and inquiry expertise, as well as to learn how to monitor and reflect on their inquiry processes.

The ThinkerTools Modeling and Simulation Tools

In contrast to the earlier version of the ThinkerTools software, in which students could only interact with prepared force-and-motion games and experiments, the new version enables students to create their own Newtonian microworlds, games, and experiments (see Figure 10.6). Using simple drawing tools, students can construct and run computer simulations. Objects, which are again called dots (the large circle shown in Figure 10.6), and barriers can be placed on the screen. Students can define and change the prop-

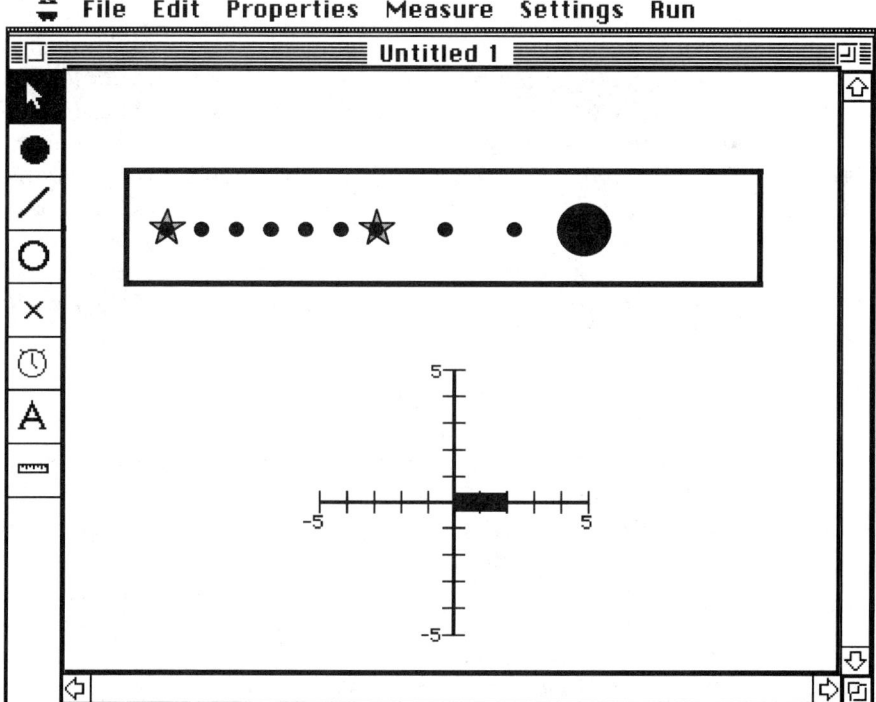

FIGURE 10.6 The ThinkerTools software provides a modeling and inquiry tool for creating and experimenting with simulations of force-and-motion phenomena.

erties of any object, such as its mass, elasticity (bouncy or fragile), and velocity. They can then apply impulses to the object to change its velocity by using the keyboard or a joystick, as in a video game. Students can thus create and experiment with a "dot-impulse model," and they can discover, for example, that when one applies an impulse in the same direction that the dot is moving, it increases the dot's velocity by 1 unit of speed. In this way, they can use simulations to discover the laws of physics and their implications.

To facilitate this inquiry process, the software includes measurement tools that allow students to make accurate observations of distances, times, and velocities easily. It also includes graphical representations of variables. For example, as in the previous version of the software, the dot leaves behind, as it moves, "dotprints" that show how far it moved in each second, and the "datacross" again shows the dot's x and y velocity components. In this version, students can also have the software keep a table or graph to record, for example, the velocity of the dot. And analytic tools such as "stepping through time" allow students to pause the simulation and to proceed time step by time step so that they can better see and analyze what is happening to the motion of the dot. In this mode, the simulation runs for a small amount of time, leaves one dotprint on the screen, and then pauses again. These analytic tools and graphical representations help students determine the underlying laws of motion. They can also be incorporated within the students' conceptual model to represent and reason about what might happen in successive time steps.

This new version of the software enables students to create experimental situations that are difficult or impossible to create in the real world. For example, they can turn friction and gravity on or off and can select different friction laws (sliding friction or gas/fluid friction). They can also vary the amount of friction or gravity to see what happens. Such experimental manipulations, in which students dramatically alter the parameters of the simulation, make it possible for them to use inquiry strategies, such as looking at extreme cases, that are hard to employ in real-world inquiry. This type of inquiry enables students to see more readily the behavioral implications of the laws of physics and to discover the underlying principles.

Instructional Approaches for Inquiry and Reflection

To embody our increased emphasis on inquiry, we developed the Thinker-Tools Inquiry Curriculum, which revolves around the new version of the ThinkerTools software. This curriculum employs a constructivist approach that focuses on inquiry and modeling. It is aimed at developing students' metacognitive knowledge—specifically, their knowledge about the nature of scientific laws and models, their knowledge about the processes of modeling and inquiry, and their ability to monitor and reflect on these processes. The pedagogical strategies include having students make their conceptual models and inquiry processes explicit, supplying instructional materials to scaffold their inquiry process, and teaching them methods for monitoring and reflecting on their progress.

Scaffolded Inquiry

The ThinkerTools Inquiry Curriculum is centered on a generic inquiry cycle, shown in Figure 10.7, derived from our earlier instructional cycle. In contrast to our earlier approach, this cycle is made explicit to students and is presented as a sequence of goals to be pursued: question, predict, experiment, model, and apply. The students start by formulating a research question. They then generate alternative hypotheses related to their question. Next, they design and carry out experiments in which they try to determine which of their hypotheses, if any, is accurate. They carry out these experiments in the context of both the computer simulation and the real world. After the students have completed their experiments, they analyze their data and try to formulate a model to characterize their findings. (An example of such a model is shown in Figure 10.1.) Once the students have developed their model, they try to apply it to different real-world situations in order to investigate its utility and its limitations. Determining the limitations of their conceptual model raises new research questions, and the students begin the inquiry cycle again.

The inquiry cycle thus starts with formulating a research question, which is perhaps the single most difficult step in scientific research. In the ThinkerTools Inquiry Curriculum, the process of formulating a research question is heavily scaffolded. The teacher begins the curriculum by tossing a bean bag around the room and asking the students to describe all of the factors that affect its motion. In this way, they see that this apparently simple motion is actually very complicated. The teacher then asks the students to think about

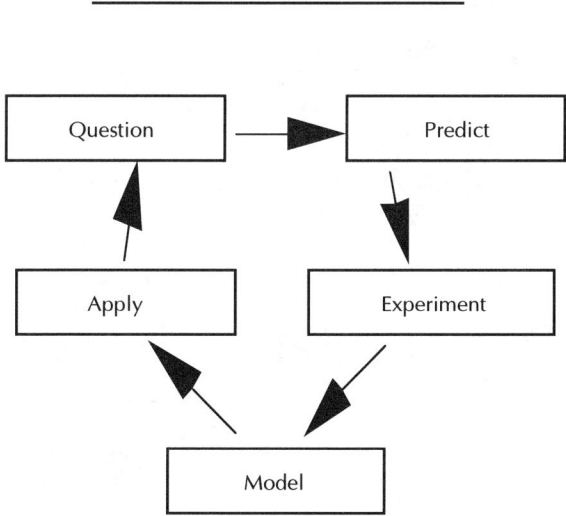

FIGURE 10.7 A portrayal of the scientific inquiry process, which students use to guide their research.

how they could simplify this situation. The research strategy is to start with simplified, idealized force-and-motion situations, such as one-dimensional motion in a world with no friction or gravity. Gradually they add complexity, such as thinking about friction, varying the mass of the object, dealing with two-dimensional motion, understanding gravity, and finally reasoning about trajectories (which is where the curriculum started with students tossing a bean bag around their classroom). This strategy of starting with simplified situations and gradually adding complexity is made explicit to the students.

The inquiry cycle is repeated with each module of the curriculum. Each time it is repeated, the conceptual models that the students are creating increase in complexity. Furthermore, the curriculum scaffolds their inquiry less

Now you will evaluate the work you just did.

Reasoning Carefully

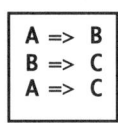	**Reasoning Carefully.** Students can reason appropriately and carefully using scientific concepts and models. For instance, they can argue whether or not a prediction or law that they or someone else has suggested fits with a scientific model. They can also show how experimental observations support or refute a model.

Circle the score that you think your work deserves.

1	2	3	4	5
not adequate		adequate		exceptional

Justify your score on the basis of your work. _____

_____

_____

Writing and Communicating Well

	**Writing and Communicating Well.** Students clearly express their ideas to each other or to an audience through writing, diagrams, and speaking. Their communication is clear enough to allow others to understand their work and reproduce their research.

Circle the score that you think your work deserves.

1	2	3	4	5
not adequate		adequate		exceptional

Justify your score on the basis of your work. _____

_____

_____

FIGURE 10.8 An example of a reflective assessment page found in the students' research books. This sample page is located at the end of the model phase of the inquiry cycle.

and less. By the end of the curriculum, the students are engaging in independent inquiry on research topics of their own choosing. To do this, they follow the inquiry cycle, which provides them with a high-level goal structure for guiding their research.

Reflective Assessment

In addition to the inquiry cycle, we introduce students to a set of criteria for monitoring and reflecting on their research. These include cognitively oriented criteria, such as "being inventive," "being systematic," and "reasoning carefully," as well as socially oriented criteria, such as "communicating well" and "teamwork." The definitions for these criteria were designed to help students understand the nature and purpose of research. The students use these criteria in a process we call reflective assessment in which they evaluate their own and each others' research (Frederiksen and Collins, 1989). For example, Figure 10.8 provides a sample reflective assessment page from the students' research books in which they use some of these criteria. In this case, "reasoning carefully" and "communicating well" are defined for the students, who then rate the research they have just completed on a five-point scale. The students then justify their ratings by explaining why their work deserves those scores.

Our hypothesis was that this reflective assessment process would help students to understand better the purpose and steps of the inquiry cycle. We also thought that having their work constantly evaluated by themselves, their peers, and their teachers would be highly motivating. Moreover, this process encourages the students to monitor and reflect on their work continually, which should improve their inquiry process. We further hypothesized that such a metacognitive reflective assessment process would be particularly important for academically disadvantaged students, because one reason why these students are low-achieving is that they lack metacognitive skills, such as monitoring and reflecting on their work. If this process is introduced and scaffolded as we illustrated, it should enable low-achieving students to learn these valuable metacognitive skills, and their performance should therefore be closer to that of high-achieving students.

Instructional Trials of the ThinkerTools Inquiry Curriculum

The ThinkerTools Inquiry Curriculum, revolving around the modeling software, the inquiry cycle, and the reflective assessment process, was implemented by three teachers in their urban classrooms. We saw these instructional trials of the curriculum as an opportunity to do a controlled study on the value of the reflective assessment process, in particular, and on the development of metacognitive skills in general. For each of the participating teachers, half of his or her classes engaged in the reflective assessment process and the other half did not. Thus all of the classes did the same Thinker-

Tools Inquiry Curriculum, but half of the classes included reflective assessment activities, such as that shown in Figure 10.8, whereas the control classes included alternative activities in which students commented on what they liked and what they did not like about the ThinkerTools curriculum.

These three teachers were teaching 12 classes in grades 7 through 9. Two of the teachers had no prior formal physics education. They were all teaching in urban situations in which their class sizes averaged almost 30 students, two-thirds of whom were minority students, and many of whom were from low-SES backgrounds. In regard to the students' achievement levels on a standardized achievement test (the Comprehensive Test of Basic Skills—CTBS), the distribution of percentile scores was almost flat, indicating that there were many low-, middle-, and high-achieving students, which is an ideal population for research purposes.

In presenting the results of these instructional trials of the ThinkerTools Inquiry Curriculum, we first focus on the students' learning of inquiry and the impact that the reflective assessment process had on that learning. Then we turn to the students' learning of physics. Because these findings are presented in depth in White and Frederiksen (1998), we will summarize only the major findings here.

The Development of Inquiry Expertise

One of our assessments of students' scientific inquiry knowledge is an inquiry test that was given both before and after the ThinkerTools Inquiry Curriculum. In this written test, the students were asked to investigate a specific research question: "What is the relationship between the weight of an object and the effect that sliding friction has on its motion?" As a first step, the students were asked to come up with alternative, competing hypotheses with regard to this question. They then had to design on paper an experiment that would determine what actually happens. Next, they had to pretend to carry out their experiment. In other words, they had to do it as a thought experiment and make up the data they thought they would get if they actually carried out their experiment. Finally, they had to analyze their made-up data to reach a conclusion and relate this conclusion to their original, competing hypotheses.

In scoring this test, the focus was entirely on the students' inquiry process. Whether or not the students' theories embodied the correct physics was regarded as totally irrelevant. Figure 10.9 presents the gain scores on this test for both low- and high-achieving students and for students in the reflective assessment and control classes. First, note that students in the reflective assessment classes had greater gains on this inquiry test. Second, note that this result was particularly strong for the low-achieving students. This is the first piece of evidence indicating that the reflective assessment process is beneficial, particularly for academically disadvantaged students who get low scores on standardized achievement tests.

When we examined this effect in more detail by looking at the gain score for each component of this test, we found that the gain scores were greatest

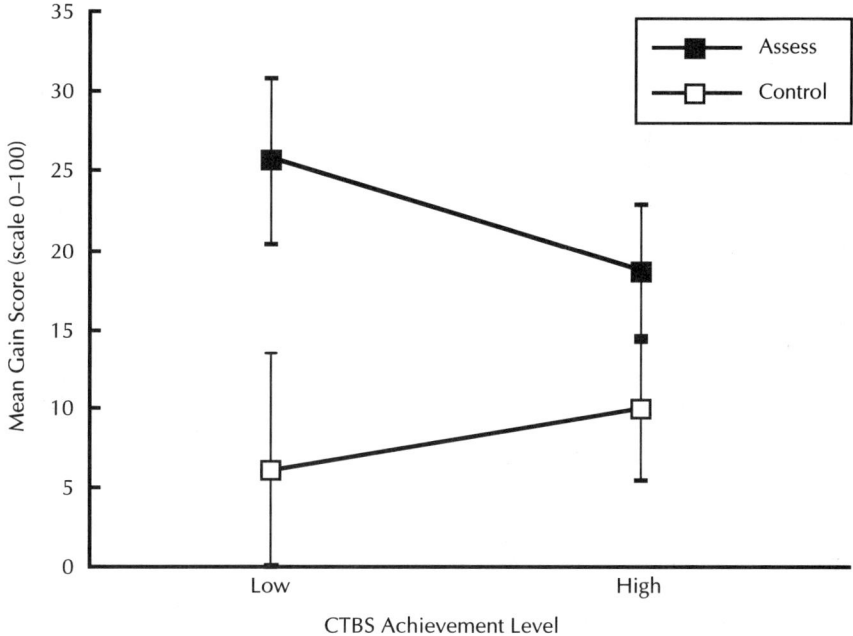

FIGURE 10.9 The mean gain scores on the inquiry test for students in the reflective assessment and control classes, plotted as a function of their achievement level.

for the more difficult aspects of the test: making up results, analyzing those made-up results, and relating them to the original hypotheses. In fact, the gain scores were greatest on this test for an attribute we call "coherence," which measures the extent to which (1) the experiments that the students designed address their hypotheses, (2) their made-up results are related to their experiments, and (3) their conclusions follow from their results, and whether they relate their conclusions to their original hypotheses. This kind of overall coherence in research is, we think, a very important indication of sophistication in scientific inquiry. It is on this coherence measure that we found the greatest difference in favor of students who engaged in the reflective assessment process.

Another measure of students' inquiry expertise is the quality of their research projects. Students carried out two projects in this course, one about half-way through the curriculum and one at the end. For the sake of brevity, we added the scores for these two projects together, as shown in Figure 10.10. These results indicate that students in the reflective assessment classes performed better on their research projects than students in the control classes. In addition, the reflective assessment process is particularly beneficial for the low-achieving students: Low-achieving students in the reflective assessment classes perform nearly as well as the high-achieving students. These findings were the same across all three teachers and all three grade levels.

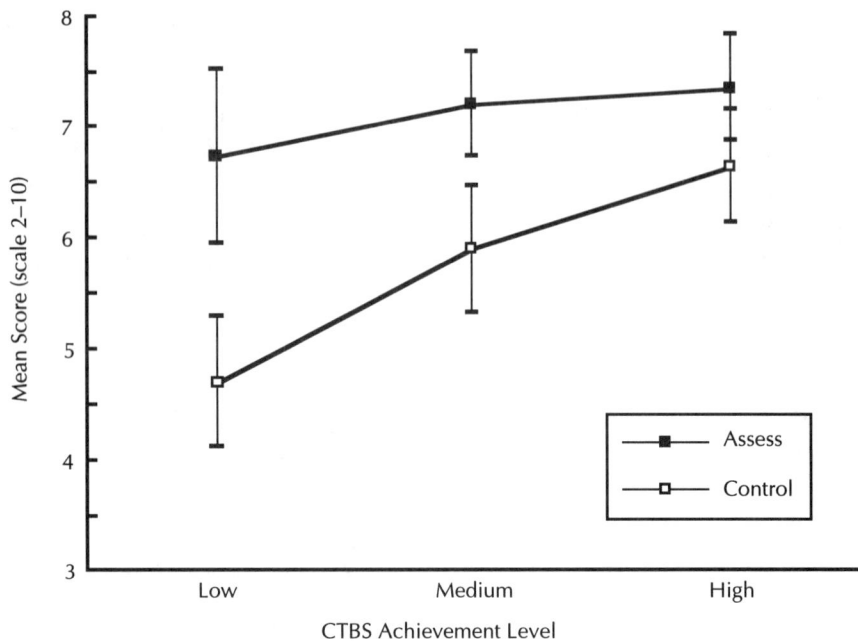

FIGURE 10.10 The mean scores on their research projects for students in the reflective assessment and control classes, plotted as a function of their achievement level.

The Development of Physics Expertise

We gave the students a physics test in which they were asked to predict and explain how forces affect an object's motion, such as in the item shown in Figure 10.4. This test was given before and after the ThinkerTools curriculum. We found significant pre-test to post-test gains. We also found that our middle-school, ThinkerTools students did better on such items than did high school physics students who were taught via traditional approaches. When we analyzed the effects of the ThinkerTools curriculum on items that represent near or far transfer in relation to contexts students had studied in the course, we found that there were significant learning effects for both the near-transfer and the far-transfer items. Together, these results show that one can teach sophisticated physics in urban, middle-school classrooms when one makes use of simulation tools combined with teaching and scaffolding the inquiry process.

Problems in the Development of Modeling Expertise

With regard to developing the students' understanding of modeling and inquiry, there are some interesting challenges. We asked students the following question in an interview at the end of the curriculum: "If you did a computer experiment and a real-world experiment and you got different results, which one would you believe and why?" We found that students have intriguingly different theories about the relationship among the computer model, their own conceptual model, and the real world. For example, some students (roughly a third of those interviewed) believe that the computer model embodies the right physics, and so their job is to discover its laws in order to create their conceptual model and make sure that their real-world experiments yield the correct results. This not unreasonable view suggests that these students understand the basic idea of a model.

Other students appear to lack such a modeling perspective and make statements such as "The computer is just a machine and so it cannot hit a ball or know any physics. What I really believe is the results of my real-world experiments because that is the only thing that is real." These students do not appreciate that the computer simulation is following laws embedded in a reasoning structure (stepping through time and using laws to calculate changes in velocity on the basis of the forces that are acting) and that versions of these laws and reasoning structures are needed for their conceptual model. To develop this understanding, it may be necessary to introduce a more explicit representation of how the computer simulation works. For example, in some of our earlier research we created simulations that could, at the students' request, illustrate such model-based reasoning by thinking out loud (White and Frederiksen, 1990).

Some students, in contrast, did appear to have the desired modeling perspective. They explained that the computer simulation is just a model and that it is an accurate model of the real world only if you include the right parameters. For example, you need to put in air resistance when there is significant air resistance, and you need to choose the correct law for resistance, such as selecting gas/fluid resistance and not sliding resistance when appropriate. Thus, if you put in the right parameters and choose the correct law for the computer model, then the computer model will be an accurate model of the real world.

To help students develop the desired modeling perspective, we created yet another version of the ThinkerTools curriculum and an augmented version of the software. This new direction for our work is in line with changing visions for science education: There is an increasing recognition that using computer-based modeling tools and learning about the process of modeling are important components of science education.

Latest ThinkerTools Research: Toward a Better Understanding of the Nature of Scientific Models and the Process of Modeling

In our latest version of the curriculum, we have focused on teaching students about the nature of scientific models and the process of modeling (Schwarz and White, 1998a, 1998b; White and Schwarz, 1997). We think this new emphasis is important for a variety of reasons. First, models and the process of modeling are fundamentally important components of the scientific endeavor. Thus students who take science courses should learn about these important processes and products of science. Second, gaining a deeper understanding of the nature of science may help students develop more fruitful epistemological beliefs. Finally, getting students to create their own computer simulation models will enable them to reify, visualize, and test their conceptual models. This may help them to develop their conceptual models more effectively as well as gain a better understanding of the process of modeling.

The ThinkerTools Modeling Software

To improve students' learning about the nature of models and modeling, we created an enhanced version of the ThinkerTools software that allows students to modify the simulation itself; it can then violate Newtonian principles and obey laws that are closer to students' alternative conceptions of force and motion. Students modify a simulation by selecting from among alternative rules that can govern the simulation's behavior. They are given three or four qualitative or semi-quantitative rules that govern a particular aspect of the simulation, such as motion in the absence of forces, the effects of friction on motion, or the relationship between mass and the effects of a force. An

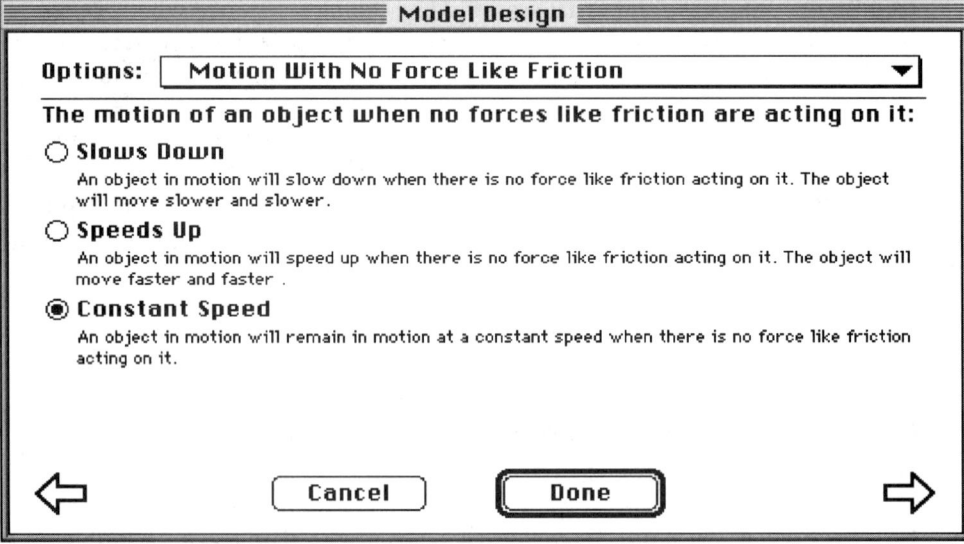

FIGURE 10.11 This screen shot illustrates how the software enables students to choose from among alternative laws of motion.

example of a set of such alternative laws is given in Figure 10.11 for the case of motion in the absence of forces.

Once students have chosen a "model design" rule, they are asked to type in a causal explanation in response to the prompt "I think this is true because. . . ." Afterwards, they can run a simulation that uses their chosen modeling rule(s). Watching the behavior of the simulation, students see the consequences of their rule(s). They can go on to create and run simulations of various phenomena using their chosen modeling rule(s). In this way, the software allows students to embody, run, and test the rules incorporated within their model.

An additional feature of the software was created to enhance students' understanding of the implications of their model and the ways in which such models "reason." When running in single-step mode, the computer articulates (using speech synthesis) the rules that are governing the motion of the dot. For example, suppose a student has chosen the rule "An object in motion will slow down when there is no force like friction acting on it" and has responded to the prompt "I think this is true because . . ." with "I have no force to keep me going." Whenever that rule is invoked by the simulation, the computer says at each time step, "Without a force like friction acting on me, I will slow down, because I have no force to keep me going." Thus, at each time step, the dot talks aloud about whichever rules are determining its motion. We designed this feature of the software to help students learn about the nature of the computer model and, more specifically, to appreciate that the simulation is governed by rules and that it embeds those rules in a reasoning structure that steps through time and applies those rules whenever their application conditions are met.

In the ThinkerTools Modeling Curriculum, students are encouraged to compare their model with the Newtonian model. The Newtonian model runs the simulation according to Newton's laws of motion. Students are never told that Newton's laws are the normative models for nonrelativistic motion—only that Isaac Newton was a famous physicist of the seventeenth century who invented important models of force and motion. Thus, from an epistemological standpoint, the Newtonian model is treated in the same way as the models students are constructing themselves. The alternative would have been to treat the Newtonian simulation as an accurate representation of real-world motion, which students use in validating their own models (as in the prior curriculum).

We created the model-design feature of the software in this simple form for several reasons. According to previous research on students' intuitive conceptions, novices think about force-and-motion phenomena primarily in qualitative terms. Therefore, creating software to imitate common intuitive conceptions should enable students to choose rules that approximate their own conceptions. The pedagogical idea is that students will be able to implement their conceptual models directly within the simulation and will then be able to see the implications of their conceptual model. We also wanted to provide a modeling structure that would enable students to gain a better understanding of the different types of modeling rules and the ways in which they are used within a model to reason about force-and-motion phenomena. Thus, expressing the rules in a qualitative (or semi-quantitative) form seemed promising as a starting point for students' conceptual development. Finally, keeping the modeling tool simple allows the students to concentrate on the models themselves; they need not spend time learning how to create a simulation by using a programming language such as Logo or Boxer (diSessa, 1993; Papert, 1980; Sherin, diSessa, and Hammer, 1993).

Instructional Approach for the ThinkerTools Modeling Curriculum

In this model-design version of the ThinkerTools curriculum, students conduct their research using a slightly revised version of the inquiry cycle, which includes six steps: question, hypothesize, investigate, analyze, model, and evaluate. As in the earlier version, they begin their research with a question, such as "What is the motion of a moving object when no forces like friction are acting on it?" They then develop alternative hypotheses about the answer to this question, such as that the object would slow down, speed up, or travel at a constant speed. Unlike the prior version of the ThinkerTools curriculum, however, students then conduct only real-world experiments in the investigation phase of the inquiry cycle. For example, in investigating motion with and without friction, the students begin by giving a plastic puck an impulse and measuring the object's speed over 1- and 2-meter distances. They then reduce the amount of friction between the puck and the floor by adding a balloon and stopper to the puck, in order to make it act like a mini-hovercraft, and repeat the experiment and measurements. Students must work especially carefully at conducting their real-world experiments and collecting data, because in contrast to the prior version of the curriculum, in which students also experimented with Newtonian computer simulations, real-world experimentation is at this point the students' only means of gathering information to create their conceptual model. (Computer experiments are not carried out until the modeling and evaluating phases of the inquiry cycle.)

Once students have gathered their data, they analyze their results. For the "motion-with-no-force" topic, students analyze the differences in speed between the first and second meters for the pucks with different amounts of friction. They then create a tentative rule to characterize their results. They next go to the computer and create a model by choosing a rule for the computer that most closely corresponds to their own rule. Then, in order to see the consequences of their rule, they run a simulation that uses their model. In the final phase of the inquiry cycle, the students evaluate their model using a set of criteria for characterizing good scientific models. Does the model accurately portray behavior? Does it embody a plausible mechanism? We think that within the model-design curriculum, it is particularly important for students to use such criteria to judge their models, because they are not constrained to working with a computer simulation that embodies the accepted Newtonian theory. Finally, as in our earlier curriculum, students evaluate their models by applying them to various situations in the real world to see their utility as well as to investigate their limitations.

In addition to having students use the software to create models, we included several classroom activities that attempted to make the forms of models and the process of modeling explicit. For example, the curriculum was introduced by having students watch and discuss a videotape of modern uses of computer models and simulations. These included computer simulations of a storm and of two galaxies colliding, an impulse-based simulation of various objects in motion, and a portion of the computer-animated movie *Toy Story*. Further, in the modeling section of the inquiry cycle, we had students read passages about the nature of models, how the computer model works, and how computer models can be useful. Finally, students engaged in classroom debates about their models. They were assigned to groups charged with creating arguments to support each model, taking into account the criteria for characterizing good models as well as considering the model's limitations.

Results of Instructional Trials of the ThinkerTools Modeling Curriculum

Overall, the results suggest that although the curriculum was fairly successful at promoting knowledge of models and modeling, as well as expertise in scientific inquiry, it was not as successful as the prior version of the curriculum at promoting conceptual understanding of Newtonian physics.

Modeling Expertise

Analyses of data from students in four classes indicate that most students developed a basic understanding of the nature of models by the end of the curriculum. For example, when we analyzed students' projects, we found that 88% of students who completed a final project included, in their project report, some sort of model that went beyond simply restating their data. Further, 91% of these models were in the form of a general-purpose rule that often included some sort of explanation. An example of a student's model is "Falling objects speed up as they fall. I think this is because gravity is always applying force on everything. . . . This rule was the closest I could find to my real-world experience. Regardless of [its] mass, the height [it was] dropped from, or [its] weight, the object will speed up until confronted with an obstacle."

A second source of evidence that students improved their understanding of the nature of models is obtained from written pre-tests and post-tests. In these modeling tests, one set of questions asks students to categorize which of a series of items they considered to be models. The items included a pencil, a globe, an equation, and a scientific rule. We found that there was a significant improvement in students' ability to identify correctly those items that could be regarded models. For example, when given the item "roughly every twenty-four hours, the sun rises in the east, and sets in the west, because the earth rotates on its axis" only 14% of students thought this was a model on the pre-test, whereas 48% of them believed it was a model on the post-test (χ^2 (1, $n = 71$) = 22.15, $p <.001$). Overall, there was improvement in students' ability to differentiate modeling items, such as a scientific rule, an equation, and the ThinkerTools simulation, from nonmodeling items, such as a pencil and a bicycle.

To illustrate further the type of understanding that students developed about the nature of models, the purpose of modeling, and the evaluation of models, we present the responses of a student answering a series of interview questions. These interviews were conducted 2½ months after the curriculum ended.

Interviewer: What do you think a scientific model means?

Student: Well, it can be a theory or rule about what you think happens in real life, or it can be a representation of something. Any representation of a real thing, like a car model, or a theory. It's a representation of the real world.

Interviewer: What is the point of trying out the different model rules on the computer?

Student: 'Cuz, then you can visualize each one instead of randomly picking one because your data says it. Like you want to be able to compare to what really happened instead of just looking at data that could be all messed up.

Interviewer: What criteria would you use to evaluate your research findings?

Student: Well, I looked if my data seemed like plausible, and I looked to see if my model could be used to predict anything really accurately.

Physics Expertise

With regard to the students' understanding of physics, analyses of our conceptual physics test show that students made only modest improvements in developing a Newtonian conceptual model. Students did demonstrate significant improvement on some questions, such as the following:

Imagine that a spaceship is coasting along in deep space. It is not near any planets or other outside forces. What will be true about the speed of the spaceship as it moves along?
    A. the speed will decrease
    B. the speed will remain the same
    C. the speed will increase

Eighty-two percent of the students chose the Newtonian answer B on the post-test, compared to only 63% on the pre-test (χ^2 (1, $n = 125$) = 13.7, $p <.001$). Overall, however, students showed only a modest improvement on the physics test, with an average post-test score of 55% correct compared to a pre-test score of 49% correct ($t_{71} = 5.27$, $p <.001$). In contrast, the prior version of the curriculum led to post-test scores of 68% correct. It should be noted that the model-design version of the curriculum did not cover as much physics as the prior curriculum. To allow for this, we also compared the effectiveness of the two curricula for the subset of items that addressed physics covered in the model-design curriculum. This comparison again indicated that the model-design version of ThinkerTools produced a smaller gain, 8.5%, than the prior version of the curriculum, where the gain was 15.3% ($t_{215} = 1.99$, $p <.05$).

Epistemological Beliefs

Students were also given an assessment that investigated their epistemological beliefs at the beginning and end of the curriculum. The test includes items that assess whether students believe scientific knowledge is coherent or composed of pieces, whether it is simple or complex, whether it is applicable or inapplicable to real life, and whether it is certain or tentative. There were significant changes in students' performance on a quarter of the items. The pattern of results for those items indicates that students changed their beliefs toward more realistic and more productive epistemologies. For example, on one such item, students were asked to decide which point of view they agreed with in the following hypothetical debate between two students:

Paul: The thing I like about science is that nothing is ever settled completely. New information could make us change old theories.

Dan: I disagree completely. Once experiments have been done and a theory has been made to explain the results of those experiments, the matter is settled.

At the end of the curriculum, students were more likely to agree with Paul, whereas at the beginning, they were more likely to agree with Dan ($t_{95} = 2.37$, $p = .02$).

Inquiry Expertise

Analyses of students' performance on our written inquiry assessment, given both as a pre- and as a post-test, indicate that the curriculum resulted in significant improvement in students' inquiry skills. As described earlier, the inquiry test presents students with a research question and asks them to generate alternative hypotheses, design an experiment, make up data, and then analyze those data and draw conclusions. Students showed significant improvement on all sections of this assessment, particularly on the analysis and conclusions sections. On the pre-test, students had a mean of 46%, whereas on the post-test they had a mean of 58% ($t_{57} = 4.4$, $p <.001$). Students who experienced the prior version of the ThinkerTools curriculum showed similar levels of performance on the post-test.

Comparison of Results

The model-design version of the ThinkerTools curriculum appears to have been as successful as our prior curriculum at developing students' inquiry skills and to have been more successful at developing their understanding of modeling. However, it was less successful at improving students' understanding of physics. The probable explanation for these differential results is that in the model-design version of the curriculum, students spend more time learning about the nature of models and the process of modeling and less time interacting with Newtonian simulations that embody the scientifically accepted model. These findings highlight the tension between using the software to discover the correct Newtonian physics and using the software to create models in order to develop a more sophisticated epistemology of science.

Discussion

Synthesizing the Various Approaches

How can we synthesize our alternative pedagogical approaches so that we can achieve our multiple pedagogical goals: enabling students to develop expertise in modeling and inquiry *and* helping students acquire a deep understanding of the subject matter? We think that it is possible to blend our various approaches by creating a curriculum that would begin by using the software in model-discovery mode and then progress to using the software in model-design mode. Alternatively, students could use the software in model-discovery mode for more difficult concepts and switch to model-design mode for easier concepts. For instance, they could interact only with Newtonian models when attempting to create laws about what happens in a world with no friction, which is counterintuitive and thus very difficult. In contrast, when they are trying to create laws for friction, which is relatively easy, they could create computer models that embody competing theories. In this way, simulations that embody the scientifically accepted Newtonian theory could be employed to facilitate the acquisition of key Newtonian conceptions that are known to require major conceptual change (such as a shift from "force causes motion" to "force causes change in motion"). Students could then engage in a more authentic modeling and inquiry process for topics that require more modest conceptual change.

There are ways of extending the use of the software as well. Students could use the model-design mode of the software in the hypothesis stage of the inquiry cycle to envision alternative theories of force and motion. They could also engage in a more explicit model-comparison process by comparing the data from their real-world experiments with those produced by various computer models. In these ways, they could take further advantage of the modeling capabilities of the software to help them explore and evaluate alternative laws and models. Such activities would also allow students to engage more fully in the processes of modeling and inquiry.

By extending and synthesizing these various ways of using the software, we could create an instructional approach that better achieves the pedagogical goals of the students' understanding the subject matter and acquiring inquiry and modeling expertise. However, our present ambitions go beyond simply blending and extending our previous approaches, because there are additional pedagogical goals that we now want to pursue. We want to enable students to progress to more abstract, mathematical ways of encoding and reasoning with scientific knowledge. We want to foster an even more effective classroom community of researchers in order to facilitate the modeling and inquiry processes. Finally, we want to enable students to create more explicit models of their inquiry and reflection processes in order to help them become expert inquirers who have knowledge of and control over their own inquiry skills. These are the foci of our latest research on the ThinkerTools Project.

Summary and Conclusion

Technological tools make possible a multitude of new and exciting approaches to science education. In this chapter, we have argued that the creation of such new approaches should be driven by pedagogical goals, which should determine the design of the computer-based tools as well as shape the creation of instructional activities and approaches. Our own work has evolved from a focus on students' conceptual models of physics toward an increasing focus on their understanding of the processes of scientific inquiry and modeling. In other words, the emphasis has shifted from understanding the subject matter to understanding the inquiry process itself. The features of our ThinkerTools software and the ways in which it is used instructionally have evolved to meet these changing goals.

In the early stages of our research, students interacted with force-and-motion microworlds that we created to make Newtonian principles as apparent as possible. We designed computer games and other experimental activities, set in the context of a progression of increasingly complex microworlds, that enabled the students' knowledge of the relevant physics to become increasingly sophisticated. Our work was highly successful in helping students to acquire relatively sophisticated conceptual models of force and motion, but it was less successful in helping students to understand the process of creating models and to engage in scientific inquiry on their own.

As our work continued and as research in the cognitive sciences also progressed, we came increasingly to recognize the importance of enabling students to learn about the processes of inquiry and modeling. We redesigned the software so that students could create their own microworlds that obey laws of their own choosing. In addition, within any microworld, students can create their own experiments, games, or other activities. And, we augmented the software to include other types of tools, such as inquiry support and reflective assessment tools that help students learn about and reflect on the inquiry process itself (White and Frederiksen, in press; White, Shimoda, and Frederiksen, in press).

As a consequence, there are now a wide variety of ways to use these software tools to meet differing pedagogical goals. In one approach, students interact with Newtonian microworlds presented to them, and their primary goal is to develop Newtonian conceptual models of force and motion. In another approach, students create their own Newtonian microworlds and study their behavior. Their goal is to construct a theory of force and motion and, in so doing, to develop, reflect on, and improve their inquiry skills. In yet another approach, students create both Newtonian and non-Newtonian microworlds, and their primary goal is to develop an understanding of scientific modeling. Here, students use computer-modeling tools to reify and investigate the implications of their own theories. They also explore alternative models that embody competing theories of force and motion. In these ways, students can engage in a variety of inquiry processes aimed at achieving differing pedagogical goals.

On the basis of this line of research, we conclude that focusing science education on the processes of scientific inquiry, modeling, and reflection has tremendous potential for enabling students to develop skills for life-long learning, problem solving, and collaboration. Students can utilize powerful tools that help them develop expertise related not only to scientific inquiry and modeling but also to learning and problem solving in general. As technology advances and our knowledge of human cognition develops, the goals for science education will continue to evolve. There is at present, for instance, an increasing recognition of the social nature of the scientific endeavor (Dunbar, 1995) and a corresponding increase in the tools available to support the social processes involved in scientific theorizing and experimentation (Pea, 1994). Similarly, in science education there is a move to create classroom communities in which social processes, such as debate and negotiation, play prominent roles as students learn via collaborative inquiry (Brown and Campione, 1996; Linn, Bell, and Hsi, in press; Scardamalia and Bereiter, 1994; White and Frederiksen, 1998). Such reformulations of our approaches to science education will become increasingly important in the twenty-first century, as new types of conceptual and social tools are developed and the nature of our society is rapidly transformed.

References

Brown, A. L., & Campione, J. C. 1996. Psychological theory and the design of innovative learning environments: On procedures, principles, and systems. In Schauble, L., & Glaser, R. (eds.), *Innovations in learning: New environments for education.* Mahwah, NJ: Lawrence Erlbaum, pp. 289–325.

diSessa, A. 1993. Toward an epistemology of physics. *Cognition and Instruction,*

10(2-3), 105-225.

Dunbar, K. 1995. How scientists really reason: Scientific reasoning in real-world laboratories. In Sternberg, R. J., & Davidson, J. E. (eds.), *The nature of insight*. Cambridge, MA: M.I.T. Press, pp. 365-395.

Frederiksen, J., & Collins, A. 1989. A systems approach to educational testing. *Educational Researcher, 18(9)*, 27-32.

Frederiksen, J., & White, B. 1998. Teaching and learning generic modeling and reasoning skills. *Journal of Interactive Learning Environments, 5*, 33-51.

Horwitz, P. 1989. Interactive simulations and their implications for science teaching. *The 1988 AETS Yearbook*.

Linn, M. C., Bell, P., & Hsi, S. In press. Using the Internet to enhance student understanding of science: The Knowledge Integration Environment. *Interactive Learning Environments*.

Papert, S. 1980. *Mindstorms: Children, computers and powerful ideas*. New York: Basic Books.

Pea, R. 1994. Seeing what we build together: Distributed multimedia learning environments for transformative communications. *Journal of the Learning Sciences, 3(3)*, 285-299.

Scardamalia, M., & Bereiter, C. 1994. Computer support for knowledge-building communities. *Journal of the Learning Sciences, 3(3)*, 265-283.

Schwarz, C., & White, B. 1998a. *Fostering middle-school students' understanding of scientific modeling*. Paper presented at the annual meeting of the American Educational Research Association, San Diego, CA.

Schwarz, C., & White, B. 1998b. *The ThinkerTools model-design software and curriculum*. Paper presented at the annual meeting of the National Association for Research in Science Teaching, San Diego, CA.

Sherin, B., diSessa, A., & Hammer, D. 1993. Dynaturtle revisited: Learning physics through collaborative design of a computer model. *Interactive Learning Environments, 3(3)*, 91-118.

White, B. 1981. *Designing computer games to facilitate learning* (Tech. Rep. No. AI-TR-619). Cambridge, MA: M.I.T. Artificial Intelligence Laboratory.

White, B. 1984. Designing computer activities to help physics students understand Newton's laws of motion. *Cognition and Instruction, 1*, 69-108.

White, B. 1989. The role of intermediate abstractions in understanding science and mathematics. *Proceedings of the Eleventh Annual Meeting of the Cognitive Science Society*. Hillsdale, NJ: Erlbaum.

White, B. 1993a. ThinkerTools: Causal models, conceptual change, and science education. *Cognition and Instruction, 10(1)*, 1-100.

White, B. 1993b. Intermediate causal models: A missing link for successful science education? In Glaser, R. (ed.), *Advances in instructional psychology, vol. 4*. Hillsdale, NJ: Lawrence Erlbaum, pp. 177-252.

White, B., & Frederiksen, J. 1990. Causal model progressions as a foundation for intelligent learning environments. *Artificial Intelligence, 24*, 99-157.

White, B., & Frederiksen, J. 1998. Inquiry, modeling, and metacognition: Making science accessible to all students. *Cognition and Instruction, 16(1)*, 3-118.

White, B., & Frederiksen, J. In press. New educational technologies and instructional approaches for facilitating scientific inquiry. In Jacobson, M., and Kozma, R. (eds.), *Learning the sciences of the 21st century: Theory, research, and the design of advanced technology learning environments*. Mahwah, NJ: Lawrence Erlbaum.

White, B., & Horwitz, P. 1988. Computer microworlds and conceptual change: A new approach to science education. In Ramsden, P. (ed.), *Improving learning: New perspectives*. London: Kogan Page, pp. 69–80.

White, B., & Schwarz, C. 1997. *Computer microworlds and scientific inquiry: Enabling students to construct conceptual models*. Paper presented at the annual meeting of the National Association for Research in Science Teaching, Chicago, IL.

White, B., Shimoda, T., & Frederiksen, J. In press. Enabling students to construct a theory of inquiry with SCI-WISE: An approach to facilitating metacognitive development. In Lajoie, S. (ed.), *Computers as cognitive tools: the Next Generation*. Mahwah, NJ: Lawrence Erlbaum.

Part 3

Toward Extended
Modeling Environments

The chapters in this section describe two very different innovations in modeling technology designed to improve science and mathematics learning. The Eisenbergs describe a modeling system for generating concrete objects as products, in order to give students experience in relating abstract representations to real phenomena in the physical world. Dede and his colleagues describe a virtual reality environment for giving students immersive experiences in simulated science worlds to improve their understanding of real phenomena not directly accessible to perception.

The Developing Scientist as Craftsperson describes work by Michael Eisenberg and Ann Eisenberg with a CAD-like program, HyperGami, that enables users to generate concrete paper sculptures from abstract mathematical descriptions. Increasingly, the day-to-day practice of science education is pervaded by the presence of computational media. Simulations, modeling tools, and virtual laboratories have become the stock in trade of the up-to-date science educator; and consequently, the young scientist is a person who spends a large proportion of his or her time in abstract and nonphysical "worlds." This move toward an increasingly "virtualized" science education has important benefits for some scientific domains and for some activities: Arguably, it is only through the simulation of especially complex systems that the student can get a sense of how such systems are capable of behaving. However, the authors argue that to relate the abstract representations in science models to real phenomena, students also need to be given more experiences with real, concrete objects. They discuss the benefits of giving science students an opportunity to derive visual and tactile pleasure from their work by producing something concrete and beautiful. The Eisenbergs seek to dissipate some of the tension between the "virtual" and "real-world" paths to learning science and mathematics. They summarize their work with the HyperGami system to ground their discussion of craftspersonship in science education. They present an overview of HyperGami and illustrate the types of constructions that have been made using the system.

Multisensory Immersion as a Modeling Environment for Learning Complex Scientific Concepts describes ScienceSpace, a set of three virtual modeling worlds designed to explore the potential utility of physical immersion and multisensory perception in enhancing science education. NewtonWorld provides an environment for investigating the kinematics and dynamics of one-dimensional motion. PaulingWorld makes possible the study of molecular structures through a variety of representations, including quantum-level phenomena. MaxwellWorld enables students to explore electrostatic forces and fields, learn about the concept of electric potential, and "discover" the nature of electric flux. Students do not employ external models of phenomena. Rather, they immerse themselves in these spatially distributed, synthetic worlds as "avatars" (graphical surrogates that serve as the virtual-world personas of the human participants), use virtual artifacts to develop knowledge, and seek to gain direct experiential intuitions about how the world operates. Virtual reality models can provide learners with three-dimensional representations; multiple perspectives and frames of reference; a multimodal interface; simultaneous visual, auditory, and haptic (touch) feedback; and new types of interaction unavailable in the real world (such as seeing through objects and teleporting). The authors' experimental results indicate that transducing data and abstract concepts into mutually reinforcing multisensory representations enhances students' understanding of scientific models. They believe that these varied aspects of multisensory immersion can provide learners with experiential metaphors and analogies that enhance their understanding of complex phenomena remote from everyday experience and can help in displacing intuitive misconceptions with alternative, more accurate mental models.

During the coming years, our concept of educational modeling and simulation will be greatly extended through further technological advances like these. Mathematical models will be used to create both real objects and virtual worlds that seem real. However real these external artifacts become, what ultimately matters for science learning is what happens inside the student's mind—not the computer simulation, and not even the underlying mathematics and science model, but the student's evolving mental model as it progresses, through empowering modeling experiences, to a deeper understanding of the world.

11

The Developing Scientist as Craftsperson

Michael Eisenberg

Ann Eisenberg

Introduction

Increasingly, the day-to-day practice of science education is pervaded by the presence of computational media. Simulations, modeling tools, and virtual laboratories have become the stock in trade of the up-to-date science educator. As a consequence, the young scientist is a person who, more and more, spends a large proportion of his or her time in abstract and nonphysical "worlds." This move toward an increasingly virtualized science education has important benefits for some scientific domains and for some activities: Perhaps only through the simulation of especially complex systems can the student get a sense of how such systems are capable of behaving. Moreover, the real, physical world constrains us as human beings—and it may constrain our scientific imaginations as well. We cannot easily experience the frictionless environments that would make many principles of Newtonian mechanics more intuitive (Chapter 10; White and Horwitz, 1987; diSessa, 1982); we do not grasp the behavior of objects moving at speeds near that of light (Horwitz, 1994); we do not see firsthand the evolution of ecosystems, a phenomenon perhaps best understood at a time scale of millennia (Dawkins, 1996). In all these cases, the building and studying of virtual worlds, simulations, and abstract models may be a crucial step in the education of the scientist.

But something is lost, too, in this move away from the physical—something pleasurable, sensual, and intellectual about the behavior of stuff. At our own university, a professor in mechanical engineering lamented that her students were increasingly arriving at college with no experience of the mechanical world, of real materials. These students, she said, have never actually sat down to fix a bicycle.

Does it matter whether students fix real bicycles, mix real chemicals, collect real butterflies, or view real stars? We believe that it does and that the advent of powerful and compelling "virtual" environments should now cause science educators to reexamine carefully the delicate relationship between computational media and real-world artifacts. Interesting hints about the role

259

of the physical world are to be gleaned from the biographies of scientists. Repeatedly, in reading about the childhood or education of famous scientists, we find that for these individuals, the presence of physical objects and the practice of "scientific handicrafts" played an important formative role. The young Stephen Hawking's bedroom has been described as "the magician's lair, the mad professor's laboratory, and the messy teenager's study all rolled into one. . . . On the sideboard stood electrical devices, the uses of which could only be guessed at, and next to those a rack of testtubes, their contents neglected and discoloured among the general confusion of odd pieces of wire, paper, glue, and metal from half-finished and forgotten projects." (White and Gribbin, 1992, p. 12); Linus Pauling learned chemistry as a young assistant in a pharmacy (Csikszentmihalyi, 1996, p. 86). Isaac Newton is reported to have tinkered with homemade mechanical devices as a youth (Bernstein, 1993, p. 162). Real-world objects, in the recollections of scientists, often seem to be associated with moments of high motivation or striking imagery. Albert Einstein, in perhaps the most famous anecdote along these lines, distinctly recalled a pivotal childhood experience in which he received a compass as a gift:

That this needle behaved in such a determined way did not at all fit into the nature of events, which could find a place in the unconscious world of concepts (effect connected with direct "touch"). I can still remember (or at least I believe I can remember) that this experience made a deep and lasting impression upon me. Something deeply hidden had to be behind things. (Bernstein, 1993, p. 161).

Similar recollections crop up in interviews with other well-known scientists. Richard Feynman, for instance, recalled working with colored floor tiles at a very young age, and he likewise mentioned an instance in which viewing a ball rolling in a wagon piqued his early curiosity about the nature of inertia (Mehra, 1994, pp. 3–5). Feynman also repaired radios and other appliances while still a youngster (Feynman, 1985). The astronomer Fred Hoyle recalled having his interest in science sparked by a chemistry set (Lightman and Brawer, 1990, p. 52). The astrophysicist Margaret Geller, in an interview (Lightman and Brawer, 1990, p. 360), recalled working with solid geometric kits as a child and added,

My father is a crystallographer. . . . He had an attraction for any kind of toy that had anything to do with geometry. . . . For example, I'd make a cube, and he'd explain to me the relationship between that and the structure of table salt. And I'd make an icosahedron, and he'd explain how you see that in the real world. . . . I would be able to visualize in 3-D. And I realize now—I've talked to lots of people in science—that very few people have that ability.

For those who enjoy scientific biography, tales like these are easy to find. But maybe they are just tales and nothing more. Biography, as the more skeptically inclined will point out, is by its very nature anecdotal evidence. And worse—biography isn't even unbiased data. The biographers are writers working for a paycheck; maybe they're looking for easy illustrations of scientific precocity, searching for punchy tales that sound like childhood epiphanies. The scientists themselves, in recollecting their childhood experiences, might be attempting to frame a coherent narrative; perhaps certain events are endowed in retrospect with an exaggerated importance so that they foreshadow later developments in much the same fashion as mystical omens do in the biographical essays of Plutarch and Suetonius.

But still, it is hard not to notice the consistent patterns among these histories, cutting as they do across boundaries of time, geography, gender, economic background, and eventual specialization. In all the cases cited—and many more in the literature—the budding interest of the young scientist seems to have been inextricably linked with the day-to-day practice of what might be called a "scientific craft." Individual stories may vary in their credibility, but we believe in the overall pattern of the biographical data. We'd go so far as to say that many, if not most, scientists need to encounter objects and experience craftspersonship as students (and probably as adult professionals, too). Because it neglects this side of human experience, a purely virtualized scientific training is incomplete and is likely to be ineffective.

Of course, we needn't view these two educational paths—one focusing on computational modeling and simulation, the other on physical objects and handicrafts—as opposites. Indeed, this chapter focuses on an attempt to dissipate some of the tension between the "virtual" and "real-world" paths to learning science and mathematics. We explore a variety of themes that have emerged as salient for us over the last several years in working on a program named *HyperGami*, a system that might be summarized as an educational CAD program for the creation of mathematical paper sculpture. In the course of developing HyperGami, working with the program, and collaborating with HyperGami students of various ages, we have become sensitized to these themes. And we have come to view them as important for science education generally.

In the remaining sections of this chapter, we summarize our work with the HyperGami system and use that work to ground our discussion of craftspersonship in science education. The second section of this chapter presents an overview of HyperGami, along with the types of constructions that we and our students have made using the system. The third, fourth, and fifth sections explore the themes that are the true focus of this chapter: the affective and social roles that scientific/mathematical objects (especially homemade objects) are capable of playing in students' lives; the pacing and "rhythm" of students' scientific activities; and the cultures and values associated with different types of physical materials in science/mathematics education. In each of these sections, moreover, we suggest ways in which computational media might be used to enhance the benefits of craft activities in science education. In the sixth and final section, we discuss several new directions of our own work and speculate about how computational tools (going well beyond HyperGami) might affect the growth of a craft culture in science education.

HyperGami: An Overview

HyperGami is a Macintosh-based software application developed by the authors in the MacScheme programming environment (S1). We have described the system at great length elsewhere (Eisenberg and Nishioka, 1997a, 1997b), so we present only a summary outline of the application here.

The basic activity in HyperGami consists of creating novel or complex paper polyhedral models and sculptures; essentially, one creates the three-dimensional shape on the computer screen, allows the software to unfold that shape into a decorable two-dimensional form, and then prints out (and constructs) the eventual model. Figure 11.1 shows the HyperGami screen in the midst of a sample scenario of just this sort. Here the user has employed the software to create a particular shape—a stretched, capped hexagonal prism. The shape is visible in the ThreeD window of the screen; the software has also "unfolded" the solid into the folding net form visible in the TwoD window of the screen. The user is now in the midst of decorating the folding net; this can be done by employing solid colors, patterns, textures, hand-drawn decorations, and so forth. HyperGami is a *programmable application* that employs an enriched version of the Scheme programming language as part of its interface (Eisenberg, 1995). The user therefore has at her disposal a complete Scheme interpreter, augmented with an extensive (and always growing) library of procedures and objects for creating and decorating HyperGami constructions. To take an especially simple corollary of this idea, in the Figure 11.1 scenario the user has employed HyperGami's turtle procedures to decorate one of the faces of her hexagonal prism with a geometric design in the spirit of Abelson and diSessa (1980).

There are many more features in the HyperGami system, but space limitations preclude our offering a more thorough discussion. We must mention several points, however, because they come up later in this chapter. First, HyperGami includes tools through which the user may transfer a decoration from the folding net to the three-dimensional view of a polyhedral object; this permits the user to predict, in some measure, what the eventual construction will look like if the current folding net is printed out and folded into three-dimensional form. Figure 11.2 shows an example. Here, the decoration from a hexagonal prism such as that shown in Figure 11.1 has been transferred to its three-dimensional rendering.

A second point is simply that paper—the basic material of HyperGami constructions—is itself an extremely rich craft medium and one whose versatility is still expanding. HyperGami creations may be printed out on standard printer paper, on glossy paper, on thick cardstock, on pretinted papers, on acetate, or on large poster-sized sheets. (We ourselves have barely begun to explore the varieties of paper available and to experiment with these papers in our constructions.)

A third point is that HyperGami includes a number of features specifically designed with the needs and problems of the paper crafter in mind. For

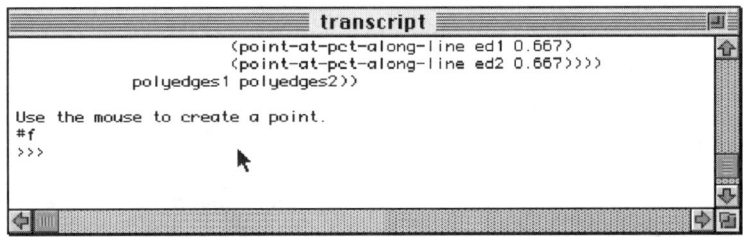

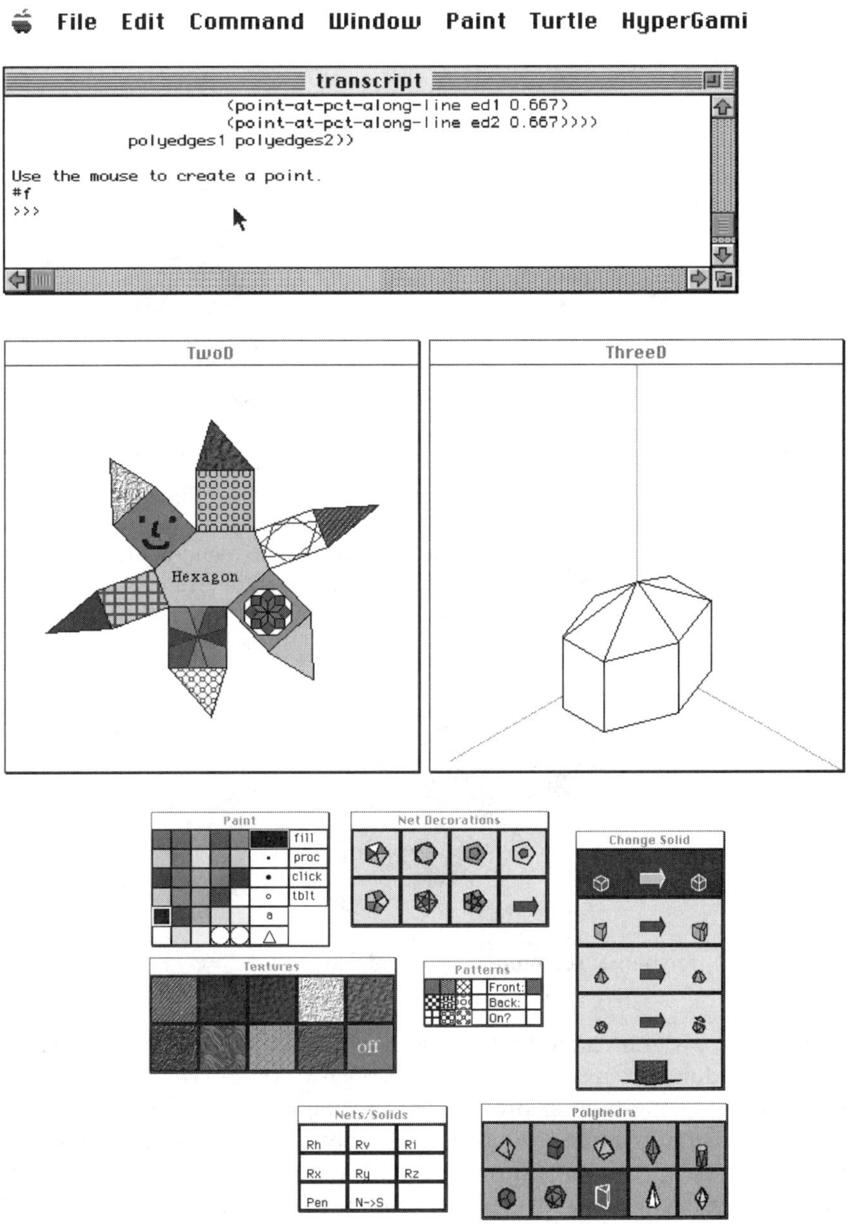

FIGURE 11.1 The HyperGami screen in the midst of a sample scenario. The transcript window at the top is the interface to the MacScheme interpreter, augmented with a large number of system-specific procedures and data objects. The shape being constructed is shown in a three-dimensional view at the right, in the ThreeD window; its unfolded version (or folding net), generated by the system, is shown at the left. In the scenario, the user has decorated the folding net using a variety of means: textures, solid colors, patterns, text, and hand-drawn and program-created design.

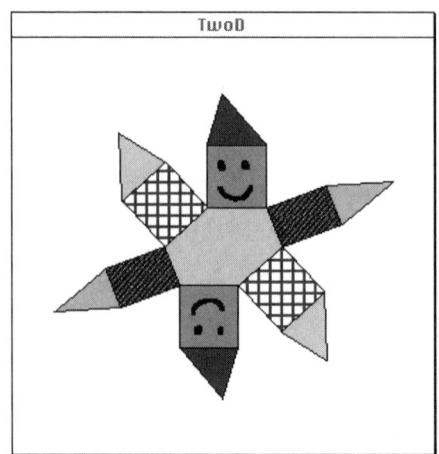 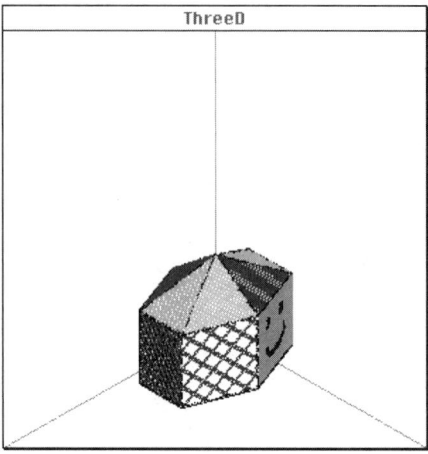

FIGURE 11.2 The user has transferred the decoration from the folding net at the left to the three-dimensional view at the right. The transfer takes place on a pixel-by-pixel basis, which makes it somewhat slow and laborious (not to mention inexact), but the view at the right does give a reasonable preview of what the eventual folded shape will look like.

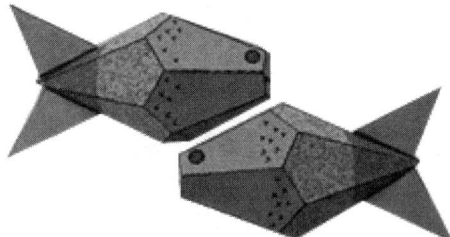

FIGURE 11.3 A polyhedral sculpture of a pair of fish.

instance, the program includes tools with which the more advanced user can "tailor" a folding net in such a way that it is easier to fold (or perhaps to decorate). Many of the built-in procedures for decorating folding nets take advantage of the geometric properties of the net's polygons (for instance, one of the example procedures furnished with the program allows the user to decorate a net polygon with "spokes" radiating from the center of the polygon to each of its vertices). Recent additions to the program permit the user to generate "tabs" on the folding nets (these greatly assist in the construction of the eventual shape), and a "surface turtle" package allows the user to move a Logo-style turtle over the entire surface of a polyhedron, "jumping" smoothly from face to face of the net as suggested by the discussion in Abelson and diSessa (1980, ch. 6).

Figures 11.3–11.9 illustrate the types of constructions that we and our students have created in HyperGami. Figures 11.3 and 11.4 depict polyhe-

FIGURE 11.4 "Tweedledee-ahedron and Tweedledum-ahedron."

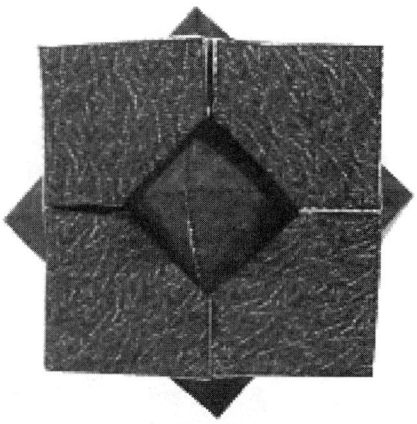

FIGURE 11.5 A "trapped octahedron" polyhedral construction by a 12-year-old boy.

dral sculptures ("orihedra") that we have created. Figure 11.3 depicts two fish created from trapezohedra and prisms. Figure 11.4 depicts two twins ("Tweedledee- and Tweedledum-ahedra") built from a variety of shapes, including the icosidodecahedron (the bodies of the figures) and the small rhombicosidodecahedron (their heads).

Figures 11.5 and 11.6 depict work done by HyperGami students. Since the development of HyperGami began, we have worked with over 50 students ranging from elementary to high school age. Typically, these students work with us as individuals or in pairs or small groups, for a period of an hour or two per week over the course of about a semester. [Eisenberg and Nishioka (1997a) and Eisenberg and Nishioka (1997b) include more details on our experiences with students.] By and large, students work toward creative projects of their own devising, such as the sculptures in Figures 11.5 and 11.6, although over the past year we have also begun the creation of more specifically curricular materials (such as exercises), mostly for use with our high school students. Figure 11.5 shows a marvelous polyhedral figure—

FIGURE 11.6 A polyhedral kangaroo designed and created by a high school senior.

FIGURE 11.7 Two "pillowhedra" made from sewn fabric.

FIGURE 11.8 A rhombic
dodecahedron of soap.

a "trapped octahedron"—in which an octahedron is shown embedded within eight pieces that collectively make up a figure rather like a surrounding cube (this piece was created by a sixth-grade boy). Figure 11.6 shows a polyhedral sculpture, composed mainly of prisms and pyramids, of a kangaroo (the body is a truncated tetrahedron) done by a twelfth-grade girl.

Finally, Figures 11.7, 11.8, and 11.9 depict work done in materials other than paper. Figure 11.7 shows a pair of "pillowhedra" created by printing out folding nets onto special paper that can then transfer its decoration to fabric; usually, this process is employed to create customized T-shirts, but it can just as easily create three-dimensional sculptures in fabric. Figure 11.8 shows a rhombic dodecahedron created from soap (using a HyperGami construction

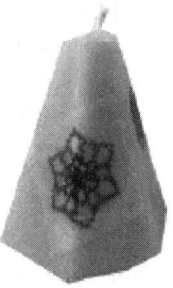

FIGURE 11.9 A truncated pyramid of wax, decorated with a turtle-graphics design.

as a mold), and Figure 11.9 shows a polyhedral wax candle (created in similar fashion). These last examples illustrate the ways in which an application such as HyperGami can lead to the construction of an immense variety of craft objects in different media.

Theme 1: The Affective Role of Objects in Science Education

In the previous section, we introduced HyperGami and presented an illustrative range of artifacts that our students and we have created with the program. As mathematical objects, HyperGami sculptures fit within a long-standing tradition of polyhedral modeling in mathematics education (see, for instance, Cundy and Rollett, 1951; Hilton and Pedersen, 1994; and Jenkins and Wild, 1985). Indeed, Margaret Geller's autobiographical anecdote, recounted earlier, suggests that such objects are effective practice materials in developing skills of spatial cognition.

For the purposes of this chapter, however, we prefer to take a broader view. Rather than focusing on the (relatively narrow) mathematical issues raised by polyhedral modeling, we wish to explore the more important and general role of computationally enriched craft activities in mathematics and science education. These issues go well beyond the specific examples of HyperGami work; they include many types of materials besides paper and other paradigms for computational media besides stand-alone applications. Nonetheless, our own experiences with HyperGami are what made these broader issues accessible to us, so we tend to use our HyperGami work as a source of illustrations.

Crafts and Affect

For us, one of the crucial aspects of our (and our students') HyperGami experience is that it has, over time, produced objects that we now enjoy having around. That's a simple observation, and at first blush it might seem almost trivial. Why should mathematics or science educators concern them-

selves with the creation of fun objects? The answer is suggested by the re-
marks from and about young scientists quoted earlier: Geller's polyhedral
models, Feynman's geometric tiles, Newton's gadgets—all seemed to provide a sensual as well as intellectual pleasure. And even more: Pleasurable
physical objects contribute to a setting, a sense of place, in which mathematics and science may be studied. Recall Hawking's room: a rich, full setting
in which a young scientist could be stimulated on a day-to-day basis.

How different this is from the physical environment of science education
as it is increasingly defined by computational media! No matter how aesthet-
ically pleasing the view from a computer screen may be, that view is only a
tiny and temporary fraction of one's setting. As computer scientists, we know
this phenomenon well: Month after month, year after year, as we toil away at
our computers, we return to our offices and see no tangible change to reflect
our work. A computer whose disk has been filled with our programs and papers looks very much like the computer on the day we purchased it. Maybe
this is why so many computer scientists' offices are decorated with screen
dumps of their work. No doubt the screen dumps are useful for a variety of
reasons, but perhaps their most important role is emotional. The screen
dumps on the wall tell the programmer that he or she is, at the end of the day,
producing something—that progress is being made. In some small measure,
the screen dumps cheer us up.

Unfortunately, the measure of cheer provided by such means remains
small. When science education and math instruction are centered in purely
virtual media, they foster relatively uninspiring physical surroundings. They
provide students with rooms, laboratories, and corridors unenlivened by the
students' own work and creations. And for many of us—perhaps most of
us—setting matters. Csikszentmihalyi *et al.* (1993), in their provocative study
of talented high school students, note that students who work within the arts
tend to have an easier time being motivated academically than those who
study math and science. Part of this motivation is provided by the studio setting in which the artists work:

[G]ifted young artists mostly work in a studio class at school. There they work by
themselves but are surrounded by peers engaged in similar activities. The drawing or
sculpture of one student is accessible to the others and therefore can be shared as it
progresses. The work itself can perhaps best be characterized as an *expressive performance.* . . . (p. 105)

For students of science and (perhaps more pointedly) mathematics, even
the most successful and enjoyable work often leaves no souvenirs, no re-
minders, no physical traces at all. To some extent this is just the way these
fields operate; solving a mathematical problem is an abstract activity whose
purpose is to leave us intellectually, not materially, enriched. But this culture
of asceticism can over time produce an unnecessary emotional strain. And it
may account for the sense of longing that occurs in the students described by
Pedersen (1988) as they encounter the polyhedral models in her office: "I

have these models in my office and students come in and beg to know how to make them. They never ask, 'What are they good for?' They know! And we know too."

Craft objects not only visually enrich the physical environment of the young mathematician and scientist, but they do so in a way that reflects the personal experience and progress of the students. In this sense, they take on biographical meaning that, say, a poster or store-bought object cannot. In the same vein, a craft object can play a role as a personal statement, as a public display, or as a gift. Repeatedly we have seen our students put their Hyper-Gami objects to just such uses. Polyhedral models and sculptures are given as presents, used as Christmas ornaments, and placed on display in the home or classroom. One student modeled her pet rooster in polyhedral form; another gave her favorite polyhedron a nickname. We ourselves have used Hyper-Gami models as thank-you notes, wedding and birthday gifts, and souvenirs of special occasions. In this way, HyperGami objects (and craft objects more generally) take on a versatile role as "social currency," and it is precisely along these day-to-day social dimensions that math and science education are typically so impoverished.

There is yet another, more intellectual role in which tangible objects excel: They symbolize and reflect a growth in skill over time. When a student creates and displays objects over the course of a semester, a year, or a period of several years, the display itself becomes a tangible reminder of the student's growing skills. Perhaps last year's mathematical mementoes are much simpler (or less polished) than this year's; perhaps last year's mechanical or electronic projects are less impressive; perhaps the student has become more adept at producing homemade scientific instruments. The crucial point is that, by contrast, the purely computational world offers little in the way of a continuous, unconsciously available record of progress. Of course a student can bring up a sequence of programs or simulations that he or she has written, but the act of bringing up that sequence is deliberate and tedious. There is little in the virtual world that is analogous to the simple process of viewing, without even meaning to, a shelf full of ever-improving craft objects.

Creating mathematical and scientific objects via handicrafts is an emotionally satisfying activity with an intellectual message: It serves to demystify the practice of math and science as a profession. A student who has created her own scientific instrumentation at home is just a little less likely to regard the laboratory with dread. A student who builds mathematical polyhedra in the classroom will be less likely to flinch when he encounters those solids in a later course in chemistry or solid geometry. A student who has built, and held, mechanical models in the garage has a sense not only of the principles but of the gritty aesthetics of engineering—the way materials bend, or heat with friction, or hold up with time, or smell when they are new. For these students/craftspeople, the scientific world is simply an outgrowth of their most homey, day-to-day activities.

Computation and the Affective Role of Crafts

How can the addition of computational media enhance the positive affective nature of scientific craftwork? Our experiences with HyperGami suggest two major answers to this question. First, by making the craftwork more personalized and expressive, computational media can nudge scientific crafts toward an activity that might more appropriately be called "scientific artwork." Rather than simply building polyhedral models, students are able to use the software to make objects with personal meaning (a dinosaur, a holiday Christmas tree, the aforementioned pet rooster). More generally, computational media allow us to rethink scientific crafts along new lines. An experiment with geometric optics can become (under the influence of the appropriate computational tool) a means for constructing artistic effects with light; a kit for building mechanical demonstrations can become a system for creating whimsical automata; a crystal-growing kit can become part of a larger application in which students design (on the computer screen) and then construct sculptures or tangible landscapes containing mineral forms.

A second answer to the question of how computation can enhance the affective role of crafts focuses on the notion of displaying or sharing objects. As we have noted, craft objects tend to be put on display or given as gifts; in either case, they are created in the anticipation of an audience other than their creator alone. Computational tools allow for craft objects to be shared, displayed, or documented in more powerful and inventive ways than were heretofore possible. In the case of HyperGami, students' works have been displayed on the World Wide Web, which gives them a huge potential audience; we ourselves have placed folding nets on the Web so that one of our sculptures could be recreated by visitors to the HyperGami website. The fact that HyperGami figures are created in a medium that is originally computational allows those figures to be duplicated, annotated, and altered in ways that are otherwise difficult. For instance, it is a relatively trivial matter to recreate the polyhedral fish sculptures of Figure 11.3 at, say, a larger scale by simply printing out the folding nets at a greater magnification. In the same vein, someone who wishes to recreate one of our sculptures with a different pattern of decoration would have little trouble simply taking the original (undecorated) folding nets and altering them; indeed, several of our younger students' projects have been of precisely this type.

Theme 2: Crafts and the Pace of Science Education

Working with craft materials takes time—often, a lot of time. Although a simple HyperGami project can generally be completed in an hour or two, a moderately complex project may easily stretch out over several evenings of construction. Of course, HyperGami is not especially unusual among craft activities in this respect. All sorts of crafts, from the creation of mosaics and stained-glass windows to the carving of wood, can at times be similarly slow-paced.

Is slow pace a problem? Set against the rhetoric of most educational research, it might seem so. Educational researchers—especially those in educational technology—often measure the success of a program by how much less time the students need to spend learning some material than they spent previously (thus one might see a claim that, because they used a particular application, students were able to pass a standardized algebra test in a third less time than they previously required). This is an impressive result, assuming that the study time is itself devoid of pleasure. If we are doing something distasteful but necessary, we certainly wish to accomplish the task in less time.

Perhaps, though, the success of science and math education should not be measured in terms of how fast we can get the ordeal over with. That metric may be appropriate for some training situations—when we need to acquire some specific skill in order to move on to more interesting work, for example—but it seems a joyless metric to apply to the educational enterprise as a whole. We might rather view the success of science and math education in terms of how much structure it is capable of bestowing on the student's leisure hours. A student who turns her room into a lab or his garage into a workshop, one who looks forward to returning to her telescope at night, one who mulls over a mathematical puzzle on the bus, and one who ponders a scientific question during a morning shower are those who have experienced a good scientific education. In contrast, a student who passes his tests in half the time and looks forward to spending that extra time in front of the television set is a problematic success story at best.

Scientific Crafts and the Rhythm of Science Education

Alfred North Whitehead, in his book *The Aims of Education* (1929), discusses the notion of a "rhythm" to the educational process:

Life is essentially periodic. It comprises daily periods, with their alternations of work and play, of activity and of sleep, and seasonal periods, which dictate our terms and our holidays. . . . These are the gross obvious periods which no one can overlook. There are also subtler periods of mental growth, with their cyclic recurrences, yet always different as we pass from cycle to cycle, though the subordinate stages are reproduced in each cycle. That is why I have chosen the term 'rhythmic,' as meaning essentially the conveyance of difference within a framework of repetition. Lack of attention to the rhythm and character of mental growth is a main source of wooden futility in education. (p. 29)

When technology is applied to science and math education, the rhythmic patterns that Whitehead describes seem, too often, to go unacknowledged. The pacing and style of many applications are reminiscent of arcade games—a mixture of constant animation, sound, and bright colors. Sometimes, of course, that sort of pace is exactly what a student might wish (we ourselves enjoy video games as much as the next person), but a relentless focus on one sort of pacing for science education leaves the student inexperienced in and uncomfortable with the slower, dreamier side of impassioned work.

We view the leisurely pace of craft activities within science education as a largely (though not uniformly) positive feature. True, a slower activity demands a longer attention span, but it also rewards that longer attention span and may thus serve over time to extend it. It distresses us to hear educational innovators simply assume that students cannot or will not concentrate—that their short attention spans are a *fait accompli*, to be accommodated to but not challenged by their tools. In our view, students' reflexive (and counter-productive) expectation of video-game pacing is a predictable and inevitable result of a culture of design that refuses to challenge itself.

Allowing for the pace of craft activities to invade science and math education is not, in other words, simply a challenge to the students' expectations; it is also a challenge to the preconceptions of the designers and educators and to the educational culture. Often, the real goal of introducing a craft activity is simply for the student to spend leisurely time contemplating the activity itself, getting the feel of the material and seeing a construction take shape over time. One would no more seek to compress the time period of this activity than to compress a performance of "Clair de Lune" into 30 seconds. And it is a healthy exercise—again in our view—for educators to assume that students will and should have periods of meditative time on their hands. Maybe that assumption is a bit of a fiction, but we think it's a fiction that can become a self-fulfilling prophecy.

The longevity—or at least the potential longevity—of craft objects is still another factor that deserves mention in this context. Craft objects, as suggested by the discussion in the previous section, have educational value that may be played out over a span of months or years. A HyperGami polyhedron on the shelf may, months later, suggest a starting point for a sculpture; or the student may suddenly notice a pattern in the shape that escaped her attention at the time the object was created. A craft object created by a fifth-grader in September, and displayed in a classroom, might be the subject of a discussion with its creator in April. Again, it is unusual for the products of classroom activity (such as workbook pages) to have any meaning or resonance for their creators more than a day or two after they are completed.

Computation and the Pacing of Scientific Craft Activities

Computational media have an important property that is useful in regulating the pacing of craft activities: They allow for a surer sense of direction and eventual success in the creation of a craft object. In our own experiences with HyperGami, we have noticed something that we have informally called the "it's-going-to-be-so-cool" effect; having begun the creation of a particular model, we can see that it is turning out well. When this effect takes hold in the midst of a construction project, it can prompt us to keep working for hours, just to see the final product.

Computational media can extend the "cool effect" of craftwork by giving the creator a relatively early view of what the finished product will (or might) be. In the case of HyperGami, the software enables the user to see what the decorated three-dimensional shape will eventually look like; sometimes it is precisely this view that reassures us that a shape is worth creating. In the absence of the sort of predictive power that computational design media offer, the craftsperson must proceed on sheer faith in the eventual product, and he may suffer intense disappointment when the result of hours of work is a product whose imperfections could have been apparent at the outset if presented by the appropriate design tool. These elements—faith, patience, and occasional disappointment—are arguably necessary elements of the crafting process. Nonetheless, having a computational tool that permits us to make sounder judgments about the eventual success of a crafting project might give the novice a less frustrating introduction to the process of crafting itself. Moreover, if well designed, such a tool should not entirely eliminate the element of risk (and those associated elements of faith, patience, and disappointment) that accompany crafting; rather, the tool would ideally enable expert crafters to raise their ambitions a bit higher.

Theme 3: Materials as Representatives of Specific Cultures

How Should We Think About Scientific Craft Materials?

Once we start considering the use of craft activities in science education, an astonishing variety of ways of interpreting materials and understanding them come to light. Materials—whether paper, Lego bricks (P2), FischerTechnik kits (P1), wooden blocks, modeling clay, yarn, balsa wood, stained glass, fabric, or myriads of others—are associated with their own particular cultures of usage. Although these cultures are rarely made explicit in discussions of scientific education, they seem to operate forcefully in the lives of students. Certain types of materials (such as wooden blocks) will be seen as appropriate for children of some ages, but not for others; a first-grade classroom will supply at least some different materials from a sixth-grade classroom.

One could devote volumes to a fine-grained critical analysis of the cultural implications of different craft materials, but it is important not to make too much of this sort of analysis, which can easily become a dry academic exercise. After all, a notion such as "the implicit culture of a material" is only an approximation. Some adults play with wooden blocks; some children work with oil paints; some little girls play with Lego bricks, some little boys with fabric. Still, it is worthwhile to explore these issues, if only to alert ourselves to potentially novel and productive ways of thinking about the design of scientific craft activities.

Our own experiences with HyperGami have suggested to us a number of dimensions along which to think about craft materials in science and mathematics education. These dimensions are not orthogonal—the placement of a certain material along one dimension will probably affect its placement along at least several others. But, for us at any rate, these dimensions offer some conceptual purchase on how to think about the overall landscape of scientific crafts.

Longevity/Permanence

Although some craft materials are especially short-lived (perhaps the most obvious examples are edible craft materials, such as chocolate), others (such as hardwood and ceramics) are meant to last. Materials such as candle wax and the paper of HyperGami constructions occupy an interesting middle ground. Paper polyhedra are hardly likely to last for decades, but they can easily last for months or years (we still have several early constructions that are over 4 years old). In some cases, additional steps may be taken to make materials last longer (for instance, we have recently begun spraying HyperGami constructions with various fixatives to prevent the colors from fading over time). Longevity is an interesting dimension in the design of scientific craft materials, because arguably those materials that can last for at least a half-year or so are capable of longer-term educational effects. Ephemeral constructions (such as the results of chemical experiments) may well have educational impact, but the creative designer of scientific crafts may think about ways of extending the longevity of such constructions. (In the case of chemical experiments, for instance, the designer might consider including in the chemistry kit certain experiments involving reactions that take place over a period of months.)

Reparability (Undoability/Redoability)

Some craft materials are designed so that constructions may be taken apart after a time. Many commercial modular materials—Lego blocks (P2), Polydrons (P3), and Zometool kits (P4), to name just a few—have this property: Once a project has been completed, it can be decomposed and its parts reused in some other project. Other craft materials—clay, paper, wood—are less likely to be used in creating "undoable" projects (a HyperGami piece, for instance, is meant to be folded once and for all). Reparability is an advantage in some ways. It means, after all, that unpromising projects can be stopped midway and that simpler projects can be used to supply parts for later ones. But reparability has other interesting consequences that are not always helpful. For one thing, "undoable" objects are, in a sense, impermanent (to recall the previous dimension), even if they are made of sturdy materials; a Lego construction is unlikely to last very long even if its individual pieces do. Undoable objects seem to inspire less emotional investment and to have less power as "social currency." This is perhaps because of the possibility of reuse: One is unlikely to give a Zometool construction as a gift, in part because one might later need the pieces of which it was made, in part because the recipient might be tempted to disassemble the gift! In any event, the reparability or undoability of craft materials has interesting implications for the uses of those materials.

Affordability

This is an obvious dimension to consider in thinking about educational crafts. Typically, the standard examples of "cheap" craft materials are yarn, paper, and water-based paints. It is worth mentioning, though, that a wide variety of newer, more specialized craft materials—suitable for science education projects—are in fact relatively affordable: "smart" materials such as "muscle wire" (Gilbertson 1993), diffraction gratings, flexible mylar mirrors, temperature-sensitive films, glow-in-the-dark paints, and so on. We have, in other writing, referred to such materials as middle-tech, somewhere on the spectrum between the obvious "high-tech" examples of digital logic and the "low-tech" examples of clay and yarn (Eisenberg and Eisenberg, 1998b). Thus, although affordability is clearly desirable in a scientific craft material, it is also a dimension along which a material's classification need not be forever fixed. After all, many of today's most commonplace materials (paper among them) were once rare and precious.

Intended Audience

Craft materials often seem designed with a specific audience in mind, at least within some particular surrounding culture. For example, the colors with which materials are made might suggest that they are intended for young children (bright primary colors). Or the fact that materials require fine motor control might suggest that they are intended for older students. Or perhaps the fact that materials (such as those in chemistry sets) have certain physical risks associated with them suggests that these materials are intended only for adults, or at least for students who have adult supervision. In our own experiences with HyperGami, we have noticed a general, age-related response to the medium of polyhedral paper sculpture: Whereas adults and young children have often expressed delight, teenagers (especially males) have often been noticeably cooler (Eisenberg and DiBiase, 1996). This has led us to rethink the types of examples that we now present to high school students. These examples now tend to de-emphasize the element of whimsy that, to teenagers, can seem perilously undignified.

Still other dimensions could be mentioned in a more thorough discussion of this kind: the portability of craft materials; their associated settings or infrastructure, and the opportunities that some materials might offer for collaborative work. Nonetheless, the four dimensions we have discussed at least suggest the lines of thought that our work with HyperGami has opened for us. These, of course, are dimensions that apply equally well to traditional and computationally enriched craft materials. In the following paragraphs, however, we note some specifically new dimensions that the advent of computation has introduced into scientific crafts.

Computation and New Ways of Thinking About Crafts

Role of Computation in Design

HyperGami, as an application, offers a clear illustration of one style of computationally enriched craft: In this style of work, the craftsperson does most of the explicit design work on the computer and then completes the construction using real-world materials. This paradigm, we believe, could profitably be extended to other sorts of scientific craft activities—geometric optics projects, crystal-growing sets, and kaleidoscope design, to name just a few. In any of these cases, one could well imagine beautiful software applications whose express purpose is to assist the student of science in creating and understanding real-world objects. And, like HyperGami, these applications could presumably expand the range of expressiveness or complexity of the craft, helping the designer create never-before-possible objects.

This is not, of course, the only way computational media can affect the design of craft objects. A software application might be used only to design part of some larger project (for example, an application might be used to decorate but not design balsa wood gliders); or an application might be used mainly to provide more elaborate or animated instructions on how to do the crafting itself (there exist CD-ROMs for teaching origami and paper airplane construction). Conceivably, software might be used in an expanded role beyond that illustrated by HyperGami. For instance, an application might be designed to assist the craftsperson not only in the design phase but during the process of physical construction itself. (One could imagine an application that helps students construct a terrarium and then later assists them in monitoring or analyzing the miniature ecosystem that they have created.)

Craft Objects with Embedded Computation

In most of the discussion thus far, we have assumed rather traditional views of computation and crafts as individual entities. That is, "computation" is provided by a large machine sitting on the desk (and probably connected to other large machines via networks), whereas "crafts" are traditional physical materials. An especially interesting direction for integrating computational media and crafts is to embed a certain degree of computation within the craft materials themselves. This is, of course, the direction Resnick and his colleagues have pursued creatively at the MIT Media Lab in developing the "programmable Lego brick" and its conceptual offspring (Resnick, 1993; Resnick et al., 1996). Likewise, the now-popular idea of "wearable computing" offers fascinating possibilities for integrating computational media and fabrics (Mann, 1997).

There are still other possibilities for integrating computation and crafts, especially those crafts that show up often in science education. One might envision, for example, a set of plastic biological models (such as those of the human heart or eye that are used in classrooms) augmented with computational elements so that students can observe phenomena (such as arrhythmia). Or the simple frameworks constructed for use with soap films (so that students can observe minimal surfaces generated by the films) might be made so that the frameworks shift or twist slightly after construction, revealing the ways in which minimal surfaces readjust under dynamic conditions.

Often, "embedded computation" within craft objects implies adding some sort of dynamic (as opposed to static) element, as in the two foregoing examples. But this needn't always be the case: Computational elements might simply be used for measuring, monitoring, or communicating. A homemade water-based barometer might be equipped with computational elements that signal an approaching storm or that simply record readings of water levels over a period of time for the purpose of later display. A homemade mobile might include a computational device whose purpose is to record and graph the movements of the mobile's arms over the course of a day.

The advent of small, light, flexible computational devices provides a richly fertile ground for experimenting with the integration of computers and crafts in novel ways. Indeed, we and our colleagues and students have used the MIT Media Lab's "cricket"—a recent, smaller and lighter version of the original programmable brick—in a wide variety of science-related projects. These include a "computationally enriched kaleidoscope" by A. Warmack, a cricket-driven dynamic color display by M. Burin, K. Johnston, and D. Olvera that employs tanks filled with water tinted in various shades, and a cricket-operated magnetic field sensor devised by T. Wrensch (Eisenberg and Eisenberg, 1998b). By thumbing through any catalog of scientific toys and crafts, or by strolling through the local science museum, one is sure to get new ideas for ways to use embedded computation within traditional scientific crafts.

Programmability/Reprogrammability/Adaptability

One final, and important, dimension worth noting in this discussion is that of programmability. Some examples of computational craftwork might employ computers as part of, say, clothing, to make it change color in different lighting conditions, but the behavior of the object under consideration is fixed by the designer and is not alterable by the user. In other cases, the user might be able to program the behavior of the object at the outset but not be able to alter the program thereafter (for instance, the user might initially specify the behavior of a programmable mobile or kaleidoscope). Or perhaps the user could reprogram the computational element in some restricted fashion but could not alter it while it was running within the constructed device. A more powerful possibility would be to allow the user to reprogram the computational elements of some craft object "on the fly," while the device is running; this would permit a user to reprogram or fine-tune, say, a programmable barometer while it was in the process of taking measurements. Finally, one might allow for the possibility of a certain level of internal adaptability in the device's programming, based perhaps on its use; for example, a computer-augmented home-built Van de Graaff generator might be constructed such as to permit higher-voltage demonstrations only after it had been successfully used in a certain number of (relatively safer) lower-voltage experiments.

New Directions in Computational Crafts

Over the last several years, we have come to derive such pleasure from integrating computation and papercrafts that it is often difficult for us to choose, like sage adults, which of the myriad lines of thought and work to pursue. Certainly, we plan more development for HyperGami itself. Recent additions to the program (besides one or two mentioned earlier in this chapter) include the implementation of loadable texture libraries, several new geometric operations on solids, and the (still experimental) development of "intelligent spatial advisors" to help users select possible customizations to perform on polyhedra (Eisenberg and Eisenberg, 1997c). In the somewhat longer term, we have begun a reimplementation of (much of) HyperGami in Java, which we hope will permit both a wider dissemination of the program and its further evaluation. And there are numerous additional research issues that we fervently wish to pursue within the narrow context of HyperGami development. These include representing the bending (and perhaps crumpling) of paper surfaces, representing additional paper-sculpture techniques (such as tearing or cutting slits in surfaces), representing sets of distinct polyhedra on the screen at one time (in the current version of HyperGami, each individual polyhedron must be constructed separately, and one does not typically view sets of polyhedra on the screen at one time).

Going beyond the specific domain of HyperGami, there is still much to do within the basic paradigm of the original program—that is, creating software applications to assist in the design of more complex or expressive scientific crafts. One might imagine an application to aid in the design of metal-ring topological puzzles, an application for the design of marionettes, an application to assist in the practice of creative glassblowing, an application for the design of new types of birdhouses, or one for the design of novel sorts of kites. Perhaps some of these ideas would fare better than others in practice, but there is a single notion behind them that is consistently worth exploring: that computational media (especially when augmented by a composable notation like a programming language) can enrich those activities that young scientists have historically found to be motivating and pleasurable.

There are other directions to pursue that transcend the basic paradigm of HyperGami-like applications. In the previous section, we alluded to new possibilities for integrating computational elements within craft objects themselves and gave a few primitive illustrations of the idea. We believe that many of the smaller, more ubiquitous pieces of scientific crafts—mirrors, motors, springs—might well be designed to include small amounts of embedded computation. (For example, one might imagine a simple "intelligent spring" that sends a signal if it is stretched beyond the limit at which it is well approximated by Hooke's law.) And there are natural ways in which the World Wide Web could augment the practice of crafts. For instance, it is quite plausible to imagine a world in which scientific craft objects routinely come with their own associated websites explaining how those objects were constructed. (Several ideas along these lines are mentioned in Eisenberg and Eisenberg, 1998a.)

It might even be possible to change the fundamental mindset with which craft objects are created. Traditionally, craft materials are not designed with educational purposes in mind. The early manufacturers of paper almost certainly never envisioned the development of mathematical papercrafts, and the first makers of soap probably never thought about the use of their invention for the study of minimal surfaces. In other words, the traditional relationship between the development of materials and the needs of science educators has been serendipitous; industry creates, and (on occasion) the world of scientific crafts catches a lucky break. Perhaps we can do better by starting from a perceived need in science education and attempting to design real-world crafts in response. Indeed—returning to the examples with which we began this paper—it may not be hopeless to design new craft activities or craft materials (perhaps with some sort of embedded computation) that illuminate concepts such as Newtonian mechanics in the absence of friction, wave/particle duality, objects moving near the speed of light, and evolutionary processes that typically take place over millennia. The real world, the world of crafts, is partly our own creation as designers, and the basic stuff of home science can itself be a target domain for innovation.

For us, these (admittedly futuristic) notions originated with paper. In designing, using, and teaching with HyperGami, we have come to feel that this system is most fruitful as an "object to think with"—a single instance of a much larger class of examples in which computers and traditional (or nontraditional) craft materials are integrated. There is something satisfying about using new technology to work within a tradition of paper geometric construction that dates back at least to Albrecht Dürer in the sixteenth century (Malkevitch 1988). And there is something satisfying in the varied pace of the HyperGami activity itself, in which abstract design on the computer screen is followed by patient and careful handling of paper in all its exquisitely tangible manifestations.

After all, it's a natural desire to employ all one's senses and cognitive powers in the course of a single project. We do not feel that a love of crafts is incompatible with technophilia, nor that an enjoyment of computer applications must detract from time spent in crafting. The world is not, or should not be at any rate, a battleground between the real and the virtual. It is instead a marvelous continuum—a source of wonders that blend and knead together the natural and the artificial, the traditional and the novel, the scientifically objective and the personally expressive, the tangible and the abstract. We anticipate a future in which ever more astonishing things will present themselves to our minds, and ever more astonishing ideas to our hands.

Acknowledgments

We are indebted to the ideas and encouragement of Hal Abelson, Robbie Berg, Fred Martin, Michael Mills, Mitchel Resnick, Brian Silverman, and Jim Spohrer, among many others. Zach Nies, Adrienne Warmack, Tom Wrensch, Andee Rubin, and Vennila Ramalingam all collaborated on the work described here. Thanks to Gerhard Fischer, Hal Eden, and the members of the Center for Lifelong Learning and Design at the University of Colorado for providing an intellectual home for our efforts. This work has been supported in part by the National Science Foundation and the Advanced Research Projects Agency under Cooperative Agreement CDA-9408607, and by NSF grants CDA-9616444 and REC-961396. The second author is supported by a fellowship from the National Physical Science Consortium, the first author by Young Investigator award IRI-9258684. Finally, we would like to thank Apple Computer, Inc. for donating the machines with which our research was conducted.

References

Abelson, H., & diSessa, A. 1980. *Turtle geometry*. Cambridge MA: M.I.T. Press.
Bernstein, J. 1993. *Cranks, quarks, and the cosmos*. New York: Basic Books.
Csikszentmihalyi, M. 1996. *Creativity*. New York: HarperCollins.
Csikszentmihalyi, M., Rathunde, K., & Whalen, S. 1993. *Talented teenagers*. Cambridge, England: Cambridge University Press.
Cundy, H. M., & Rollett, A. P. 1951. *Mathematical Models*. London: Oxford University Press.
Dawkins, R. 1996. *The blind watchmaker*. New York: Norton.
diSessa, A. 1982. Unlearning Aristotelian physics: A study of knowledge-based learning. *Cognitive Science, 6*, 37-75.
Eisenberg, M. 1995. Programmable applications: Interpreter meets interface. *SIGCHI Bulletin, 27:2*, 68-83.
Eisenberg, M., & DiBiase, J. 1996. Mathematical manipulatives as designed artifacts: The cognitive, affective, and technological dimensions. *Proceedings of the International Conference on the Learning Sciences, 1996*, Chicago, pp. 44-51.
Eisenberg, M., & Nishioka, A. 1997a. Orihedra: Mathematical sculptures in paper. *International Journal of Computers for Mathematical Learning, 1(3)*, 225-261.
Eisenberg, M. & Nishioka, A. 1997. Creating polyhedral models by computer. *Journal of Computers in Mathematics and Science Teaching, 16(4)*, 477-511.
Eisenberg, M., & Eisenberg, A. 1998a. Shop class for the next millennium. *Journal of Interactive Media in Education*.
Eisenberg, M., & Eisenberg, A. 1999. Middle tech: Blurring the division between high and low tech in education. In A. Druin, ed., The Design of Children's Technology, San Francisco: Morgan Kaufmann, 244-273.
Eisenberg, M., & Eisenberg, A. 1998c. Designing real-time software "advisors" for 3d spatial operations. In preparation.
Feynman R. 1985. *"Surely you're joking, Mr. Feynman!"* New York: Bantam Books.
Gilbertson, R. 1993. *Muscle wires project book*. San Rael, CA: Mondo-Tronics.

Hilton, P., & Pedersen, J. 1994. *Build your own polyhedra*. Menlo Park, CA: Addison-Wesley.

Horwitz, P., Taylor, E. F., & Barowy, W. 1994. Teaching special relativity with a computer. *Computers in Physics*, 8, 92-97.

Jenkins, G., & Wild, A. 1985. *Making Shapes, vols. 1, 2, and 3*. Diss, England: Tarquin.

Lightman, A., & Brawer, R. 1990. *Origins*. Cambridge, MA: Harvard University Press.

Malkevitch, J. 1988. Milestones in the history of polyhedra. In Senechal, M., and Fleck, G. (eds.), *Shaping space: A polyhedral approach*. Boston: Birkhäuser, pp. 80-92.

Mann, S. 1997. Wearable computing: A first step toward personal imaging. *IEEE Computer*, *30(2)*, 25-32.

Mehra, J. 1994. *The beat of a different drum: The life and science of Richard Feynman*. Oxford, England: Oxford University Press.

Pedersen, J. 1988. "Why study polyhedra?" In Senechal, M., and Fleck, G. (eds.), *Shaping space: A polyhedral approach*. Boston: Birkhäuser, pp. 133-147.

Resnick, M. 1993. Behavior construction kits. *Communications of the ACM*, *36(7)*, 64-71.

Resnick, M., Martin, F., Sargent, R., & Silverman, B. 1996. Programmable bricks: Toys to think with. *IBM Systems Journal*, *35(3)*, 443-452.

White, M., & Gribbin, J. 1992. *Stephen Hawking: A life in science*. New York: Dutton.

White, B., & Horwitz, P. 1987. ThinkerTools: Enabling children to understand physical laws. Report No. 6470, BBN Laboratories.

Whitehead, A. N. 1929. *The aims of education*. New York: Mentor Books (printed in 1949).

Software

S1: LightShip Software. *MacScheme*. Palo Alto, CA.

Products

P1: FischerTechnik , Waldachtal, Germany.

P2: Lego Systems, Inc. Billund, Denmark.

P3: Cuisenaire Co. of America. *Polydrons*. White Plains, NY.

P4: BioCrystal, Inc. *ZomeTool*. Boulder, CO.

12

Multisensory Immersion as a Modeling Environment for Learning Complex Scientific Concepts

Chris Dede

Marilyn C. Salzman

R. Bowen Loftin

Debra Sprague

The power of technology to change one's intellectual viewpoint is one of its greatest contributions, not merely to knowledge, but to something even more important: understanding . . . it goes beyond the limits of human perception.

Arthur C. Clark, Technology and the Limits of Knowledge

In every aspect of our knowledge-based society, fluency in understanding complex information spaces is an increasingly crucial skill (Dede and Lewis, 1995). In research and industry, many processes depend on people utilizing complicated representations of information (Rieber, 1994). Increasingly, workers must navigate complex information spaces to locate data they need, must find patterns in information for problem solving, and must use sophisticated representations of information to communicate their ideas (Kohn, 1994; Studt, 1995). Further, to make informed decisions about public-policy issues such as global warming and environmental contamination, citizens must comprehend the strengths and limitations of scientific models based on multivariate interactions. In many academic areas, students' success now depends on their ability to envision and manipulate abstract multidimensional information spaces (Gordin and Pea, 1995). Fields in which students struggle with mastering these types of representations include mathematics, science, engineering, statistics, and finance.

Research on learning scientific concepts yields insights into why understanding complex information spaces is difficult. Many scientific domains deal with abstract and multidimensional phenomena that people have difficulty comprehending. Mastery of abstract scientific concepts requires that students build flexible and runnable mental models (Redish, 1993). Frequently, these scientific models describe phenomena for which students have no real-life referents (Halloun and Hestenes, 1985a) and incorporate invisible factors and abstractions (Chi, Feltovich, and Glaser, 1991; White, 1993). Students learning science need to be able to sift through complex information spaces, identifying what is important and what is not and recognizing critical patterns and relationships. Learners may need to translate among reference frames, to describe the dynamics of a model over time in order to predict how changes in one factor influence other factors, and to reason qualitatively about physical processes (McDermott, 1991).

Developing effective pedagogical strategies and simulation technologies for teaching complex science concepts presents a substantial challenge for educational researchers and instructional designers. Despite the utilization of new teaching approaches, tools, and technologies, students struggle with abstractions in science. They not only enter their courses with gaps and inaccuracies in their conceptual understanding of the material but also often leave with unaltered misconceptions (Halloun and Hestenes, 1985b; Reif and Larkin, 1991). Students' lack of real-life referents for intangible phenomena, coupled with an inability to reify ("perceptualize") abstract models, is an important aspect of this problem. In an effort to help students comprehend abstract information spaces, finding ways to utilize our biologically innate ability to make sense of physical space and perceptual phenomena seems a promising approach.

Using Models and Simulations to Convey Complex Scientific Concepts

Guided inquiry experiences using scientific models that reveal the shortcomings of learners' current conceptual frameworks can help wean students from erroneous beliefs. Before the formal representations that scientists use are introduced, these models can develop learners' abilities to understand intuitively how the natural world functions. Fostering in students the ability to predict qualitatively the behavior of phenomena under investigation is a valuable foundation for teaching them to manipulate quantitative formulas. Also, students are not empty vessels to be filled with theories; they have firmly held, often erroneous beliefs about how reality operates. Model-based instruction can help learners' existing mental models evolve into more accurate conceptions of reality.

To date, uses of information technology to apply these pedagogical principles have centered on creating computational tools and two-dimensional virtual representations that students can manipulate to complement their memory and intelligence in constructing more accurate mental models. Perkins (1991) classifies the types of "constructivist" paraphernalia instantiated via information technology as information banks, symbol pads, construction kits, phenomenaria, and task managers. Transitional objects (such as Logo's "turtle") are used to facilitate translating personal experience into abstract symbols (Papert, 1988; Fosnot, 1992). Thus, technology-enhanced constructivist learning currently focuses on how representations and tools can be used to mediate interactions among learners and natural or social phenomena.

However, high-performance computing and communications capabilities are opening up new possibilities in modeling scientific phenomena (Dede, 1995). Like Alice walking through the looking glass, the virtual reality interface enables learners to immerse themselves in distributed, synthetic environments. They can become "avatars" (computer-graphics representations that serve as personas of human participants in the virtual world) who collaborate in inquiry-based "learning by doing" and use virtual artifacts to construct knowledge. The key features that virtual reality (VR) adds to modeling as a means of constructivist learning are

- *Immersion:* Learners develop the subjective impression that they are participating in a "world" that is comprehensive and realistic enough to induce the willing suspension of disbelief (Heeter, 1992; Witmer and Singer, 1994). By engaging students in learning activities, immersion may make important concepts and relationships more salient and memorable, helping learners to build more accurate mental models. Also, inside a head-mounted display, the learner's attention is focused on the virtual environment and is not subject to the distractions present in many other types of educational environments.
- *Multiple three-dimensional representations and frames of reference:* Spatial metaphors can enhance the meaningfulness of data and provide qualitative insights (Erickson, 1993). Enabling students to interact with spatial representations from various frames of reference may deepen learning by providing different and complementary insights (Arthur, Hancock, and Chrysler, 1994).
- *Multisensory cues:* Via high-end VR interfaces, students can interpret visual, auditory, and haptic (tactile) displays to gather information, while using their proprioceptive system to navigate and control objects in the synthetic environment. This potentially deepens learning and recall (Psotka, 1996).
- *Motivation:* Learners are intrigued by interactions with well-designed immersive "worlds," which encourages them to devote more time and concentration to a task (Bricken and Byrne, 1993).
- *Telepresence:* Geographically remote learners can experience a simultaneous sense of presence in a shared virtual environment (Loftin, 1997).

By using a VR interface, instructional designers can not only display how a model can aid in interpreting a scientific phenomenon but can also enable learners (1) to experience being part of the phenomenon and (2) to participate in a shared virtual context within which the meaning of this experience is socially constructed.

The Potential of Multisensory Immersion for Learning Scientific Concepts

The virtual reality interface has the potential to complement existing approaches to science instruction. By themselves becoming part of a phenomenon (a student becomes, for example, a point-mass undergoing collisions in an immersive virtual environment without gravity or friction), learners gain direct experiential intuitions about how the natural world operates. In particular, good instructional design can make those aspects of virtual environments that are useful in understanding scientific principles salient to learners' senses. For example, in two-dimensional Newtonian microworlds, students often ignore objects' velocities, focusing instead on position. In our comparable immersive environment, NewtonWorld, learners "inside" a moving object are themselves moving; this three-dimensional, personalized frame of reference directs attention to velocity as a variable. In NewtonWorld, we heightened this saliency by using multisensory cues to convey multiple, simultaneous representations of relative speeds. As another example of the power of "perceptualization," learners who struggled with the concepts underlying our vector-field-based immersive environment, MaxwellWorld, reported that representations providing redundant data simultaneously through visual, auditory, and haptic stimuli aided their comprehension. Transducing data and abstract concepts (such as energy) into mutually reinforcing multisensory representations may be an important means of enhancing understanding of scientific models.

In addition, researchers are documenting that the social construction of knowledge among students in a shared, text-based virtual environment makes possible innovative, powerful types of collaborative learning. As we discuss later, adding immersive, multisensory representations to these textual "worlds" might increase communicative and educational effectiveness. Overall, we believe that various aspects of multisensory immersion, when applied to scientific models, can provide learners with experiential metaphors and analogies that (1) help them understand complex phenomena that are remote from their everyday experience (relativity, quantum mechanics) and (2) help displace "common-sense" misconceptions with alternative, more accurate mental models.

Challenges in Using Virtual Reality for Learning

In spite of its potential benefits, many barriers intrinsic to current virtual reality technology can block students' mastery of scientific concepts. These challenges to educational design are as follows:

- Virtual reality's physical interface is cumbersome (Krueger, 1991). Head-mounted displays, cables, 3-D mice, and computerized clothing all can interfere with interaction, motivation, and learning.

- Display resolution is inversely proportional to field of view. A corresponding trade-off exists between display complexity and image delay (Piantanida, Boman, and Gille, 1993). The low resolution of current VR displays limits the fidelity of the synthetic environment and prevents virtual controls from being clearly labeled.
- VR systems have limited tracking ability with delayed responses (Kalawsky, 1993).
- Providing highly localized 3-D auditory cues is challenging because of the unique configuration of each person's ears. Also, some users have difficulty localizing 3-D sounds (Wenzel, 1992).
- Haptic feedback is extremely limited and expensive. Typically, only a single type of haptic feedback can be provided by computerized clothing; for example, one glove may provide heat as a sensory signal but cannot simultaneously provide pressure. In addition, using computerized clothing for output can interfere with accurate input on users' motions.
- Virtual environments require users to switch their attention among the different senses for various tasks (Erickson, 1993). To walk, users must pay attention to their haptic orientation; to fly, they must ignore their haptic sense and focus on visual cues. Also, as Stuart and Thomas (1991) describe, multisensory inputs can result in unintended sensations (such as nausea due to simulator sickness) and unanticipated perceptions (such as being aware of virtual motion but feeling stationary in the real world).
- Users often feel lost in VR environments (Bricken and Byrne, 1993). Accurately perceiving one's location in the virtual context is essential to both usability and learning.
- The magical (unique to the virtual world) and literal (mirroring reality) features of VR can interact, reducing the usability of the interface (Smith, 1987). Also, some researchers have demonstrated that realism can detract from learning rather than enhance it (Wickens, 1992).

As virtual reality technology evolves, some of the challenges to educational design will be overcome. At present, however, achieving the potential of immersive, synthetic worlds to enhance learning requires transcending these interface barriers through careful attention to usability issues.

Another class of potential problems with the use of immersive virtual worlds for education is the danger of introducing new or unanticipated misconceptions that result from the limited nature of the "magic" possible via this medium. For example, learners will not feel their sense of personal physical weight alter, even when the gravity field in the artificial reality they have created is set to zero. The cognitive dissonance that this mismatch creates, which reflects conflicting sensory signals, may create both physiological problems (such as simulator sickness) and false intellectual generalizations. One aspect of our research is examining the extent to which manipulating learners' visual, auditory, and tactile cues may induce subtle types of misconceptions about physical phenomena. The medium (virtual reality) should not detract from the message (learning scientific principles).

The Virtual Worlds of ScienceSpace

ScienceSpace is a collection of virtual worlds that we designed to explore the potential utility of physical immersion and multisensory perception to enhance science education (Dede, Salzman, and Loftin, 1996). ScienceSpace now consists of three worlds—NewtonWorld, MaxwellWorld, and Pauling-World—in various states of maturity. All three worlds are built using a polygonal geometry. Colored, shaded polygons and textures are used to produce detailed objects. These objects are linked together and given behaviors through the use of NASA-developed software (VR-Tool) that defines the virtual worlds and connects them to underlying physical simulations. Interactivity is achieved through the linkage of external devices (such as a head-mounted display) using this same software. Finally, graphics-rendering, collision detection, and lighting models are provided by other NASA-developed software.

Our hardware architecture includes a Silicon Graphics Onyx Reality Engine2 four-processor graphics workstation, Polhemus magnetic tracking systems (with a 3Ball or stylus), and a Virtual Research VR4 head-mounted display. One Polhemus tracker is in the 3Ball or stylus that the participant holds in one hand; a second is mounted on a fixture and held in the other hand; and a third is mounted on the head-mounted display. The hand holding the 3Ball or stylus is represented in the virtual world as a hand with the index finger extended (aligned with the user's hand). Attached to the second tracker is a menu system. Sound is produced by a Silicon Graphics Indy workstation and delivered via head-mounted display headphones and external speakers. Vibrations are delivered to a subject's torso by a "vest" with embedded subwoofers. This interface enables us to immerse students in 3-D virtual worlds using the visual, auditory, and haptic senses. Students use a virtual hand (controlled by the 3Ball), menus, and direct manipulation to perform tasks in these immersive virtual environments.

NewtonWorld

NewtonWorld provides an environment for investigating the kinematics and dynamics of one-dimensional motion. In NewtonWorld, students spend time in and around an activity area, which is an open "corridor" created by colonnades on each side and a wall at each end (see Figure 12.1). Students interact with NewtonWorld using a "virtual hand" and a menu system, which they access by selecting the small 3-Ball icon in the upper left corner of the head-mounted display. Students can launch and catch balls of various masses and can "beam" (teleport) from the ball to cameras strategically placed around the corridor. The balls move in one dimension along the corridor, rebounding when they collide with each other or the walls. Equal spacing of the columns and lines on the floor of the corridor aid learners in judging distance

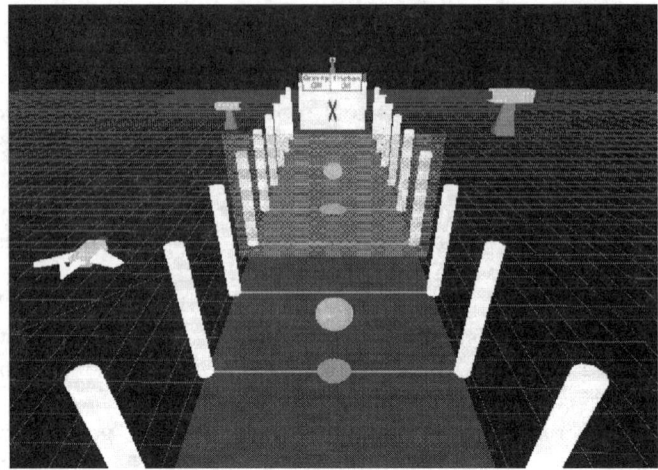

FIGURE 12.1 Above the corridor, showing cameras, balls with shadows, and the far wall.

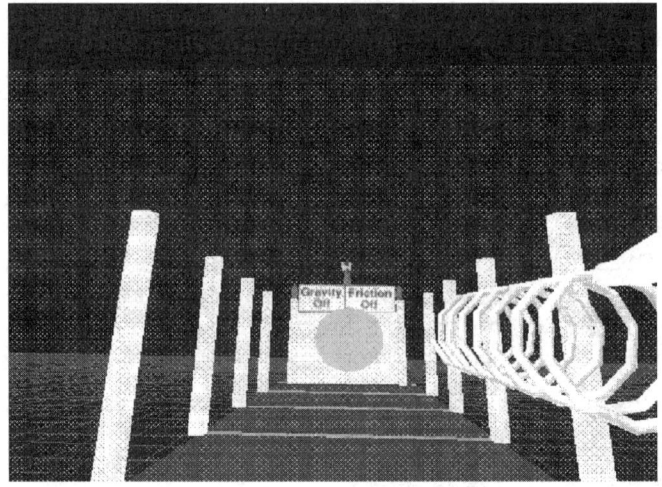

FIGURE 12.2 After launch, illustrating the spring-based launching mechanism.

and speed. Signs on the walls indicate the presence or absence of gravity and friction.

Multisensory cues help students experience phenomena and direct their attention to important factors such as mass, velocity, and energy. For example, potential energy is made salient through tactile and visual cues, and velocity is represented by auditory and visual cues. Currently, the presence of potential energy before launch is represented by a tightly coiled spring, as well as via vibrations in the vest. As the ball is launched (Figure 12.2) and po-

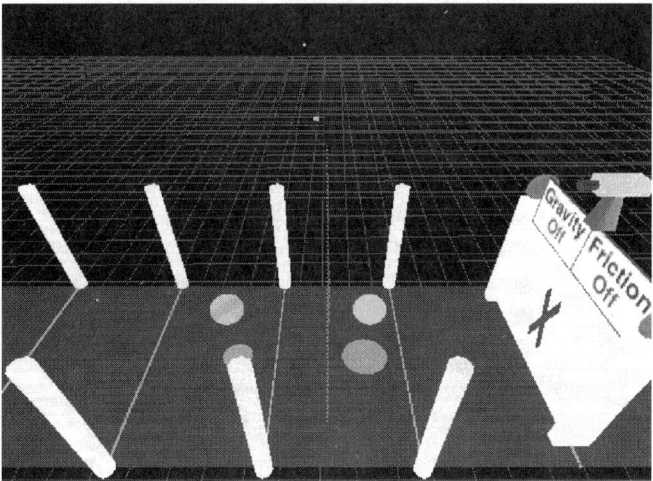

FIGURE 12.3 A collision seen from the center-of-mass reference frame.

tential energy becomes kinetic energy, the spring uncoils and the energy vibrations cease. The balls now begin to cast shadows whose areas are directly proportional to the amount of kinetic energy associated with each ball. On impact, when kinetic energy is instantly changed to potential energy and then back to kinetic energy again, the shadows disappear and the vest briefly vibrates. To help students judge the velocities of the balls relative to one another, the columns light and chime as the balls pass.

Additionally, we provide multiple representations of phenomena by allowing students to assume the sensory perspectives of various objects in the world. For example, students can become one of the balls in the corridor, a camera attached to the center of mass of the bouncing balls (Figure 12.3), or a movable camera hovering above the corridor. Figure 12.4 shows a collision seen from just outside one colonnade. These features aid learners in understanding the scientific models underlying Newton's three laws, potential and kinetic energy, and conservation of momentum and energy.

NewtonWorld was the first virtual environment we built, so its current interface does not incorporate the sophisticated features we developed in designing MaxwellWorld and PaulingWorld. Accordingly, we are redesigning NewtonWorld to take advantage of these new capabilities. Figures 12.5 and 12.6 show sketches illustrating our redesign, which is presently under construction. New features include a "scoreboard" to help learners relate qualitative and quantitative representations, an improved interface based on a "roadway" metaphor, three levels of interaction that support progressively more complex types of learning activities, and the inclusion of perfectly elastic and perfectly inelastic collisions. Dede, Salzman, Loftin, and Ash (in preparation) provides additional details of our design strategies and early research results for NewtonWorld.

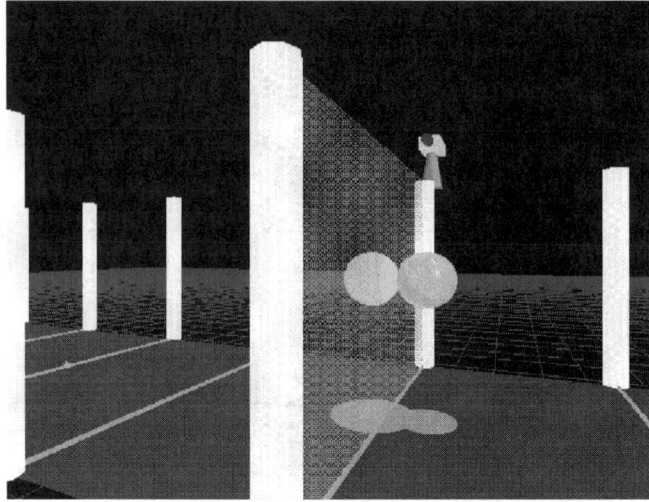

FIGURE 12.4 A collision seen from just outside a colonnade.

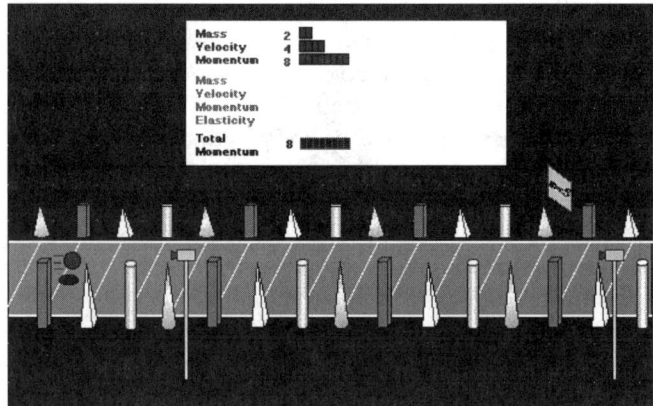

FIGURE 12.5 Level 1 of redesigned NewtonWorld, showing "scoreboard" and "roadway."

PaulingWorld

PaulingWorld makes possible the study of molecular structures through a variety of representations, including quantum-level phenomena. PaulingWorld is in its early stages of development. Learners can view, navigate through, superimpose, and manipulate five different molecular representations: wire-frame, backbone, ball-and-stick, amino acid, and space-filling models. See Figures 12.7 through 12.10 for examples of these models. To design the immersive multisensory representations and underlying scientific models that

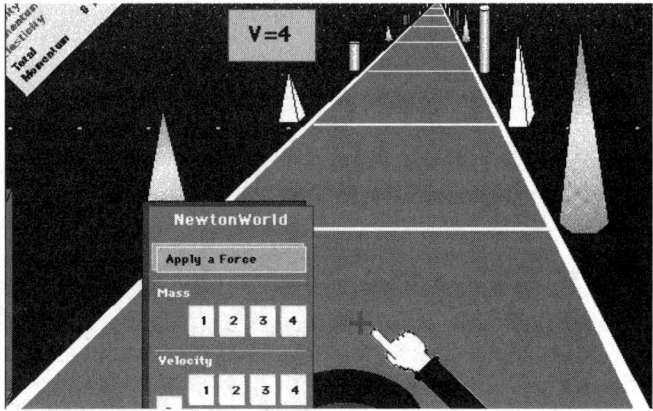

FIGURE 12.6 Within the "roadway" view.

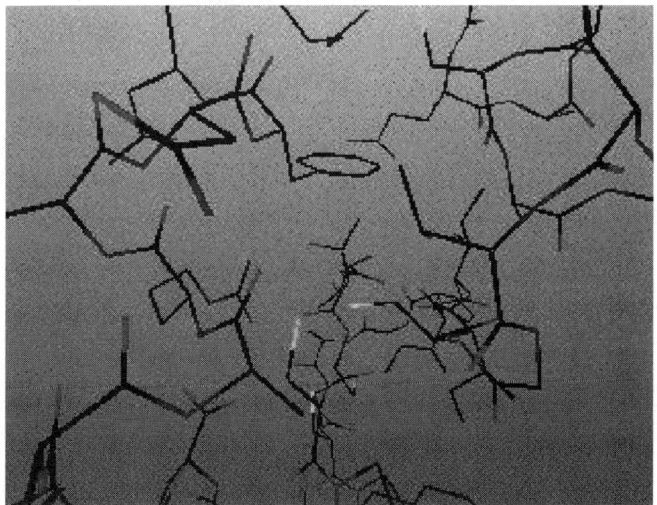

FIGURE 12.7 Wireframe model.

we will use for quantum-mechanical bonding phenomena, we are working with an NSF-funded project, "Quantum Science Across the Disciplines," led by Peter Garik at Boston University (http://qsad.bu.edu/).

MaxwellWorld

Although we discuss examples from all three of our virtual worlds, this chapter focuses on our design and evaluations of MaxwellWorld as an illustration of how models based on multisensory immersion can aid in the learning of complex scientific concepts. To date, we have collected more research data

FIGURE 12.8 Backbone model.

FIGURE 12.9 Ball-and-stick model with some amino acids.

on learning in MaxwellWorld than in our other virtual environments, and MaxwellWorld also illustrates some particularly interesting applications of scientific modeling to education.

MaxwellWorld allows students to explore electrostatic forces and fields, learn about the concept of electric potential, and "discover" the nature of electric flux. The fieldspace in this virtual world occupies a cube approximately 1 meter on a side, with Cartesian axes displayed for convenient reference. The small size of the world produces large parallax when viewed from nearby, making its three-dimensional nature quite apparent.

FIGURE 12.10 Space-filling model.

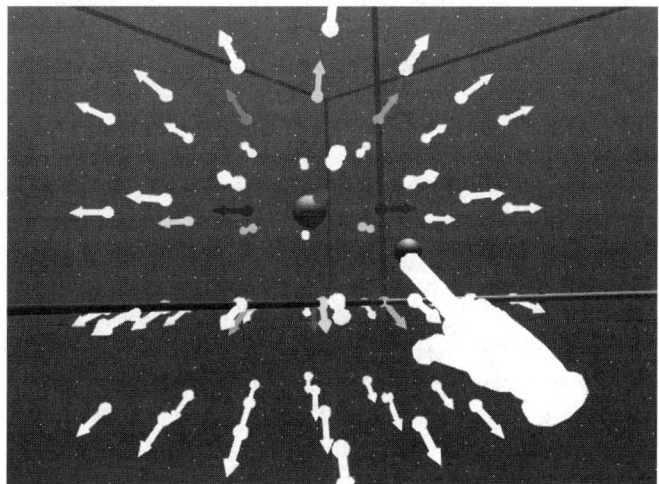

FIGURE 12.11 User exploring a field with test charges and field lines.

Students use a virtual hand, menu, direct manipulation, and navigation to interact with this world (see Figure 12.11). The virtual hand is attached to the 3Ball, which is held in one hand. The menu is attached to the tracker held by the other hand. Attaching the menu to the user's other hand allows students to remove the menu from their field of view, while keeping it immediately accessible. Students select menu items by holding up the menu with one hand, pointing to the menu option with the virtual hand, and depressing the 3Ball button (see Figure 12.12). Thus menu selection in MaxwellWorld is similar to menu selection on two-dimensional interfaces in which users manipulate the menu with a cursor controlled by a mouse. MaxwellWorld also utilizes direct

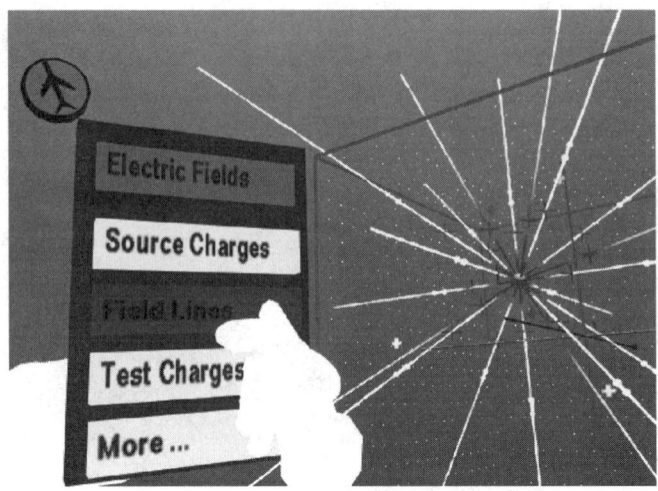

FIGURE 12.12 Activating the menu via the virtual hand.

manipulation. Once users have selected objects from the menu, they can place them in the world, move them around, and delete them. Finally, users can change their location by selecting the navigation mode via the menu, pointing the virtual hand in the desired direction, and depressing the 3Ball button.

Our decision about which vector field phenomena and representations to incorporate into MaxwellWorld was based on the advice of our domain expert, Dr. Edward Redish from the University of Maryland. Using the virtual hand, students can place both positive and negative charges of various relative magnitudes into the world. Once a charge configuration is established, learners can instantiate, observe, and interactively control model-based scientific representations of the force on a positive test charge, electric field lines, potentials, surfaces of equipotential, and lines of electric flux through surfaces. For example, a small, positive test charge can be attached to the tip of the virtual hand. A force meter associated with the charge then depicts both the magnitude and the direction of the force of the test charge (and hence the electric field) at any point in the workspace. A series of test charges can be "dropped" and used to visualize the nature of the electric field throughout a region. In our most recent version of MaxwellWorld, learners can also release a test charge and watch its dynamics as it moves through the fieldspace (Figure 12.13) and then "become" the test charge and travel with it as it moves through the electric field.

In like manner, an electric field line can be attached to the virtual hand. A student can then move his or her hand to any point in the workspace and see the line of force extending through that point. MaxwellWorld can also display many electric field lines to give students a view of the field produced by a charge configuration (Figure 12.11). In another mode of operation, the tip of the virtual hand becomes an electric "potential" meter that, through

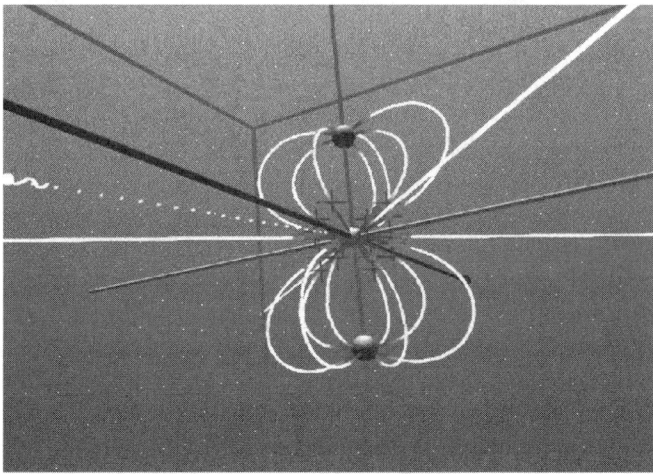

FIGURE 12.13 Dipole with moving test charge.

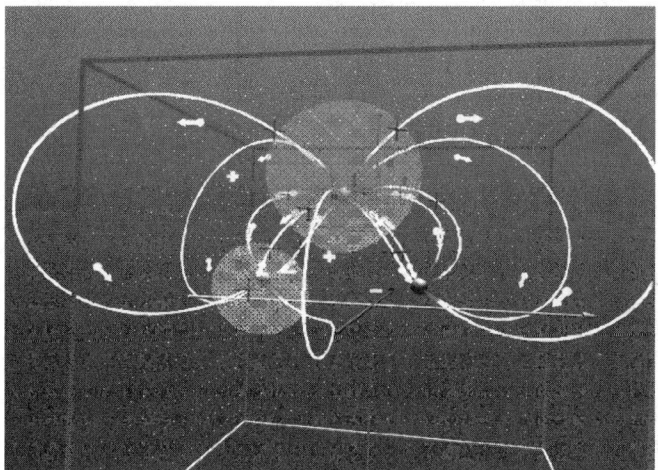

FIGURE 12.14 Tripole with equipotential surface.

a simple color map and a "=" or "−" sign on the finger tip, allows students to explore the distribution of potential in the fieldspace. By producing and manipulating equipotential surfaces, learners can watch how the shapes of these surfaces change in various portions of the fieldspace (Figure 12.14). By default, the surfaces are colored to indicate the magnitude of the potential across the surface; however, the student can also choose to view the electric forces as they vary across the surface. This activity helps students distinguish between the concepts of electric force and potential.

Via the production of a "Gaussian" surface, the flux of the electric field through that surface can be visually measured. Gaussian surfaces can be placed anywhere in the workspace by using the virtual hand to anchor the sphere; the radius (small, medium, or large) is selected from the menu. This representation enables students to explore flux through a variety of surfaces when these are placed at various points in the field. All these capabilities combine to make it possible to represent many aspects of the complex scientific models underlying vector field phenomena.

Conducting Research on ScienceSpace

We have developed elaborate, customized assessment methodologies for evaluating the usability and learnability of our ScienceSpace worlds (Salzman, Dede, and Loftin, 1995). Although infrequent, potential side effects such as "simulator sickness" necessitate the inclusion of special questions and protections to ensure users' comfort. Moreover, because each person evolves a unique psychomotor approach to interacting with the three-dimensional nature of physical space, individuals appear to have much more varied responses to 3-D, multimodal interfaces than to the standard 2-D graphical user interface with menus, windows, and mouse. Evaluating the multisensory dimensions of an immersive virtual world adds another level of complexity to the assessment process.

Thus, portions of our protocols center on calibrating and customizing the virtual world's interface to that particular learner. Throughout the sessions, we carefully monitor the learning process and record student comments and insights. We also videotape the hours of time we spend with each subject so that we can study these records for additional insights. Finally, our protocols are designed so as to help us capture various aspects of the learner's experience, in addition to assessing educational outcomes. By focusing on the students' experience as well as on their learning, we gain insights that guide the refinement of the user interface and help us understand how to leverage VR's features for modeling science.

The following is a summary of the four issues our protocols are designed to assess.

- *The VR experience*. The VR experience can be characterized along several dimensions. We have focused on usability, simulator sickness, meaningfulness of our models and representations, and motivation. Our most recent addition has been the inclusion of questions to assess how immersed students feel in the modeling environment.
- *Learning*. Here our goal is to determine whether and how students progress through learning tasks within the virtual environment, to assess their mastery of concepts at both the descriptive and the causal levels (discussed later), and to assess whether their learning can be generalized to other domain-specific problems.
- *The VR experience vs. learning*. We want to understand the relationship between the VR experience and learning and to identify when the two may conflict.

- *Educational utility*. We hope to be able to demonstrate that the system is a better (or worse) teaching tool than other pedagogical strategies by comparing the quality and efficiency of learning among alternatives that vary in cost, instructional design, and pedagogical strategy.

This careful evaluation strategy is generating data from which we are gaining insights into how multisensory immersion can enhance learning, as well as how virtual reality's usability can be enhanced. Many of the strategies underlying these assessment methodologies and instruments are also generalizable to a wide range of synthetic environments beyond VR.

MaxwellWorld: Formative Evaluation

In the summer of 1995, we assessed our initial version of MaxwellWorld as a tool for remediating misconceptions about electric fields and teaching concepts with which students are unfamiliar. During the sessions, we administered one to three lessons on the construction and exploration of electric fields (electric force, superposition, test charges, field lines); electric potential (potential and kinetic energy, potential difference, work, potential vs. force); and the concept of flux through open and closed surfaces, leading up to Gauss's law.

Our observations during these sessions, students' predictions and comments, usability questionnaires, interview feedback, and pre- and post-test knowledge assessments helped us determine whether this early version of MaxwellWorld aided students in correcting any of their preexisting misconceptions and in learning underlying scientific concepts with which they were not familiar. These experiences were also valuable in developing modifications to MaxwellWorld to enhance learning outcomes.

The findings that follow are based on 14 high school students and 4 college students who participated in these evaluations. Thirteen of the 14 high school students had recently completed their senior year; 1 student had recently completed his junior year. All students had completed at least a course in high school physics. Each session lasted for approximately 2 hours. Students were scheduled on consecutive days for the first two sessions, and the third session was conducted approximately 2 weeks later.

All of the students enjoyed learning about electric fields in MaxwellWorld. When asked about their general reactions to MaxwellWorld, a majority of the students commented that they felt it was a more effective way to learn about electric fields than either textbooks or lectures. Students cited the three-dimensional representations, the interactivity, the ability to navigate to multiple perspectives, and the use of color as characteristics of MaxwellWorld that were important to their learning experiences.

Evaluations before and after lessons showed that lessons in MaxwellWorld deepened the students' understanding of the distribution of forces in an electric field and enhanced their comprehension of the scientific models that interrelated representations such as test charge traces and field lines. Manipulating models of the vector fields in three dimensions appeared to play an important role in their learning. Several students who were unable to describe the distribution of forces in any electric field before using MaxwellWorld gave clear descriptions during the post-test interviews and demonstrations. Also, manipulating field lines and traces in three dimensions helped students visualize the distribution of force. As an illustration, one student expected field lines to radiate from a single charge along a flat plane and was surprised to see that they radiated in three dimensions. Another student expected to see field lines cross but found that this could not occur.

Although this initial version of MaxwellWorld helped students qualitatively understand three-dimensional (3-D) superposition, students had difficulty applying superposition when solving post-test problems. Learners appeared to understand the concept of superposition during the lessons and particularly enjoyed the demonstrations of superposition (moving the source charges dynamically changes the traces and field lines); they often alluded to this during the post-testing. However, many of them had difficulty applying superposition to post-test demonstrations and sketches, a result that indicates the need to refine our modeling and instruction.

This early version of MaxwellWorld extended traditional 2-D representations to include (1) the third dimension; (2) the ability to manipulate representations as a means of understanding the dynamics of electrostatic models; and (3) two-color schemes to measure and distinguish the magnitude of the force on, and the potential experienced by, test charges, field lines, and equipotential surfaces. These representational capabilities helped students to deepen their understanding of physics concepts and models. The post-test outcomes showed that students were able to learn about flux through open and closed surfaces using MaxwellWorld. All students performed well during post-testing, demonstrating an understanding of important and difficult-to-master concepts such as Gauss's law, field vs. flux, and directional flux.

Although only four of the students used MaxwellWorld to learn about electric potential, all of them demonstrated that they could visualize the distribution of potential for basic charge arrangements, interpret the meaning of a distribution of potential, identify and interpret equipotential surfaces, relate potential difference and work, and describe some of the differences between electric force and electric potential. All were particularly surprised to see (1) 3-D representations of equipotential surfaces, particularly in the case of a bipole (two charges of the same size and magnitude), and (2) the varying nature of forces over an equipotential surface.

We observed significant individual differences in the students' abilities to work in the 3-D environment and with 3-D controls, as well as in their susceptibility to the symptoms of simulator sickness (eye strain, headaches, dizziness, and nausea). Some students learned rapidly to use the menus, manipulate objects, and navigate, but others required guidance throughout the sessions. Most students experienced nothing more than slight eyestrain; however, two students experienced moderate dizziness and slight nausea during the first session and consequently did not return for the second session. No student complained of any symptoms during the first 30–45 minutes of the lesson, an outcome that reinforced our strategy of using multiple, short learning experiences.

These "lessons learned" from an early formative evaluation of Maxwell-World are consistent with evaluative data collected on our other Science-Space worlds (Salzman, Dede, and Loftin, 1995). To wit,

- Enabling students to experience phenomena from multiple perspectives appears to help them understand complex scientific concepts and models.
- Multisensory cues appear to engage learners, direct their attention to important behaviors and relationships, help them understand new sensory perspectives, prevent errors through feedback cues, and enhance ease of use.
- Simulator sickness and system usability pose potential threats to the learning process.
- Talk-aloud protocols employing a cycle of prediction-observation-comparison (White, 1993 and Chapter 10) are highly effective for administering lessons and for monitoring usability and learning in VR modeling environments.

Our early evaluations of MaxwellWorld indicate that using this type of scientific modeling helped students to deepen their understanding of electric fields and electric potential and to correct misconceptions about these phenomena. However, these studies did not establish the extent to which students' learning was due to (1) the method of instruction (the lessons), (2) scientific models and representations that could have been used equally well in a conventional 2-D modeling environment, or (3) the unique features of multisensory immersion in virtual reality.

MaxwellWorld: Multisensory Immersion vs. Conventional 2-D Representations

In January 1996, we initiated an extended study designed to accomplish two goals: (1) compare learning and usability outcomes from MaxwellWorld to those from a highly regarded and widely used two-dimensional microworld, EM Field™, which covers similar material, and (2) assess the usability and learnability of an enhanced version of MaxwellWorld with additional modeling and representational capabilities suggested by results from the initial formative evaluation described above.

The first stage of this study compared MaxwellWorld and EM Field on the extent to which representational aspects of these simulations influenced learning outcomes. EM Field runs on standard desktop computers and presents learners with 2-D representations of electric fields and electric potential, using quantitative values to indicate strength (Trowbridge and Sherwood, 1994). To make the two learning environments comparable, we removed some of MaxwellWorld's more powerful features and designed lessons to utilize only those features of MaxwellWorld for which EM Field had a counterpart; this limited version of MaxwellWorld we designated MW_L. With these conditions constraining the functionality of the VR environment, the

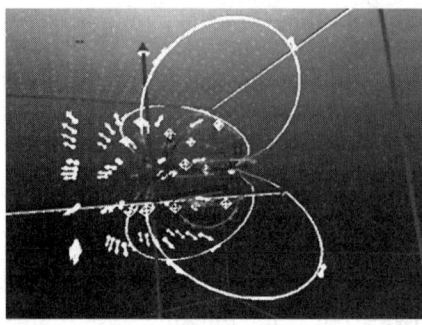 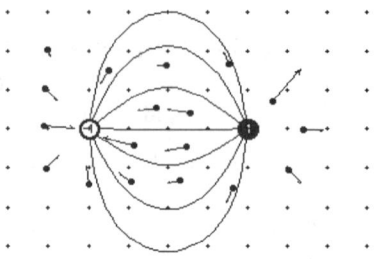

FIGURE 12.15 A dipole with field lines and test charge in MW_L (left) and EM Field (right).

primary differences between the simulations were representational dimensionality (EM Field's 2-D vs. MW_L's 3-D) and type (EM Field's quantitative vs. MW_L's qualitative). See Figure 12.15.

In the second stage of the study, we utilized MaxwellWorld's full range of capabilities (including multisensory input) to determine what value these features added to the learning experience. Through the pre-test for stage two, we also examined the extent to which students, after a period of 5 months, retained mental models learned in either environment. Through this two-stage approach, we hoped to separate the relative contributions of 3-D representation and multisensory stimulation as instrumental to the learning potential of virtual reality.

Initial Hypotheses

Our initial hypotheses for this two-phase study were as follows:

Learning. Learning can occur along three dimensions. First, there is *conceptual understanding*—students' ability to define key concepts and describe interrelationships among significant representations in the scientific model. Second, there is *2-D understanding*—students' abilities to create and interpret 2-D representations of the phenomena (the ability to illustrate concepts on paper). Third is *3-D understanding*—students' abilities to create and interpret full 3-D representations of the phenomena (this reflects their ability to visualize the true three-dimensional nature of the concept). We expected students to learn along all three dimensions while completing lessons in either MW_L or EM Field.

Learning in EM Field vs. MW_L. Our previous experience with VR learning environments indicated that 3-D simulations are likely to facilitate the construction of more thorough and accurate mental models of intrinsically three-dimensional phenomena. Therefore, we hypothesized that students who used MW_L would perform better on conceptual questions than those who used EM Field. Additionally, students in earlier studies of MaxwellWorld had demonstrated the ability to represent phenomena using 2-D sketches after working in the 3-D simulation. However, because students learning in MW_L would need to translate 3-D information into 2-D information for the tests, we did not expect their performance to exceed the performance of students working in a 2-D environment, who would not need to perform this translation. Therefore, we hypothesized that students who used MW_L would not perform significantly worse on questions requiring two-dimensional understanding than those who use EM Field. Finally, we felt that working in 3-D would result in a better three-dimensional understanding of the phenomena than working in 2-D. Therefore, we hypothesized that students who used MW_L would demonstrate significantly better three-dimensional understanding than those who used EM Field.

Learning in the Full Version of MaxwellWorld. In phase two of the study, with the constraints on MaxwellWorld's performance removed, we hypothesized that students would identify the multisensory representations as valuable for their learning.

Retention. We hypothesized that, over a 5-month period, students would experience significantly greater retention of learning in MW_L than in EM Field.

We also identified several additional factors that would probably influence the learning experience and outcomes:

Simulator Sickness. Our work with virtual realities suggested that many students would experience mild symptoms of simulator sickness, particularly eye strain, at some point during their use of the system. We hypothesized that MW_L and MaxwellWorld students would experience significantly more such symptoms than EM Field students. We further expected simulator sickness to interfere with learning.

Nature of the Learning Experience. Inherent to any human–computer interaction are the subjective experiences of usability, motivation, and ability to understand the representations. We expected students to find MW_L and MaxwellWorld more motivating and meaningful than EM Field. We hypothesized that such greater motivation might result in increased learning (through such factors as increased student attention and concentration). We also expected students to find MW_L and MaxwellWorld more difficult to use than EM Field. However, because careful design of the interface and lessons had greatly reduced usability problems in MaxwellWorld, we hypothesized that these interface challenges characteristic of VR would not interfere with learning.

Stage One of the Comparison Study

The first stage of this study was completed by 14 high school students (12 males and 2 females). All students had 1½ years of high school physics and were recruited from a physics class in a local high school. (The gender disparity in the sample population reflects the fact that relatively few women take high school physics.) Students' performance in their science and math classes varied; their grades ranged from A through C (A's and B's were the norm). Although students were advanced in their knowledge of physics relative to the typical high school population, the pre-test they took at the start of this study indicated that most remembered little about electric fields and electric potential (this confirms the limits of conventional approaches to teaching this type of scientific material). Students participated in two 2-hour learning experiences, completing lessons in either EM Field or MW_L. The lessons focused on electric fields and electric potential, mirroring concepts covered in our initial formative evaluation of MaxwellWorld.

Procedure and Materials

Evaluations were conducted in our virtual reality lab, where we videotaped and logged student–administrator interactions. Male and female students were assigned randomly to one of two groups: EM Field and MW_L. Both groups were equivalent in terms of their science background, and both groups of students participated in two sessions. Before the sessions, students were asked to complete Recruitment questionnaires. During Session 1, students completed a Background questionnaire and a Pre-lesson test, and then they completed Lesson 1. During Session 2, students completed Lesson 2, which was followed by a Post-lesson test, Experience questionnaire, and Interview. Immediately following each lesson, students also completed a Simulator Sickness questionnaire. Lessons required approximately 1 hour 15 minutes to complete, and sessions lasted for approximately 2 hours. Students were given a break-time approximately half-way through each lesson.

Background Questionnaire and Recruitment Questionnaires. These questionnaires were used to characterize our sample for this study. They elicited information about the students' demographic characteristics, educational backgrounds, attitudes toward learning science, and any history of motion sickness.

Lessons. Lessons were constructed so that the informational content and learning activities were the same for both groups. Lesson 1 focused on the construction and exploration of electric fields, Lesson 2 on electric potential and its relationship to the electric field. These lessons were administered orally by the test administrator. Learning activities in the lessons relied on a cycle of "predict-observe-compare." This served two purposes: to help us gauge the students' understanding and progress, and to prime them for the upcoming activity. Each successive learning activity built on the previous activities, increasing both in level of complexity and in the integration of information that was necessary.

Pre- and Post-lesson Tests. We used two versions of a Pre-/Post-lesson test to assess learning. Half of the students in each group were randomly assigned to receive version A for the pre-test and version B for the post-test, and vice versa. These tests examined three dimensions of understanding for each concept: conceptual understanding (ability to define concepts), two-dimensional understanding (ability to create and interpret 2-D representations), and three-dimensional understanding (ability to create and interpret 3-D representations). The first two sections were administered on paper. The third section was administered orally, and students used physical 3-D manipulatives to demonstrate their understanding.

Experience Questionnaire. This questionnaire was used to assess the nature of the learning experience. It consisted of a series of 7-point anchored rating scales related to usability, motivation, and ability to understand the representations. Here is an example: "Using the menu system was . . . [very difficult −3] . . . to . . . [+3 very easy]."

Simulator Sickness. The Simulator Sickness questionnaire, SSQ (Kennedy, Lane, Berbaum, and Lilienthal, 1993), consisted of a series of 4-point ratings of symptoms associated with simulator sickness. It can be analyzed to yield oculomotor, disorientation, nausea, and overall simulator sickness scores.

Interviews and Qualitative Data. To help us understand the nature of the statistical outcomes, as well as to diagnose strengths and weaknesses of EM Field and MaxwellWorld and of our lessons, we collected the following qualitative data: students' predictions and observations throughout the lessons; their comments, likes, and dislikes; their suggestions for improvement; and their reflections on the learning process.

Analyses and Results

Stage one of our study yielded the following outcomes:

Learning. As anticipated, students learned as a result of completing the lessons with either MW_L or EM Field. Students were better able than before to define concepts, describe concepts in 2-D, and demonstrate concepts in 3-D.

Learning in MW_L vs. EM Field. MW_L students were better able to define concepts than EM Field students. Also, MW_L students were not any worse than EM Field students at sketching concepts in 2-D. A closer examination of the sketches shows that although MW_L students performed better on the force sketches, they performed worse on the sketches related to potential, which resulted in total sketch scores that were similar for the two groups. Finally, MW_L students were better able than EM Field students to demonstrate concepts and their underlying scientific models in 3-D. Despite the inherent three-dimensionality of the demonstration exercises (as well as our use of the terms *surface* and *plane* in the lessons), EM Field students typically confined their answers to a single plane; drew lines when describing equipotential surfaces; and used terms such as *circle*, *oval*, and *line*. In fact, only one of the seven students in the EM Field group described the phenomena in a three-dimensional manner. In contrast, MW_L students described the space using 3-D gestures and referred to equipotential surfaces by using terms such as *sphere* and *plane*.

Simulator Sickness Scores. As we anticipated, MaxwellWorld's immersive VR environment induced more symptoms associated with simulator sickness than did EM Field's monitor-based 2-D environment. In MW_L, overall simulator sickness scores tend to be slightly higher on day 1 than on day 2. This may have been due to adaptation to the VR environment. In a result consistent with our earlier research findings, there appear to be large individual differences in the way students react to the VR environment. We found that simulator sickness scores did *not* significantly predict learning outcomes. Though it constituted a minor nuisance, simulator sickness did not interfere with mastering the material.

The Nature of the Learning Experience. Students rated MW_L as more motivating than EM Field. Ideally, we would like to see the ratings for motivation even higher. However, we suspect that the intensity and the controlled nature of the lessons may prevent students from feeling extremely motivated during the learning experience. We found that motivation scores did not significantly predict learning outcomes, and motivation alone did not account for the differences in learning in each group.

Students found using the various features of MW_L significantly more difficult than using EM Field. Further, ratings for the ability to understand MW_L's representations were slightly, though not significantly, higher than the ratings for EM Field. The variability in ratings was greater for EM Field than for MW_L, which suggests that there were more individual differences in ability to extract information from the EM Field representations than from the MW_L representations.

Student Comments. Students' comments provide further insight into the nature of the learning experience. Overall, students described MW_L as easy to use, interesting, and informative. They especially liked the three-dimensional representations, the ability to see phenomena from multiple perspectives, and the interactivity of the system. MW_L students found using the 3Ball and virtual hand somewhat challenging and indicated that the responsiveness of MW_L was problematic at times. Students described EM Field as very easy to use but somewhat boring. They found the simplicity of its graphics both a strength and a weakness. Additionally, more MW_L students than EM Field students indicated that they found it easy to remain attentive during sessions.

Stage Two of the Comparison Study

Procedure and Materials

During stage two, we examined the "value added" by the full power of MaxwellWorld's multisensory representations. Seven EM Field and MW_L students returned for stage two, conducted approximately 5 months after stage one. All students experienced the full power of MaxwellWorld, receiving an additional lesson (built upon the concepts taught in earlier lessons) that relied on multisensory cues to supplement the visual representations. The auditory and haptic representations that were used provided, simultaneously, information redundant to that expressed through the visual sensory channel. We also assessed stage-one retention at the beginning of stage two. (The retention test was an abbreviated version of the Post-lesson test used in stage one.)

Analyses and Results

No statistically significant differences in retention outcomes were observed. However, with only seven participants, this stage of the study had very low power. Our limited data suggest that with a larger number of subjects, retention of 3-D understanding might be significantly higher for MaxwellWorld participants than for EM Field participants.

Data for stage two did yield insights into the value of multisensory representations. Students learned from visual and multisensory representations used in the lesson and demonstrated significantly better understanding of concepts, 2-D sketches, and 3-D demos after the lesson than before it. Ratings concerning multisensory representations (haptic and sound), post-lesson understanding, and student comments all suggest that learners who experienced difficulty with the scientific concepts found that multisensory representations helped them understand more than purely visual representations did.

Summary of This Comparison Study

Both stages lend support to the thesis that immersive 3-D multisensory representations can help students develop more accurate and causal mental models than 2-D representations. Learning outcomes for stage one show that MW_L learners were better able than EM Field learners to understand the space as a whole, recognize symmetries in the field, and relate individual visual representations (test charge traces, field lines, and equipotential surfaces) to the electric field and electric potential. MW_L students appeared to visualize the phenomena in 3-D, whereas EM Field students did not.

Subjective ratings for stage one yielded converging evidence that the virtual worlds' representational differences were responsible for differences in learning. In stage one, students rated the representations used in MW_L as easier to understand than the representations used in EM Field. Second, differences in learning could not be attributed solely to motivation (which was higher in MW_L than in EM Field). Additionally, MW_L students learned more even though they experienced more usability problems and simulator sickness. Finally, during interviews, students cited MW_L's immersive 3-D representations as one of its key strengths. In stage two, the enhancement of visual representations with multisensory cues appeared to enhance learning, especially for students who had trouble grasping the concepts.

Outcomes of this study support the following findings related to modeling scientific concepts:

• Virtual modeling experiences such as those provided by EM Field and MW_L should be integrated with initial instruction to avoid the forming of misconceptions that are difficult to remediate later. Although students in both the EM Field and the MW_L group demonstrated a better overall understanding of the topics on the post-test than on the pre-test, some students with a moderate knowledge of electrostatics at pre-test benefited less than students who demonstrated little or no knowledge at pre-test. In addition, some of the more advanced students who had misconceptions appeared to have a hard time overcoming them despite experiences in the virtual worlds.

- Immersive 3-D multisensory representations such as those used in MW_L may facilitate the students' development of comprehensive and runnable mental models more effectively than the 2-D representations of EM Field. Learning outcomes, subjective ratings, and comments from both stages one and two all provide evidence that supports this finding.
- Although no new types of misconceptions were introduced by conducting the learning experiences in an immersive environment, students have a number of misconceptions about electrostatic phenomena, and some of them are difficult to remediate. Working with the students yielded insights into the nature of their preexisting misconceptions. For example, learners have a strong tendency to think of charges in an electric field independently, and they have trouble describing the nature of superpositional fields and potential for sets of charges. Experiences in both MW_L and EM Field clearly helped students to think about this issue, but the students still had some difficulty understanding regions between sets of charges. In addition, field line representations are notoriously difficult to comprehend. Even after use of EM Field or MW_L, several students continued to have misconceptions about the meaning of field lines, although most learners gained a greater understanding of this representational formalism. At the conclusion of the lessons in both systems, some students did not fully understand how the electric field influences charged objects and the interrelationship between potential and force. Modeling environments and activities must be carefully designed to try to avoid such shortcomings.

Although the subject population is small, the results of this study suggest that the three-dimensional nature of VR is an aid to learning and that the virtual reality experience can be more meaningful and motivating for students than comparable 2-D microworlds. Given that many capabilities of Maxwell-World were suppressed in this study, these findings are a promising indication of the potential of immersive scientific models to enhance educational outcomes.

Next Steps in Our Virtual Reality Research

Over the next 2 years, we plan to extend our current research on the Science-Space worlds along several dimensions. We will conduct tests on Maxwell-World (on immersive frames of reference and on multiple sensory channels) to examine the contribution of "perceptualization" to scientific model-based learning. Using the revised version of NewtonWorld, we also intend to examine how, by facilitating innovative types of student collaborations, virtual reality may enhance the nature of social constructivist learning. These three planned studies are described in more detail below.

Further, to examine challenges in curriculum integration and in classroom implementation, we will move our VR worlds out of laboratory environments into precollege settings. In public-school classrooms (fifth grade for NewtonWorld, twelfth grade for MaxwellWorld), we plan to integrate VR experiences into science instruction. By this means, we hope to find out whether our laboratory results on learning and usability hold up in the more complex environment of schools and to determine how students and teachers adapt VR environments to their needs and interests.

Understanding Frames of Reference as a Means of "Perceptualization"

We believe that transforming current scientific visualization tools into "perceptualization" experiences could augment their power to boost learning. We have documented that adding multisensory perceptual information helped students who were struggling to understand the complex scientific models underlying NewtonWorld and MaxwellWorld. Providing experiences that leverage human pattern recognition capabilities in three-dimensional space, such as shifting among various frames of reference (points of view), also extends the perceptual nature of a visualization. These enhanced visualization (or perceptualization) techniques facilitate student experiences that increase the saliency and memorability of abstract scientific models, potentially enhancing the learning process.

By using frames of reference in virtual reality, we can provide learners with experiences that they would otherwise have to imagine. For example, we can enable students to become part of a phenomenon and experience it directly. Alternatively, we can let learners step back from the phenomenon to get a global view of what is happening. One frame of reference may make salient information that learners might not notice in another frame of reference. Further, multiple frames of reference might help students fill in gaps in their knowledge and become more flexible in their thinking.

Although there are numerous frames of references, many can be classified as one of two types: exocentric or egocentric (McCormick, 1995; Wickens and Baker, 1995). See Figure 12.16. In our MaxwellWorld study on frames of references and perceptualization, the two concepts learners will be asked to master are the distribution of force in electric fields and the motion of test charges through electric fields. Comprehending distribution depends more heavily on global judgments than on local judgments, whereas understanding motion requires more local judgments than global judgments. We will examine how the egocentric frames of reference, the exocentric frames of refer-

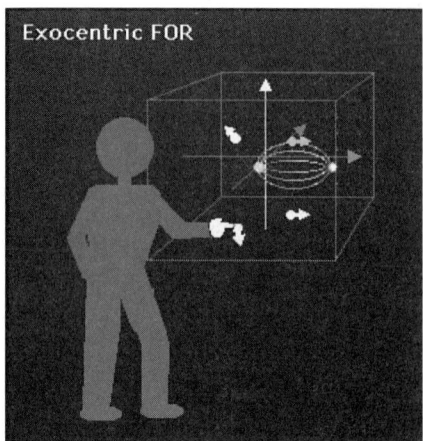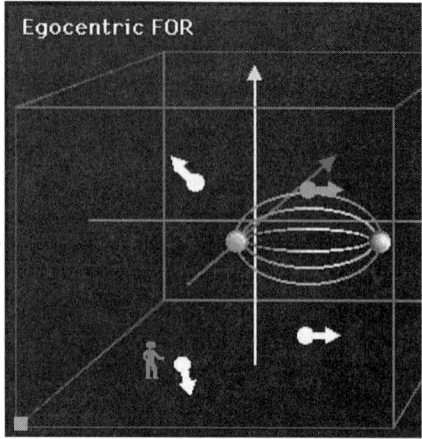

FIGURE 12.16 Exocentric and egocentric frames of reference in MaxwellWorld.

ence, and the be-centric frames of reference (utilization frames of reference of both types) shape mastery of these two types of material.

Mastery of scientific models can be assessed on two levels: descriptive and causal. If an individual can describe what he or she is examining and identify patterns in the data, his or her mastery can be characterized as *descriptive*. If that person can further interpret the meaning of the patterns and manipulate the information for problem solving, his or her mastery is *causal* as well. Causal mastery reflects a deeper understanding of the information; it is what we seek to accomplish in teaching learners about scientific models. We will examine both descriptive and causal mastery, as they are reflected in performance on frame of reference learning tasks, as a means of providing insights into the strengths and weaknesses of frames of references.

Multisensory Cues and Perceptualization

Through a study of visual, auditory, and haptic (touch/pressure) sensory cues, we plan to extend our explorations on how multisensory immersion influences learning. Various sensory modalities can provide similar, mutually confirming input. Or, through each sensory channel conveying different data, they can increase the amount of information conveyed to the learner. Little is known about what level of redundancy in sensory input is optimal for learning or about how much information learners can process without sensory overload. Moreover, each sense uniquely shapes the data it presents (for example, perceived volume and directionality of sound are nonlinear, vary with the pitch of the input, and are idiosyncratic to each person). Such complex considerations affect any decision about which sensory channel to use in presenting information to learners. Virtual reality provides a good research environment for exploring these design issues, as well as for exploring how multisensory immersion shapes collaborative learning.

Immersive Collaborative Learning as a Means of Enhancing Social Constructivism

As a near-term research initiative in our ScienceSpace worlds, we will investigate the effectiveness of collaborative learning situations in which three students in the same location rotate roles among (1) interacting with the world via the head-mounted display, (2) serving as external guide, and (3) participating as a reflective observer. We also plan to experiment with collaborative learning among distributed learners inhabiting a shared virtual context. The student would act and collaborate, not as himself or herself, but behind the mask of an "avatar": a surrogate persona in the virtual world. Loftin (1997) has already demonstrated that two users can simultaneously manipulate a shared immersive environment by using a communications bandwidth as low as that of a standard ISDN telephone line. By adapting military-developed distributed simulation technology, we could scale up to many users in a shared, interactive virtual world.

Collaboration among learners' avatars in shared synthetic environments may support a wide range of pedagogical strategies (such as peer teaching, Vygotskian tutoring, and apprenticeship). In addition, adding a social dimension makes technology-based educational applications more intriguing to those students who are most motivated to learn when intellectual content is couched in a social setting. In virtual environments, however, interpersonal dynamics provide leverage for learning activities in a manner rather different from typical face-to-face collaborative encounters (Dede, 1995). We believe that our ScienceSpace worlds offer an intriguing context for extending such work on social constructivism in virtual environments. Physical immersion and multisensory stimulation may intensify many of the psychological phenomena noted above, and "psychosocial saliency" may be an interesting counterpart to perceptual saliency in enhancing learning. Important issues to be addressed include the relative value of providing learners with graphically generated bodies and the degree to which the "fidelity" of this graphical representation affects learning and interaction (here fidelity is not simply visual fidelity but also the matching of real body motions to the animation of the graphical body).

Lessons Learned to Date on Learnability and Usability in Virtual Reality

What generalizations can we make about model-based science learning from our research to date with immersive multisensory virtual environments? On the basis of lessons learned from all our ScienceSpace worlds, we are developing design heuristics, assessment methodologies, and insights, some of which are applicable to a range of educational-modeling environments beyond virtual reality.

Learning and Knowledge Representation

Our goal is to develop an overarching theory of how learning difficult, abstract material can be strongly enhanced by scientific models instantiated via multisensory immersion and perceptualization. The following are illustrative themes applicable across all the virtual worlds we have created.

- Multisensory cues can engage learners, direct their attention to important behaviors and relationships, prevent interaction errors through feedback cues, and enhance perceived ease of use.
- The introduction of new representations and perspectives can help students gain insights that help them correct misconceptions formed through traditional instruction (many representations used by physicists are misleading for learners), as well as aiding learners in developing accurate mental models. Our research indicates that qualitative representations (such as the shadows that show kinetic energy in NewtonWorld) can increase saliency for crucial features of both phenomena and traditional representations.
- Three-dimensional representations seem to help learners understand phenomena that pervade physical space. Being immersed in a 3-D environment is also motivating for learners.
- Learner motivation is high in virtual reality environments, even when novelty effects wear off. The inclusion of interactivity, constructivist pedagogy, and challenge, curiosity, fantasy, and beauty (Malone and Lepper, 1984) all seem to augment student interest and involvement.
- Initial experiences in working with students and teachers suggest that it may be possible to achieve collaborative learning by having two or more students work together and take turns "guiding the interaction," "recording observations," and "experiencing activities" in the VR environment. Extending this to collaboration among multiple learners located in a shared synthetic environment may further augment learning outcomes, as may features (such as a "Hall of Fame") that provide social recognition for learner achievements.
- In addition to pre-test and post-test assessments of learning, continuous evaluation of progress through lessons is critical to diagnosing the strengths and weaknesses of the virtual worlds. We have found that talk-aloud protocols employing a cycle of prediction-observation-comparison are highly effective for monitoring usability and learning.

We believe these early results are clear evidence that VR has the potential to facilitate certain types of scientific-model-based learning more effectively than any other pedagogical modalities.

Challenges in Current Virtual Reality Interfaces

We have found the following usability issues to be characteristic of virtual reality interfaces.

- Limitations of the physical design and optics in today's head-mounted displays may cause discomfort for users. Because the visual display is an integral part of interaction and communication of information in these learning environments, these limitations presently reduce usability and learning. Delays in VR system response time can also be a factor with complex environments. Both of these problems are becoming less serious as hardware technology advances.
- Immersion does present some challenges for lesson administration (for example, students in the head-mounted display are not able to access written instructions or to complete written questions). We have found that oral interaction works well.
- Students exhibit marked individual differences in interaction style, ability to interact with the 3-D environment, and susceptibility to simulator sickness.
- To help learners utilize educational virtual worlds, calibrating the display and virtual controls for each individual is important. Additionally, monitoring and systematically measuring simulator sickness is vital, because malaise signals interface problems and also can explain why a student is having trouble with certain learning activities.
- Spreading lessons over multiple VR sessions appears to be more effective than covering many topics in a single session. For example, although students began to challenge their misconceptions during a single 3-hour NewtonWorld session, many had trouble synthesizing their learning during post-testing. We believe that factors such as fatigue and cognitive overload in mastering the interface influenced these outcomes. In contrast, we completed our MaxwellWorld evaluations over several sessions, tackling fewer topics during each session and dedicating less time per session to pre-testing or post-testing. Reviews and post-tests demonstrated that students were better able to retain and integrate information over multiple lessons.

In our judgment, none of these interface challenges precludes developing compelling learning experiences in virtual reality.

Implications of Our Work for Using Models in Science Education

The results of our research can inform larger debates within the science education community on the best practice in using models and simulations to help students learn complex scientific concepts. Some of the issues actively discussed among researchers studying the utility of models for learning science are listed here (Feurzeig, 1997). After each topic, we present our beliefs about what ScienceSpace research contributes to resolution of the issue.

The tension between computer-based modeling activities, and real-world observation and laboratory experimentation.

The Debate. In interacting with a model, learners are manipulating a representation of reality that can simplify complex scientific concepts and their interrelationships. However, unless carefully designed, models can oversimplify reality in a manner that later makes deeper understanding of phenomena harder to attain. Still, models that go beyond simulation to allow learners to change underlying variables and relationships—to illustrate how the idealized phenomenon functions by altering it in ways not possible in reality—can make possible a kind of meta-understanding that cannot be achieved via real-world experimentation. Yet real-world phenomena are more "real" to learners: more believable, more fully sensory. On the other hand, some complex scientific concepts (such as relativity and quantum mechanics) involve intangible phenomena unobservable in the everyday macroscopic settings to which learners have access. Models are the only means by which students form nonabstract impressions of such phenomena. Given these relative strengths and limits, what should be the pedagogical balance between interacting with models and experiencing reality itself?

Our Contribution. Models based on multisensory immersion give learners experiences closer to the perceptual aspects of reality than any other simulation medium. Our research suggests that virtual reality is a potentially powerful means of bridging the gap between models and real-world experimentation by combining strengths of each: the sensorial, immersive involvement of real-world experiences and the emphasis on crucial variables for understanding that models can provide (in our work, through perceptual saliency). In our research so far, we have not found that carefully designed "almost real" models induce new types of learner misconceptions. However, we do believe transitional learning experiences that gradually remove the affordances of models to reveal the full complexity and confusion of reality are important for generalizability and transferability of learning. The best pedagogical strategy may involve beginning with real-world experiments to show the complexity and counterintuitive nature of phenomena, then using models to simplify the situation and to enhance comprehension via interactive representations, and finally combining and extending the models to show how the complexity of real-world behavior emerges from a multiplicity of simultaneous underlying causes.

The tension between modeling in science research and modeling in science education.

The Debate. This issue concerns the differences between modeling by experts and modeling by novices—in particular, between the modeling tools used by scientists and those used by precollege students. Some researchers claim that under the guidance of professionals, typical students (especially at the secondary school level) can learn scientific concepts by using the same models and supercomputing facilities used by research scientists. Others insist that all but the brightest high school students need specially designed modeling tools and applications to introduce them to model-based inquiry.

Our Contribution. In our design of representations for virtual reality, we have noted that part of the difficulty in mastering complex scientific concepts is the misleading representational formalisms and terminology that have emerged historically in science and now are entrenched as standard professional notation. Students come to us with misconceptions that appear to be linked to these traditional representations. We find that, despite our best efforts to compensate for the shortcomings of these formalisms, students sometimes remain confused about how to relate conventional representations to reality and how to use standard scientific terminology to convey their ideas. Two examples from electrostatics illustrate this point.

First, from their prior physics instruction, many of our learners in Maxwell-World have initial misconceptions linked to the "field line" representation. For experts, field lines are a quick way of ascertaining the direction of a vector field along a series of points. However, novices understandably develop several intuitive misconceptions through analogical reasoning: Field lines illustrate the path an untethered test charge would take through the field, the force does not vary from point to point along the field line, field lines can cross, and so forth. Additionally, learners often have difficulty relating field lines to another common representation of force, test charge traces. In MaxwellWorld, we attempt to overcome the shortcomings of the traditional field line representation by adding several enhancements. First, field lines are colored according to the strength of the force along them, which helps students visualize how the force varies from point to point. Second, our "enhanced" field lines can be continuously manipulated in 3-D. By grabbing a point on a field line and moving it, students can see how characteristics of the field line (both the shape and the strength of the field along it) change from point to point, and they can verify that field lines will never cross. Finally, by releasing a test charge on a field line, learners can see that the test charge moves along the field line only when the line does not curve.

Another example of a problematic representation is the "equipotential surface," which indicates a set of points across which a test charge's electric potential (or energy) would remain constant. In 2-D, this surface appears to be a line, so students have trouble distinguishing equipotential surfaces from field lines. Further, the standard formalism for equipotential surfaces does not convey information about the magnitude of the surface's potential. Nor does this representation help students relate the concepts of potential and force on the surface (this is also a problem with field lines). Consequently, students have trouble remembering which representation tells them about electric field (or force) and which tells them about electric potential (or energy). For example, we have observed a number of students describing field lines when asked to describe equipotential surfaces, and vice versa. At a deeper level, students have trouble distinguishing the concept of electric field (or force) from electric potential (or energy). For example, when students are asked whether the force on a test charge will vary or remain constant as they move it along an equipotential surface in a complex field, they most commonly predict that it will remain constant. We have enhanced the equipotential surfaces displayed in MaxwellWorld in an effort to compensate for these shortcomings of the standard formalism.

In general, these traditional scientific representations have one thing in common: They fail to make salient information that may be obvious to the expert, but not to the novice. The missing data often are crucial in providing the foundation for understanding how these models represent reality. Our approach has been to enhance traditional representations, adding new information and offering learners an opportunity to investigate the interrelationships among them. However, we have sometimes found ourselves limited in the extent to which we can build on these conventional formalisms, and even our enhanced versions are subject to some of the same misinterpretations.

We believe that researchers in the modeling community need to investigate the strengths and limitations of both enhanced and unique representations that are less subject to misinterpretation than those scientific formalisms that have emerged historically, before the availability of visualization tools. As we have found in our work, new notational systems may enable students to learn the underlying scientific concepts more readily. Unfortunately, learning with models based on new representations does not intrinsically convey the standard formalisms used by scientists to represent concepts. Therefore, our research suggests that until the scientific community is willing to replace historic formalisms with alternative, equally accurate representations more readliy comprehended by novices, many students will need specially designed modeling applications that focus on making salient otherwise "cognitively opaque" notational systems.

The tension between computer visualization of a model's output behavior and computer visualization of a model's structure and component processes.

The Debate. Computational modeling programs often employ visual representations of the model's behavior—animated displays of the outputs generated in the course of running the model. (Indeed, many researchers use computational models primarily for obtaining visualizations of model behavior, and *modeling* is thought of as almost synonymous with *visualization*.) Typically, scientists who conduct computational modeling research with sophisticated visualization facilities (at supercomputer centers, for instance) are content with programs that visualize a model's output behavior ("data visualization") but not its internal structure and component processes. Researchers disagree about whether this "output only" approach to visualization should be followed in science education.

Our Contribution. As we noted earlier, mastery of scientific models can be assessed on two levels: descriptive and causal. Descriptive mastery indicates that an individual remembers representations and their behavioral interrelationships; causal mastery shows a deeper understanding about what these descriptive dynamics imply about the nature of reality. In our evaluations of multisensory immersion's educational utility, we are careful to define causal mastery as the true goal and are not overly impressed when students exhibit descriptive mastery (even though the ability to describe a phenomenon's dynamics is a richer type of learning than presentational instruction typically achieves). Our experiences with educational modeling suggest that both inculcating causal understanding in students and measuring their attainment of this capability would be far more difficult with "output only" models than with the structure-and-processes instructional design we use, which allows real-time manipulation of causal factors in order to observe secondary effects.

The tension between learning to use models and learning to design and build models.

The Debate. Clearly, it is important for students to learn to use preconstructed models, but researchers differ as to how much (and how) pupils can learn to design and build their own models. Some argue that if students don't learn how to create models in classroom settings, how can we expect them to develop fluency at model building in workplace contexts? Further, constructivist learning theorists argue that students can comprehend much about model-based inquiry from engagement in the process of building models and simulations—indeed, that the process of designing and building models is an essential part of learning to use models as investigative tools and of understanding the strengths and limitations of models as a means for representing reality.

Our Contribution. For the very difficult scientific concepts on which our research is based—material that warrants the "sledgehammer" power of multisensory immersion to enhance learning—designing appropriate representations, interactive interfaces, and educational experiences is very challenging. That naive students could rapidly construct meaningful models of these complex phenomena is unlikely, however well designed the authoring tools they utilize. Our studies suggest that—at least for this type of counterintuitive, abstruse material—the use of preconfigured models for guided inquiry is much more efficient and probably just as effective as learners creating models from scratch.

Beyond these issues that are currently the subject of lively debate, we wish to mention a weakness that plagues most present approaches to model-based science learning: the lab-like nature of the learner's experiences. Controlled manipulations of a phenomenon, as in a scientific laboratory setting, are vital for understanding its nature, yet they are unmotivating to learners who are not already interested in science. Beginning with more playful and game-like exploration is important for motivating most students, and ending with these types of activities probably also boosts the transferability and generalizability of learning. At this point, our ScienceSpace worlds are as subject to this criticism as most other science-based educational models, yet we believe a major strength of multisensory immersion will be its capacity to support playful exploration in fantastical settings. As we develop our worlds, we plan to incorporate activities that support game-like competitions, invite exploration of curiously configured, beautiful environments (for example, Mandelbrot spaces), and contextualize scientific phenomena within an "edutainment" context (such as MaxwellWorld-like field spaces within the warp engines in a StarTrek virtual environment). We believe one of the most important challenges for model-based instructional design will be making these science learning environments more motivating and intriguing without weakening their educational value.

Conclusions drawn from an incipient set of studies on virtual reality as a modeling medium certainly do not provide definitive, generalizable answers about model-based instructional design. However, our experiences and research results provide a different perspective on the strengths and limitations of model-based learning, and further studies to explore the potential power of multisensory immersion certainly seem indicated.

Conclusion

The virtual reality interface has the potential to complement existing approaches to model-based instruction in science. An overarching goal in our ScienceSpace research is to develop a theory of how multisensory immersion aids learning. In our virtual worlds, we can simultaneously provide learners with 3-D representations; multiple perspectives and frames of reference; a multimodal interface; simultaneous visual, auditory, and haptic feedback; and types of interaction unavailable in the real world (such as seeing through objects, flying like Superman, and teleporting). With careful design, these capabilities can combine to create a profound sense of motivation and concentration conducive to mastering complex, abstract material.

By themselves becoming part of phenomena, learners gain direct experiential intuitions about how the natural world operates. Instructional design can make those aspects of virtual environments that promote understanding of scientific principles salient to learners' senses, and multisensory cues can heighten this saliency. Our experimental results indicate that transducing data and abstract concepts into mutually reinforcing multisensory representations is a valuable means of enhancing understanding of scientific models. Providing experiences that leverage human pattern recognition capabilities in three-dimensional space, such as shifting among various frames of reference (points of view), also extends the perceptual nature of a visualization. In addition, the social construction of knowledge among students immersed in a shared virtual environment may support innovative, powerful types of collaborative learning.

Overall, we believe that multisensory immersion, when applied to scientific models, can provide learners with experiential metaphors and analogies that help them understand complex phenomena remote from their everyday experience and can help displace intuitive misconceptions with alternative, more accurate mental models. Studying this new type of learning experience to chart its strengths and limitations is an important frontier for cognitive science research, scientific modeling, and constructivist pedagogy.

Beyond its implications for model-based learning of science, we believe that our research illuminates larger issues related to students' understanding complex information spaces. In every aspect of our knowledge-based society, fluency in utilizing complicated representations of information is an increasingly crucial skill. Comprehending models that include sophisticated interrelationships, such as nonlinearities and feedback loops, is important not only for scientists but also for workers and citizens. Such complex behaviors are typical of many crucial phenomena in modern civilization, and our well-being vitally depends on understanding the strengths and limitations of the decision-making models we create of those situations. Inculcating in students model-assessment skills such as sensitivity analysis is not simply a way of meeting discipline-based science standards as educational outcomes; these are survival skills necessary for our time, just as irrigation and planting skills were for agricultural economies. The educational standards of the near future will probably reflect a focus that transcends knowledge of various isolated disciplines and emphasizes integrated skills central to twenty-first-century work and citizenship. Model-based learning has much to contribute as we strive to understand how best to conceptualize and achieve this next generation of educational standards.

Acknowledgments

This work is supported by NSF's Applications of Advanced Technology Program, Grants RED-93-53320 and 95-55682, and by NASA through a grant (NAG 9-713) and through access to equipment and computer software. The authors gratefully acknowledge the aid of Katy Ash, Brenda Bannan, Craig Calhoun, Jim Chen, Erik Geisler, Jeff Hoblit, Pat Hyde, Deirdre McGlynn, Edward (Joe) Redish, Saba Rofchaei, Chen Shui, and Susan Trickett.

References

Arthur, E.J., Hancock, P.A., & Chrysler, S.T. 1994. Spatial orientation in virtual worlds. In *Proceedings of the 37th Annual Meeting of the Human Factors and Ergonomics Society*. Santa Monica, CA: Human Factors Society, pp. 328–332.

Bricken, M., & Byrne, C. M. 1993. Summer students in virtual reality. In Wexelblat, A. (ed.), *Virtual reality: Applications and exploration*. New York: Academic Press, pp. 199–218.

Chi, M.T.H., Feltovich, P.J., & Glaser, R. 1991. Categorization and representation of physics problems by experts and novices. *Cognitive Science*, 5, 121–152.

Clark, A.C. 1973. Technology and the limits of knowledge. In Boorstin, D. (ed.), *Technology and the frontiers of knowledge*. New York: Doubleday.

Dede, C. 1995. The evolution of constructivist learning environments: Immersion in distributed, virtual worlds. *Educational Technology 35(5)*, 46–52.

Dede, C., and Lewis, M. 1995. *Assessment of emerging educational technologies that might assist and enhance school-to-work transitions.* Washington, DC: National Technical Information Service.

Dede, C., Salzman, M., & Loftin, B. 1996. ScienceSpace: Virtual realities for learning complex and abstract scientific concepts. In *Proceedings of IEEE Virtual Reality Annual International Symposium 1996.* New York: IEEE Press, pp. 246-253.

Dede, C., Salzman, M., Loftin, B., & Ash, K. (in preparation). Using virtual reality technology to convey abstract scientific concepts. In Jacobson, M., & Kozma, R. (eds.), *Learning the sciences of the 21st century: Reseach, design, and implementation of advanced technological learning environments.* Hillsdale, NJ: Lawrence Erlbaum.

Ellis, S.R., Tharp, G.K., Grunwald, A.J., & Smith, S. 1991. Exocentric judgments in real environments and stereoscopic displays. In *Proceedings of the 35th Annual Meeting of the Human Factors Society.* Santa Monica, CA: Human Factors Society, pp. 1442-1446.

Erickson, T. 1993. Artificial realities as data visualization environments. In Wexelblat, A. (ed.), *Virtual reality: Applications and explorations.* New York: Academic Press, pp. 1-22.

Feurzeig, W. 1997. Personal communication to the authors. Cambridge, MA: BBN Labs.

Fosnot, C. 1992. Constructing constructivism. In Duffy, T.M., & Jonassen, D.H. (eds.), *Constructivism and the technology of instruction: A conversation.* Hillsdale, NJ: Lawrence Erlbaum, pp. 167-176.

Gordin, D.N., & Pea, R.D. 1995. Prospects for scientific visualization as an educational technology. *The Journal of the Learning Sciences, 4(3),* 249-279.

Halloun, I.A., & Hestenes, D. 1985a. Common sense concepts about motion. *American Journal of Physics, 53,* 1056-1065.

Halloun, I.A., & Hestenes, D. 1985b. The initial knowledge state of college students. *American Journal of Physics, 53,* 1043-1055.

Heeter, C. 1992. Being there: The subjective experience of presence. *Presence: Teleoperators and Virtual environments, 1(1),* 262-271.

Kalawsky, R.S. 1993. *The science of virtual reality and virtual environments.* New York: Addison-Wesley.

Kennedy, R.S., Lane, N.E., Berbaum, K.S., & Lilienthal, M.G. 1993. Simulator sickness questionnaire: An enhanced method for quantifying simulator sickness. *The International Journal of Aviation Psychology, 3(3),* 203-220.

Kohn, M. 1994. Is this the end of abstract thought? *New Scientist,* September 17, 37-39.

Krueger, M. 1991. *Artificial reality II.* New York: Addison-Wesley.

Loftin, R.B. 1997. Hands across the Atlantic. *IEEE Computer Graphics & Applications, 17(2),* 78-79.

Malone, T.W., & Lepper, M.R. 1984. Making learning fun: A taxonomy of intrinsic motivations for learning. In Snow, R.E., & Farr, M.J. (eds.), *Aptitude, learning and instruction.* Hillsdale, NJ: Lawrence Erlbaum.

McCormick, E.P. 1995. *Virtual reality features of frames of reference and display dimensionality with stereopsis: Their effects on scientific visualization.* Unpublished master's thesis, University of Illinois at Urbana-Champaign, Urbana, Illinois.

McDermott, L.C. 1991. Millikan lecture 1990: What we teach and what is learned — closing the gap. *American Journal of Physics, 59,* 301-315.

Papert, S. 1988. The conservation of Piaget: The computer as grist for the constructivist mill. In Foreman, G., & Pufall, P.B. (eds.), *Constructivism in the computer age.* Hillsdale, NJ: Lawrence Erlbaum, pp. 3-13.

Perkins, D. 1991. Technology meets constructivism: Do they make a marriage? *Educational Technology 31(5)*, 18-23.

Piantanida, T., Boman, D.K., & Gille, J. 1993, Human perceptual issues and virtual reality. *Virtual Reality Systems, 1(1)*, 43-52.

Psotka, J. 1996. Immersive training systems: Virtual reality and education and training. *Instructional Science 23(5-6)*, 405-423.

Redish, E. 1993. The implications of cognitive studies for teaching physics. *American Journal of Physics, 62(9)*, 796-803.

Regian, J.W., Shebilske, W., & Monk, J. 1992. A preliminary empirical evaluation of virtual reality as a training tool for visual-spatial tasks. *Journal of Communication 42*, 136-149.

Reif, F., & Larkin, J. 1991. Cognition in scientific and everyday domains: Comparison and learning implications. *Journal of Research in Science Teaching 28*, 743-760.

Rieber, L.P. 1994. Visualization as an aid to problem-solving: Examples from history. In *Proceedings of Selected Research and Development Presentations at the 1994 National Convention of the Association for Educational Communications and Technology*. Washington, DC: AECT, pp. 1018-1023.

Salzman, M.C., Dede, C., & Loftin, R.B. 1995. Learner-centered design of sensorily immersive microworlds using a virtual reality interface. In J. Greer (ed.), *Proceedings of the 7th International Conference on Artificial Intelligence and Education*. Charlottesville, VA: Association for the Advancement of Computers in Education, pp. 554-564.

Smith, R.B. (1987). Experiences with the alternate reality kit: An example of the tension between literalism and magic. In *Proceedings of CHI+GI 1987*. New York: Association for Computing Machinery, pp. 324-333.

Stuart, R., & Thomas, J.C. 1991. The implications of education in cyberspace. *Multimedia Review 2*, 17-27.

Studt, T. 1995. Visualization revolution adding new scientific viewpoints. *R & D Computers & Software*, October 14-16.

Trowbridge, D., & Sherwood, B. 1994. EM Field. Raleigh, NC: Physics Academic Software.

Wenzel, E.M. 1992. Localization in virtual acoustic displays. *Presence, 1(1)*, 80-107.

White, B. 1993. ThinkerTools: Causal models, conceptual change, and science education. *Cognition and Instruction 10* 1-100.

Wickens, C. 1992. Virtual reality and education. *IEEE Spectrum*, 842-847.

Wickens, C.D., & Baker, P. 1995. Cognitive issues in virtual reality. In Barfield, W., & Furness, T. (eds.), *Virtual environments and advanced interface design*. New York: Oxford Press.

Witmer, B.B., & Singer, M.J. 1994. *Measuring presence in virtual environments* (ARI Tech Report No. 1014). Alexandria, VA: U.S. Army Research Institute for the Behavioral and Social Sciences.

Conclusion

Introducing Modeling
into the Curriculum

Wally Feurzeig

Nancy Roberts

The integration of modeling and simulation into mathematics, science, and the social sciences succeeds or fails for the same reasons most other educational strategies succeed or fail. To learn effectively, students must be invested in the topic under study. It helps a great deal if students believe the subject is of importance or interest to them personally. It is easier to draw in marginally interested students by compelling, hands-on activities than by lectures and assigned readings. As Spitulnik, Krajcik, and Soloway state in Chapter 3 of this book, "Model building becomes a powerful activity for engaging students in doing and thinking about science. Science is no longer something that is read about in a book, but rather becomes an activity through which phenomena are studied, manipulated, sometimes controlled and perhaps even acted upon."

The studies here suggest that students' attention is more easily captured and that more in-depth learning takes place if students build and work with their own models rather than working solely with "expert-built" models or observing the use of models in demonstrations. As Ogborn (Chapter 1) notes, "We consistently find, over a range of ages, that students' own models are initially simpler than those they can cope with when models are presented to them, but that improvements made to their own models are much more interesting, and lead to more complex models, than are changes they make to models provided for them." When a teacher provides a model to students, she has to make the model world meaningful to them (Horwitz, Chapter 8; Barowy and Roberts, Chapter 9; White and Schwarz, Chapter 10). When students create and extend their own models, they model a world as they understand it. Their understanding improves as their models improve (Ogborn, Chapter 1; Spitulnik, Krajcik, and Soloway; Chapter 3; Wilensky, Chapter 7).

Several authors suggest a strategy involving students' use of expert models as well as models of their own making. For students whose mental models of the phenomena of interest are not strong, an expert model may be the only place to start. After exploration and experiment with the model across a variety of simulations, a student's mental model of the domain becomes stronger and he can develop the ability first to modify the given model and then to move toward building his own models. As Ogborn puts it, "This suggests a pedagogic strategy, using both exploratory and expressive modes. Exploratory use of the system can introduce systems of some complexity, whilst expressive use encourages further thinking and abstraction."

Another strategy—one that is more akin to the way models are developed professionally—can be employed in the classroom. A model-building team is formed comprising of both modeling experts and subject-matter experts in all the areas relevant to the problem domain under study. The modeling experts have the high-level tool-building skills, whereas the subject-area experts have in-depth knowledge and a strong mental model of the problem area. The same kinds of teams can be created in the science classroom. Some students will have a strong understanding of the problem being studied, some will have strong abilities to translate these ideas into the language of the chosen tool, and some, perhaps because of their insecurity in both areas, can be the question askers and model testers, who try a wide variety of simulation runs to see whether the model behaves as expected. This is the approach that Spitulnik, Krajcik, and Soloway described in their case study of Jamie, Lauren, and Rachel (Chapter 3).

Whichever strategy for introducing modeling is embraced, people at the NSF modeling conferences that gave rise to this book, along with its authors, agree that

- Computational modeling can dramatically enliven science and mathematics education; it can be designed and used in a way that engages students in active investigation; and it can offer students compelling learning experiences that enhance scientific understanding.
- Modeling should be introduced into the science curriculum at elementary school levels. The ubiquity of modeling and simulation tools in the sciences and professions underscores the importance of the early introduction of modeling.
- Modeling is a unique and valuable vehicle for understanding and better coping with complexity in real-world problems.

Schools ought to be able to give all students the opportunity to gain a firm scientific foundation for understanding and interpreting the world. Current inequities in the quality of science education, and administrative practices such as ability grouping and tracking, can lead to unequal learning opportunities and foster the creation of a scientific elite. Disenfranchised students are effectively prohibited from active and thoughtful participation in many practical, social, and political aspects of life that involve science. Their access to higher levels of science study and their entry into science- and technology-oriented careers is severely restricted (Shymansky and Kyle, 1992).

Students of all abilities and at all grade levels should be able to

- make sense of the world in a deeper and more comprehensive way than is currently possible.

- move fluently among different kinds of modeling paradigms—including models with different graphical, mathematical, and physical representations.
- become more personally and creatively engaged in the process of constructing knowledge.

Children can be— and should be—engaged in the process of designing, building, and experimenting with increasingly realistic and expressive models in many domains of mathematics and science. Modeling ideas and activities should come to play a regular and central role throughout the science curriculum.

Reference

Shymansky, J., and Kyle, W. 1992. Overview: Science curriculum reform. *Journal of Research in Science Teaching, 29(8),* 745–747.

Appendix A

Websites of Contributing Authors

Authors' Websites	Associated Software
Chapter 2:	STELLA® (http://www.hps-inc.com/)
http://www.teleport.com/~sguthrie/cc-stadus.html	
Chapter 3: *http://www.cogitomedia.com/*	Model-It
Chapter 4: *http://copernicus.bbn.com/fm/*	Function Machines
Chapter 5:	StarLogo
http://starlogo.www.media.mit.edu/people/starlogo/ind ex.html	
Chapter 7: *http://www.ccl.tufts.edu/cm/*	Gas Lab (StarLogoT)
Chapter 8: *http://genscope. concord.org/*	GenScope
Chapter 10: *http://thinkertools.berkeley.edu:7019/*	ThinkerTools
Chapter 11:	HyperGami and Java Gami
http://www.cs.colorado.edu/~eisenbea/hypergami/	
http://www.cs.colorado.edu/~eisenbea/javagami/	
Chapter 12: *http://www.virtual.gmu.edu/*	ScienceSpace

Appendix B

Contents of Enclosed CD-ROM

Chapter 1: WorldMaker (PC) and LinkIt (Mac and PC) and student models
Chapter 2: STELLA run-time version (Mac and PC) and models
Chapter 3: Model-It (Mac) and student models
Chapter 4: Function Machines (Mac) and student models
Chapter 5: StarLogo (Mac and PC) and student models
Chapter 7: StarLogoT (Mac) and GasLab and figures (Mac and PC)
Chapter 8: GenScope (Mac) and figures (Mac and PC)
Chapter 9: Movies showing students engaging in model-related discussions (Mac and PC)
Chapter 10: ThinkerTools (Mac)
Chapter 11: Examples of HyperGami products (Mac and PC) and JavaGami (Mac)
Chapter 12: Movies showing virtual reality environments of ScienceSpace (Mac and PC)

Some CD drives may have difficulty in reading or working with the CD. All files are available for download from http://www.springer-ny.com/supplements/feurzeig/.

Index

2-D modeling, *vs.* multisensory immersion, 299–306, 300f

Abstract topics, WorldMaker models of, 20–21, 21f
Accelerated motion, 54–55, 54f, 56f
Acceleration, constant, 53, 53f
Accumulations, 172
Action
  local, 14
  in thinking, 8–9
Adaptive models, 190–193
Affect, crafts and, 267–270
  computation and, 270
Affordances, 185–189, 186f–187f
Aggregate behavior models, 140
Aggregate modeling, 171–172
Aitkins, Peter, 19
Analysis by synthesis, 130
Approaches, pedagogical. *See also* specific approaches, e.g., ThinkerTools
  synthesis of, 252–253
Arthur, Brian, 133
Arts, student motivation in, 268
Assessment, 194–195

Backput iteration, 97, 97f
Backwards modeling, 170–171
Behavior, environment on, 135
Biology, WorldMaker models of, 15–17, 16f
Biology classes, CC-STADUS in, 60

Calculus, reform, 60–61

Cardiovascular system model, 215–217, 215f
Causal diagramming, 220
Causal-loop diagrams, 9, 10f
Causal models, intermediate, 227–229, 228f, 229f
CC-STADUS
  in biology classes, 60
  cross-curricular models in, 43–45, 44f
  documentation in, 45, 47
  future plans for
    long-term, 67–68
    near, 65–67
  history of, 38–40
  implementation of, 40–43
    follow-up of, 48–49
    revisions to, 48–49
  intent of, 40
  in mathematics
    implementation of, 63
    student models in, 63–65, 64f, 65f
    system dynamics of, 61–62, 62f
  in physics, 52–59, 53f–59f
    for accelerated motion, 54–55, 54f, 56f–57f
    for decay models, 58f, 59
    for ping-pong ball falling model, 55, 57f–58f
    for reinforcing traditional teaching, 55
    for rocket model, 56, 58f
  student response to, 49–51
  teacher receptiveness to, 51–52
  verification and validation of, 47

CC-STADUS (*cont.*)
 whole-class environment in, 57, 59
CC-SUSTAIN, 66–67
Cell automation, 14
Cellular automata, 119
 models of, 140
Centralized mindset, 114–115, 116–118
Change, rates of, 27f, 28, 28f
Chaos, mathematical, 106–107
Chemical change, WorldMaker models
 of, 17, 18f
Chemical equilibrium, simple, 17, 18f
Climate change unit. *See* Global climate
 change unit
Cognitive conflict, 220
Complex concepts
 learning about, 282–283
 models/simulations for, 283–286
Complexity, modeling of, 100–107,
 105f, 106f
Computation
 affective role of crafts and, 270
 pacing of scientific craft and, 272–
 273
Computation, parallel, 107–112
 data blocking in, 108–112, 109f,
 111f
 deadly embrace in, 108
 race conditions in, 107, 108f
Computational crafts. *See* Craft(s);
 HyperGami
Computational models. *See also*
 Model(s); specific models, e.g.,
 StarLogo
 true, 115–116, 131, 172–173
Computational science, 179
Computer-based manipulatives, 183–
 184
Computers, in science education, 5–6
Concrete experience, 174
Concrete operational thinking, 7–8
Concreteness, 173–174
Conflict, cognitive, 220
Connected Probability project, 152
 extensible models in, 169–170
 gas-in-a-box model in, 153–160,
 153f, 154f, 157f, 159f
 stages of pedagogy of, 170
Constructionists, 167, 167t
Constructivism, social, multisensory
 immersion for, 308

Constructivist learning, technology-
 enhanced, 283–284
Constructivists, 167
Content-rich model, 43–45, 44f
Copy epistemology, 217–218
Craft, scientific, 261
 culture and, 273–275
 educational pace and, 270–273
 educational rhythm and, 271–272
 pacing and, computation and, 272–
 273
Craft objects
 affordability of, 275
 audience for, 275
 longevity of, 272, 274
 performance of, 274
 reparability (undoability/redoability)
 of, 274–275
Crafts
 affect and, 267–270
 computation and, 270
 computational (*See* HyperGami)
Cross-curricular models, in CC-STADUS,
 43–45, 44f
Cross-Curricular Systems Thinking and
 Dynamics Using STELLA (CC-
 STADUS). *See* CC-STADUS
Cross-Curricular Systems Using STELLA
 (CC-SUSTAIN), 66–67
Culture, materials as, 273–275
Curriculum-rich models, 45

Data blocking, 108–112, 109f, 111f
Deadly embrace, 108
Decay, first-order, 143
Decay chains, reactive, 19–20
Decentralized design, 119
Decentralized modeling, 2–3, 114–136.
 See also Decentralized thinking
 interest in, 114–115
 learning with, 114–115
 with StarLogo, 119–131 (*See also*
 StarLogo)
Decentralized thinking, 2–3, 131–135
 centralized mindset *vs.*, 114–115,
 116–118
 emergent objects in, 134–135
 environment in, 135
 interest in, 114–115
 levels and, 134
 order from randomness in, 133–134

positive feedback in, 132–133
resistance to, 114
in StarLogo, 119–131 (*See also* Star-
Logo)
tools for, 118–121, 121f
Demonstration modeling, 167–168
Design, of teaching models. *See* Model
design
Diagramming, causal, 220
Diagrams, causal-loop, 9, 10f
Diffusion, of randomly moving parti-
cles, 17–20, 20f
Documentation, of CC-STADUS, 45, 47
Dynamic systems modeling, 151

Ecology, WorldMaker models of, 15–17,
16f
Education, science
computers in, 5–6
mathematics in, 5
Einstein, Albert, 260
Elements, primitive modeling, 168–169
Emergent objects, 134–135
Empirical testing, 211
Environment, on behavior, 135
Epistemology, copy, 217–218
Equilibrium, thermal, 19, 20f
Essentials
bus example of, 24
of problems, 23–24
Evaluation, rational, 210–211
Evaporation, 19
Events, in thinking, 7–8
Evidence, *vs.* theory, 216
Evolution phase, 232
Existence proof, 214
Experience(s)
concrete, 174
minimally abstracted, 203
Exploratory modeling, 171
Explorer. *See* Modeling, as inquiry
Explorer: Cardiovascular Model, 215–
217, 215f
Extensible, 161–166, 163f, 164f
Extensible models, 161–166, 163f,
164f, 169–170

Farmers and pests model, 12–13, 13f,
16–17
Feedback, positive, 132–133

Feynman, Richard, 260
Figuring, 203
First-order decay, 143
Flows, 39
Formal modeling, 173–174
Formal operations thinking, 8
Formalism, 173–174
Formalization phase, 232
Forrester, Jay, 39
Forward modeling, 171
Frame democracy, 185
Function machines, 95–112
classroom investigations of, 100–
104, 102f–103f
iteration and recursion in, 98–100,
99f–100f
mathematics models from, 100–104,
101f–103f
modeling complexity in, 104–107,
105f–106f
parallel computation in, 107–112
data blocking in, 108–112, 109f,
111f
deadly embrace in, 108
race conditions in, 107, 108f
Function Machines, website of, 323
Function machines language, 95–98,
96f–98f

Gas-in-a-box model
creation of, 153–160, 153f, 154f,
157f, 159f
use of, 161–166, 163f, 164f
Gas Particle Collision Exploration Environment (GPCEE), 152
GasLab, 151–176
dynamic systems modeling and, 151
in Connected Probability project,
152
gas-in-a-box model in, 153–160,
153f, 154f, 157f, 159f
in modeling pedagogy, 166–174
(*See also* Modeling)
aggregate *vs.* object-based, 171–
173
concreteness *vs.* formalism in,
173–174
construction *vs.* use of, 166–168,
167t
languages in, 169–170

GasLab
  in modeling pedagogy (*cont.*)
    phenomena-based *vs.* exploratory,
      170-171
    size in, 168-169
    symbolization forms in, 171
    tool kit for, 161-166, 163f, 164f
  website of, 323
Geller, Margaret, 260
GenScope, 186-189, 186f, 187f
  redesign of, 189-190, 190f, 191f
  scripted version of, 192
  website of, 323
Global climate change unit, 71-91
  account of, 73-74
  conclusions on, 90
  content understanding in, 75-83
    objects and factors in, 76-79, 78t
    purpose of, 75-76
    relationships and explanations in,
      76-83
    relationships between factors in,
      79-83, 80t, 83f
  context of, 71
  Global Model Planner in, 73
  goals of, 72-73, 72t
  inquiry understanding in, 83-88
    constructing/evaluating argument
      in, 87
    model construction in, 84-86, 85f
    problem definition in, 83-84
  map of, 74f
  nature of science understanding in,
    88-90
    evaluation of models in, 89-90
    purpose of, 88-89
  projects and models in, 71-72
Global Model Planner, 73
Gravitational field, 18

Harmonic oscillator, 30-32, 31f, 33f
Hawking, Stephen, 260
Hoyle, Fred, 260
HyperGami, 261-279
  affective role of objects in, 267-270
  basic activity in, 262, 262f
  computation and crafts in
    adaptability of, 277-278
    in design, 276
    embedded, 276-277
    new directions in, 278-279

programmability/reprogrammabil-
    ity of, 277-278
  constructions with, 264-266, 264f-
      267f
    by designers, 264-265, 264f-265f
    by students, 265-267, 265f-267f
  crafts in
    affect and, 267-270
    pace of education and, 270-273
    rhythm of education and, 271-272
  features of, 262-267, 264f
  materials and culture in, 273-275
  as math objects, 267
  overview of, 262-267
  website of, 323

Ideology problem, 20-21, 21f
Immersion, multisensory, 282-317. *See
    also* Multisensory immersion
Inquiry
  guided, 207-217, 208f, 210f, 215f
  modeling as (*See* Modeling, as in-
      quiry)
  scaffolding of, 239-241, 239f, 240f
Inquiry cycle, 239-241, 239f
Instructional cycle, four-phase, 231-233
Intermediate causal models, 227-229,
    228f, 229f
IQON, 35-37
IRF (Initiate, Response, Follow-
    up/Feedback), 203-207, 204f
Iteration
  backput, 97, 97f
  in function machines, 98-100, 99f

Johnson, Mark, 7

Keller, Evelyn Fox, 117
Kuhn, 216

Lakoff, George, 7
Langton, Christopher, 14
Languages, 95
  function machines, 95-98, 96f-98f
  in GasLab, 169-170
  modeling
    aggregate, 172
    general-purpose *vs.* content do-
        main, 169-170
  object-object translation (OOTL), 3,
      5, 138-148 (*See also* Object-

object translation language
     (OOTL))
Scheme, 172
  in HyperGami, 261
  STELLA as, 172
Law, Nancy, 12, 13–14
Learning
  content of, 115
  through modeling, 114–115 (*See
     also* Model(s); Modeling)
Leiser, David, 117–118
LinkIt, 10, 10f, 24–37
  design of, 34
  examples of, 27–33
     harmonic oscillator, 30–32, 31f,
        33f
     for multiplication, 28–29, 29f
     population growth in, 29–30, 30f,
        31f
     rain forest, 24–27, 26f
     rates of change with, 27f, 28, 28f
     thermostat, 32–33, 33f
  limitations of, 34–35
  student use of, 35–37
  variables and links in, 24–25
Links, modeling with, 24–37. *See also*
     LinkIt
Logic, Piaget on, 8
Logistic machines, 105–106, 106f
Logo, 119
  vs. StarLogo, 119–121
Lotka-Volterra equations, 130, 131,
     173–174

Majenjo Daro model, 43–44, 44f
Manipulatives
  computer-based, 183–184
  scriptable, 191–193
Maragoudaki, Eleni, 12–14, 13f
Math education
  critiques of, 38–39
  tradition in, 38
Mathematics, 5
  CC-STADUS in
     implementation of, 63
     student models in, 63–65, 64f, 65f
     system dynamics of, 61–62, 62f
  high school, system dynamics in,
     60–65, 62f, 64f
  modeling in, 6
  visual modeling tool for, 95–112

(*See also* Visual modeling tool,
     mathematics)
Maxwell-Boltzmann distribution law,
     153–160, 153f, 154f, 157f, 159f,
     164–165
MaxwellWorld, 285, 291–296, 293f–
     295f
  formative evaluation of, 297–299
  multisensory immersion *vs.* 2-D in,
     299–306, 300f
Mayr, Ernst, 117, 180
Microworlds, computer, 229–231, 230f
Middle-out instructional approach,
     231–233
Mindsets
  centralized, 114–115
  decentralized (*See* Decentralized
     thinking)
  tools and media in, 135–136
Model(s), 182
  adaptive, 190–193
  aggregate behavior, 140
  of cardiovascular system, 215–217,
     215f
  cellular automata, 140
  complex concepts in, 283–286
  construction of, encouraging, 115
  content-rich, 43–45, 44f
  curriculum-rich, 45
  design of (*See* Model design)
  extensible, 161–166, 163f, 164f,
     169–170
  gas-in-a-box
     creation of, 153–160, 153f, 154f,
        157f, 159f
     use of, 161–166, 163f, 164f
  generality of, 173
  intermediate causal, 227–229, 228f,
     229f
  Majenjo Daro, 43–44, 44f
  in mathematics, 6
  multisensory (*See* Multisensory im-
     mersion; ScienceSpace; Virtual
     reality)
  nature of, 217–219
  output *vs.* structure and processes
     in, 314–315
  PERS, 44–45
  Population Ecology, 203–207, 204f
  Rulers packet, 45, 46f
  simulation *vs.* real world in, 312

Model(s) (*cont.*)
   system dynamics, 140
   teaching, 181-182
   for theorizing, 6
   true computational, 115-116, 131,
      172-173
   using *vs.* designing/building, 315-
      316
   varieties of, 180-181
   Waves, 200-203, 200f
Model design, 182-196
   adaptive, 109-193
   affordances in, 185-189, 186f-187f
   assessment in, 194-195
   in GenScope, 186-189, 186f, 187f
      redesign of, 189-190, 190f, 191f
   objects and manipulations in, 183-
      184
   purpose in, 182
   real world examples of, 193-194
   redesign of, 189-190, 190f, 191f
   representations in, 184-185
   as rule-based systems, 183
   semantics in, 183
   vs. use, 315-316
Model-It, 70-71, 131
   website of, 323
Modelers, training of. *See* CC-STADUS;
   CC-SUSTAIN
Modeling
   activity levels in, 145-146
   aggregate *vs.* object-based, 171-173
   backwards, 170-171
   concreteness *vs.* formalism in, 173-
      174
   construction *vs.* use of, 166-168,
      167t
   decentralized, 114-136
     interest in, 114-115
     learning with, 114-115
     with StarLogo, 119-131 (*See also*
      StarLogo)
   demonstration, 167-168
   of dynamic systems, 151
   forward, 171
   GasLab for, 166-174, 167t
   introduction into curriculum of,
      320-323
   languages in, 169-170
   by learners, 198

   multisensory immersion in, 282-
      317 (*See also* Multisensory im-
      mersion)
   pedagogy of, 166-174, 167t
   phenomena-based *vs.* exploratory,
      170-171
   power and relevance of, 146
   purpose of, 91-93
   in research *vs.* education, 312-314
   size in, 168-169
   symbolization in, 171
   understanding and, 146-148, 146f-
      147f
   virtual reality for, 284
   vs. laboratory experiment, 145-148,
      146f-147f
Modeling, as inquiry
   adult role in, 221-222
   approach to, 199-200
   classroom experiments on, 203-
      207, 204f
   discussion of, 219-221
   guided inquiry in, 207-217, 208f,
      210f, 215f
   issues in, 198-199
   Nature of Models interview in, 217-
      219
   open-ended exploring as, 200-203,
      200f
   purpose of, 197-222
Modeling languages. *See also* specific
   languages, e.g., StarLogo
   aggregate, 172
   general-purpose *vs.* content domain,
      169-170
Modeling tools, 1-4. *See also* specific
   tools, e.g., WorldMaker
Motion, accelerated, 54-55, 54f, 56f
Motivation, student, in arts *vs.*
   math/science, 268
Motivation phase, 231-232
MultiLogo, 118
Multiplication, in LinkIt, 28-29, 29f
Multisensory immersion, 282-317. *See*
   also ScienceSpace; Virtual reality
   frames of reference in, 307-308,
      308f
   implications of work on, 311-316
   learning and knowledge representa-
      tion in, 310

lessons learned on, 309-311
MaxwellWorld in, 291-296, 293f-
    295f (*See also* MaxwellWorld)
model output *vs.* structure and pro-
    cesses in, 314-315
model use *vs.* design/building in,
    315-316
modeling in research *vs.* education
    and, 312-314
NewtonWorld in, 287-289, 288f-
    291f
PaulingWorld in, 290-291, 291f-
    293f
perceptualization from cues in, 308
potential of, 285
research on, 296-297
    future, 306-307
simulation *vs.* real world in, 312
in social constructivism, 308
virtual worlds of, 287-296
vs. conventional 2-D, study of, 299-
    306, 300f
    initial hypothesis in, 300-301
    stage one of, 301-304
    stage two of, 304-305
    summary of, 305-306

Nature of Models interview, 217-219
Network, two-machine, 96, 96f
Newton, Isaac, 260
NewtonWorld, 285, 287-289, 288f-
    291f

Object-based modeling, 171-173
Object-object translation language
    (OOTL), 3, 5, 138-148
    applications of, 139
    conceptualization of events in, 140-
        141
    in driving StarLogo engine, 144-
        145, 145f
    equations in, 141
    intent of, 138-139
    model-based inquiry with, 141-144,
        142f-143f
    modeling *vs.* lab experiment and,
        145-148, 146f-147f
    models in, 139-140
    objects in, 141
    with other modeling tools, 145

visual representations in, 139
Objects
    affective role of, 267-270
    emergent, 134-135
    modeling with, 10-11, 11f
    in thinking, 7-8
    transitional, 283
Open-ended exploring, 200-203, 200f

p-prims, 203
Pacing, of scientific craft, 270-273
    computation and, 272-273
Paley, William, 117
Papercrafts, mathematical. *See* Craft(s);
        HyperGami
Parallel computation, 107-112
    data blocking in, 108-112, 109f,
        111f
    deadly embrace in, 108
    race conditions in, 107, 108f
Pauling, Linus, 260
PaulingWorld, 290-291, 291f-293f
Pedagogical approaches. *See also*
        specific approaches, e.g.,
        ThinkerTools
    synthesis of, 252-253
PERS model, 44-45
Personal connections, 116
Phenomena-based modeling, 170-171
Physical change, WorldMaker models
        of, 17-20, 20f
Physics, 39
Physics classes, CC-STADUS in, 52-59,
        53f-59f
    for accelerated motion, 54-55, 54f,
        56f-57f
    for constant velocity and constant
        acceleration, 53-54, 53f
    for decay models, 58f, 59
    for ping-pong ball falling model, 55,
        57f-58f
    to reinforce traditional teaching, 55
    for rocket model, 56, 58f
Piaget, Jean, on thinking, 7-8
Pond organisms, simple, 15-16, 16f
Population Ecology model, 203-207,
        204f
Population growth, in LinkIt, 29-30,
        30f, 31f
Positive feedback, 132-133

Programming languages, 95. *See also* specific languages, e.g., Function machines language
Proof, existence, 214

Quantitative reasoning, 9

Rabbits and grass model, 11f, 12, 16-17
Race conditions, 107, 108f
Rain forest
    causal loop diagram of, 10f
    in LinkIt, 24-27, 26f
Rainfall, 19
Randomness, order from, 133-134
Rates of change, 27f, 28
Rational evaluation, 210-211
Reactive decay chains, 19-20
Reasoning
    formal *vs.* concrete, 8
    quantitative, 9
    rules or laws of, 8
    semi-quantitative, 8-9
Recursion, in function machines, 99-100, 99f
Redesign, of models, 189-190, 190f, 191f
Reflective assessment, 240f, 241
RelLab, 185
Repeat machine, 98-99, 99f
Representations, 184-185
Rethinking, in LinkIt, 25-26
Rhythm, of science education, 271-272
Rulers packet models, 45, 46f
Rules, modeling with, 10-11, 11f

Scaffolding, 192-193
Scheme programming language, Hyper-Gami, 261
Science education
    critiques of, 38-39
    pace of, 270-273
    rhythm of, 271-272
    tradition in, 38
ScienceSpace, 282-317
    challenges with, 285-286
    implications of work on, 311-316
    learning and knowledge representation in, 310
    lessons learned on, 309-311
    MaxwellWorld in, 291-296, 293f-295f

formative evaluation of, 297-299
multisensory immersion *vs.* 2D in, 299-306, 300f
models and simulations theory in, 283-285
NewtonWorld in, 287-289, 288f-291f
PaulingWorld in, 290-291, 291f-293f
potential for, 285
research on
    conducting, 296-297
    future, 306-309, 308f
website of, 323
Scientific craft, 261
Scientific understanding, models for, 70-93
    global climate change unit, 71-91 (*See also* Global climate change unit)
    Model-It, 70-71
    purpose of, 91-93
Scientists, childhood experiences of, 260-261
Scriptable manipulatives, 191-193
Scripts, 192-193
Semi-quantitative reasoning, 8-9
Simplification, in WorldMaker, 23-24
Simulation(s), 182
    complex concepts via, 283-286
    introduction of, 320-323
    in ThinkerTools, 237-238, 237f
    vs. real world, 312
    vs. stimulation, 116
Slime mold cells
    research on, 117
    StarLogo simulation of, 120, 121f
Socio-cognitive dissonance, 220
StarLogo, 119-121, 121f
    decentralized thinking in, 131-135
    emergent objects in, 134-135
    environment in, 135
    levels and, 134
    modeling systems in, 130-131
    object-object translation language in, 144-145, 145f
    order from randomness in, 133-134
    positive feedback in, 132-133
    student projects with, 121-131
        rabbits and grass, 127-131

termites and wood chips, 125–127, 126f
traffic jams, 122–125, 123f, 124f
student reaction to, 131–132
website of, 323
STELLA, 131. *See also* CC-STADUS; CC-SUSTAIN
as aggregate modeling language, 172
in mathematics, system dynamics of, 61–62, 62f
student use of, 35
website of, 323
Stimulation, *vs.* simulation, 116
Symbolization, new forms of, 171
SYM*BOWL, 66
Syntonic knowledge, 131
System dynamics
in high school mathematics, 60–65, 62f, 64f
models of, 140
teacher view of, 43, 51–52
System modelers, training of. *See* CC-STADUS; CC-SUSTAIN

Teaching models, 181–182
Theorizing, models for, 6
Theory, *vs.* evidence, 216
Thermal equilibrium, 19, 20f
Thermostat, 32–33, 33f
ThinkerTools, 226–254
computer microworlds in, 229–231, 230f
intermediate causal models in, 227–229, 228f, 229f
middle-out instructional approach in, 231–233
instructional trials of, 233–236, 234f, 235f
research on, early, 227–229, 228f, 229f
research on, latest, 245–252
instructional approach in, 248–249
instructional trials in, 249–252
modeling software in, 246–248, 246f
research on, recent, 236–245
inquiry curriculum in, 241–245, 243f, 244f
inquiry in, 239–241, 239f, 240f

metacognitive knowledge in, 238
modeling and simulation tools in, 237–238, 237f
reflective assessment in, 240f, 241
website of, 323
ThinkerTools Inquiry Curriculum, trials of, 241–245, 243f, 244f
ThinkerTools Modeling Curriculum
approach in, 248–249
trials of, 249–252
Thinking
action in, 8–9
concrete operational, 7–8
formal operations, 8
objects and events in, 7–8
stages of, 7
substance of, 7
Training, of system modelers. *See* CC-STADUS; CC-SUSTAIN
Transfer phase, 232–233
Transitional objects, 283
2-D modeling, *vs.* multisensory immersion, 299–306, 300f

Understanding, models for, 2, 70–93
global climate change unit, 71–91
(*See also* Global climate change unit)
Model-It, 70–71
purpose of, 91–93

VANTS (Virtual ANTS), 14
Variables, modeling with, 24–37. *See also* LinkIt
Velocity, constant, 53, 53f
Virtual reality. *See also* Multisensory immersion; ScienceSpace
challenges with, 285–286
future research on, 306–307
for modeling, 284
of ScienceSpace, 287–296
MaxwellWorld in, 291–296, 293f–295f
NewtonWorld in, 287–289, 288f–291f
PaulingWorld in, 290–291, 291f–293f
Virtual reality interfaces. *See also* MaxwellWorld; specific interfaces, e.g., ScienceSpace
challenges with, 310

Visual modeling tool, mathematics,
    95–112
  classroom investigations of, 100–
    104, 101f–103f
  function machine in, 95
    iteration and recursion of, 98–100,
      99f–100f
  function machines language in, 95–
    98, 96f–98f
  modeling complexity in, 104–107,
    105f–106f
  parallel computation in, 107–112
    data blocking in, 108–112, 109f,
      111f
    deadly embrace in, 108
    race conditions in, 107, 108f
von Neumann, John, 14
"von Neumann bottleneck," 107

Watchmaker argument, 117
Waves model, 200–203, 200f
Websites, 323

Whitehead, Alfred North, 271
WorldMaker, 10–24
  description of, 14–15
  explosion of ideas with, 23
  limitations of, 22
  modeling with objects and rules in,
    10–11, 11f
  with older students, 23
  results and possibilities with, 22–24
  science models with, 15–21
    of abstract topics, 20–21, 21f
    of biology and ecology, 15–17, 16f
    of chemical change, 17, 18f
    of physical change, 17–20, 20f
  simplification in, 23–24
  student use of, 12
  with young children, 12–14, 13f
  with younger students, 22–23

Zone of proximity development
    (zoped), 220

HIGH PERFORMANCE SYSTEMS, INC. SOFTWARE LICENSE AGREEMENT

Before opening this envelope, please review the following terms and conditions of this Agreement carefully. This is a legal agreement between you and High Performance Systems, Inc. The terms of this Agreement govern your use of this software. Opening this envelope or use of the enclosed materials will constitute your acceptance of the terms and conditions of this Agreement.

1.   Grant of License.

In consideration of payment of the license fee, which is part of the price you paid for this software package (referred to in this Agreement as the "Software"), High Performance Systems, Inc., as Licensor, grants to you, as Licensee, a non-exclusive right to use and display this copy of the Software on a single computer (i.e., a single CPU) only at one location at any time. To "use" the Software means that the Software is either loaded in the temporary memory (i.e., RAM) of a computer or installed on the permanent memory of a computer (i.e. hard disk, CD ROM, etc.). You may use and display this copy of the Software on as many computers for which you have a license (as is indicated on your invoice and/or packing slip). You may install the Software on a common storage device shared by multiple computers, provided that if you have more computers having access to the common storage device than the number of licensed copies of the Software, you must have some software mechanism which locks-out any concurrent users in excess of the number of licensed copies of the Software (an additional license is not needed for the one copy of Software stored on the common storage device accessed by multiple computers).

2.   Ownership of Software.

As Licensee, you own the magnetic or other physical media on which the Software is originally or subsequently recorded or fixed, but High Performance Systems, Inc. retains title and ownership of the Software, both as originally recorded and all subsequent copies made of the Software regardless of the form or media in or on which the original or copies may exist. This license is not a sale of the original Software or any copy.

3.   Copy Restrictions.

The Software and the accompanying written materials are protected by U.S. Copyright laws. Unauthorized copying of the Software, including Software that has been modified, merged, or included with other software or of the original written material is expressly forbidden. You may be held legally responsible for any copyright infringement that is caused or encouraged by your failure to abide by the terms of this Agreement. Subject to these restrictions, you may make one (1) copy of the Software solely for back-up purposes provided such back-up copy contains the same proprietary notices as appear in this Software.

4.   Use Restrictions.

As the Licensee, you may physically transfer the Software from one computer to another provided that the Software is used on only one computer at a time. You may not distribute copies of the Software or the accompanying written materials to others. You may not modify, adapt, translate, reverse engineer, decompile, disassemble, or create derivative works based on the Software. You may not modify, adapt, translate or create derivative works based on the written materials without the prior written consent of High Performance Systems, Inc.

5.   Transfer Restrictions.

This Software is licensed to only you, the Licensee, and may not be transferred to anyone else without the prior written consent of High Performance Systems, Inc. Any authorized transferee of the Software shall be bound by the terms and conditions of this Agreement. In no event may you transfer, assign, rent, lease, sell or otherwise dispose of the Software on a temporary or permanent basis except as expressly provided herein.

6.   Termination.

This license is effective until terminated. This license will terminate automatically without notice from High Performance Systems, Inc. if you fail to comply with any provision of this license. Upon termination you shall destroy the written materials and all copies of the Software, including modified copies, if any.

7.   Update Policy.

High Performance Systems, Inc. may create from time to time, updated versions of the Software. At its option, High Performance Systems, Inc. will make such updates available to Licensee and transferees who have returned the Registration Card which accompanies this software package.

8.   Disclaimer of Warranty and Limited Warranty.

THE SOFTWARE AND ACCOMPANYING WRITTEN MATERIALS ARE PROVIDED "AS IS" WITHOUT WARRANTY OF ANY KIND, EXPRESS OR IMPLIED OF ANY KIND, AND HIGH PERFORMANCE SYSTEMS, INC. SPECIFICALLY DISCLAIMS THE WARRANTIES OF FITNESS FOR A PARTICULAR PURPOSE AND MERCHANTABILITY.

However, High Performance Systems, Inc. warrants to the original Licensee that the disk(s)/CD on which the Software is recorded is free from defects in materials and workmanship under normal use and service for a period of ninety (90) days from the date of delivery as evidenced by a copy of the receipt of purchase. Further, High Performance Systems, Inc. hereby limits the duration of any implied warranty(ies) on the disk/CD to the period stated above. Some jurisdictions may not allow limitations on duration of an implied warranty, so the above limitation may not apply to you.

THE ABOVE ARE THE ONLY WARRANTIES OF ANY KIND, EITHER EXPRESS OR IMPLIED, THAT ARE MADE BY HIGH PERFORMANCE SYSTEMS, INC. ON THE SOFTWARE. NO ORAL OR WRITTEN INFORMATION OR ADVICE GIVEN BY HIGH PERFORMANCE SYSTEMS, INC., ITS DEALERS, DISTRIBUTORS, AGENTS OR EMPLOYEES SHALL CREATE A WARRANTY OR IN ANY WAY INCREASE THE SCOPE OF THIS WARRANTY, AND YOU MAY NOT RELY UPON SUCH INFORMATION OR ADVICE. THIS WARRANTY GIVES YOU SPECIFIC LEGAL RIGHTS. YOU MAY HAVE OTHER RIGHTS, WHICH VARY ACCORDING TO JURISDICTION.

9.   Limitations of Remedies.

NEITHER HIGH PERFORMANCE SYSTEMS, INC. NOR ANYONE ELSE WHO HAS BEEN INVOLVED IN THE CREATION, PRODUCTION OR DELIVERY OF THE SOFTWARE SHALL BE LIABLE FOR ANY DIRECT, INDIRECT, CONSEQUENTIAL, OR INCIDENTAL DAMAGE (INCLUDING DAMAGE FOR LOSS OF BUSINESS PROFIT, BUSINESS INTERRUPTION, LOSS OF DATA, AND THE LIKE) ARISING OUT OF THE USE OF OR INABILITY TO USE THE SOFTWARE EVEN IF HIGH PERFORMANCE SYSTEMS, INC. HAS BEEN ADVISED OF THE POSSIBILITY OF SUCH DAMAGE. AS SOME JURISDICTION MAY NOT ALLOW THE EXCLUSION OR LIMITATION OF LIABILITY FOR CONSEQUENTIAL OR INCIDENTAL DAMAGE.

High Performance Systems', Inc. entire liability and your exclusive remedy as to the disk(s)/CD shall be replacement of the defective disk/CD. If failure of any disk/CD has resulted from accident, abuse or misapplication, High Performance Systems, Inc. shall have responsibility to replace the disk/CD. Any replacement disk/CD will be warranted for the remainder of the original warranty period or thirty (30) days, whichever is longer.

10.   Miscellaneous.

This Agreement shall be governed by the laws of the State of New Hampshire and you agree to submit to personal jurisdiction in the State of New Hampshire. This Agreement constitutes the complete and exclusive statement of the terms of the Agreement between you and High Performance Systems, Inc. It supersedes and replaces any previous written or oral agreements and communications relating to this Software. If for any reason a court of competent jurisdiction finds any provision of this Agreement, or portion thereof, to be unenforceable, that provision of the Agreement shall be enforced to the maximum extent permissible so as to effect the intent of the parties, and the remainder of this Agreement shall continue in full force and effect.